Multiplets of Transition-Metal Ions in Crystals

This is Volume 33 in
PURE AND APPLIED PHYSICS
A Series of Monographs and Textbooks
Consulting Editors: H. S. W. MASSEY AND KEITH A. BRUECKNER
A complete list of titles in this series appears at the end of this volume.

Multiplets of Transition-Metal Ions in Crystals

Satoru Sugano
Institute for Solid State Physics
University of Tokyo
Minato-Ku, Tokyo
Japan

Yukito Tanabe
Department of Applied Physics
University of Tokyo
Bunkyo-Ku, Tokyo
Japan

Hiroshi Kamimura
Department of Physics
University of Tokyo
Bunkyo-Ku, Tokyo
Japan

Academic Press *New York and London* **1970**

ACADEMIC PRESS, INC.
111 Fifth Avenue, New York, New York 10003

United Kingdom Edition published by
ACADEMIC PRESS, INC. (LONDON) LTD.
Berkeley Square House, London W1X 6BA

LIBRARY OF CONGRESS CATALOG CARD NUMBER: 79-107572

PRINTED IN THE UNITED STATES OF AMERICA

Contents

VI. Low-Symmetry Fields

VII. Spin-Orbit Interaction

VIII. Fine Structure of Multiplets

IX. Interaction between Electron and Nuclear Vibration

X. Molecular Orbital and Heitler–London Theories

Appendix I.	**Character Tables for the Thirty-Two Double Point Groups, $\bar{G}$**	280
Appendix II.	**Tables of Clebsch–Gordan Coefficients, $\langle \Gamma_1\gamma_1\Gamma_2\gamma_2 \mid \Gamma\gamma \rangle$, with Cubic Bases**	286
Appendix III.	**Wigner Coefficients $\langle j_1m_1j_2m_2 \mid jm \rangle$**	289
Appendix IV.	**Matrix Elements of Coulomb Interaction**	294
Appendix V.	**Complementary States in the (t_2, e) Shell**	302
Appendix VI.	**Tables of Clebsch–Gordan Coefficients with Trigonal Bases, $\langle \Gamma_1M_1\Gamma_2M_2 \mid \Gamma M \rangle = \langle \Gamma M \mid \Gamma_1M_1\Gamma_2M_2 \rangle^*$**	305
Appendix VII.	**Tables of Reduced Matrices of Spin-Orbit Interaction**	308
Appendix VIII.	**Calculation of $\langle aS\Gamma \parallel L \parallel a'S\Gamma' \rangle$**	323
Appendix IX.	**Symmetric and Antisymmetric Product Representations**	325
	Subject Index	327

Preface

The purpose of this book is to introduce graduate students as well as research physicists, chemists, and electronic engineers to the essence of the theory of multiplets of transition-metal ions in crystals, more simply known as ligand field theory. The reader is assumed to be familiar with the fundamentals of quantum mechanics and, in particular, with the theory of atomic spectra. This book may be used as a textbook for a full-year course for graduate students.

Much effort has been made to present the material simply and clearly without a sacrifice of depth. For clearness illustrative examples are always given for each topic. The book is written in a self-sustaining form; consequently it contains only a minimum number of references. The reader who wants to know individual works related to the theory and its application should refer to the books cited at the end of the Introduction.

Throughout this book the formulation is based on the strong field scheme, and no use will be found of the weak field scheme. The exclusive use of one of the two schemes has been based on the hope that it would increase the readability of the book. A further discussion of this choice will be found in the Introduction.

In order to keep a suitable balance among the chapters, several important topics have been omitted which should properly be discussed in Chapter IX. They include the broadening and shift of zero-phonon lines, and effects of tunneling between Jahn–Teller distortions.

It is our great pleasure to express sincere thanks to Professor Masao Kotani for guiding us to this field. We are also indebted to Professors C. J. Ballhausen, C. K. Jørgensen, J. de Heer, and P. O. D. Offenhartz for criticism and comments on the manuscript, and to Drs. R. G. Shulman, J. H. Gallagher, and P. M. Maas for helpful discussions. One of the authors (S. S.) is grateful to the National Science Foundation for providing him with the opportunity of preparing the preliminary manuscript at the University of Colorado.

Finally we should like to thank Miss E. Hidaka for her typing of the manuscript and the staff of Academic Press for their cooperation.

Acknowledgments

The authors are indebted to the following for permission to use published and unpublished figures in the text: T. Kushida; H. C. Longuett-Higgins; A. Misu; M. H. L. Pryce; J. W. Stout; American Institute of Physics; Physical Society of Japan; Royal Society (London).

Multiplets of Transition-Metal Ions in Crystals

INTRODUCTION

It is well known that the spectral lines of atoms having many electrons are classified into multiplets which are assigned to transitions between terms. Each term that consists of almost degenerate discrete energy levels is specified by SL. Here we are speaking of the case in which the spin-orbit interaction is relatively small. For example, the lower terms of a Cr^{3+} ion, which has three electrons outside the closed shell, are known as follows:

Terms	Term energies (cm^{-1})
^4F	0
^4P	~14,200
^2P	~14,200
^2G	~15,200
^2D	~20,400
^2H	~21,200
^2F	~36,700

These terms are the quantum states in which three outer-shell electrons are accommodated in the $3d$ atomic orbitals. Energy separation of these terms are due to the Coulomb interaction between these outer-shell electrons. Optical transitions between these terms are seen in gaseous Cr^{3+} ions.

When a Cr^{3+} ion is incorporated as an impurity in a white sapphire, an Al_2O_3 crystal, the crystal exhibits a beautiful red or pink color. The absorption spectrum of this crystal called ruby, in the visible spectral region, is illustrated in Fig. 5.9 on p. 118. Inorganic complex salts involving, for example, $[Cr(H_2O)_6]^{3+}$ molecular ions and antiferromagnetic crystals Cr_2O_3 also show qualitatively similar absorption spectra in the visible region. As naturally expected, these absorption spectra in crystals are quite different from the spectrum of gaseous Cr^{3+} ions.

However, in the past ten years, it has been established that the spectral lines and bands in the insulating crystals involving *d*-electrons, such as those described above, are also classified into multiplets which are assigned to transitions between terms. The terms in this case have been found to be the quantum states in which *d*-electrons are accommodated in some orbitals relatively localized around the transition-metal elements. Naturally, this theory for the crystal spectra is similar to the theory of atomic multiplets, but it differs from it in that atoms in crystals do not have spherical symmetry but approximately cubic or tetrahedral symmetry in many cases, because of their surroundings. This difference brings a new aspect to the theory of multiplets. We call this theory of crystal multiplets *ligand field theory*.

The original form of the ligand field theory can be found in the crystalline field theory developed by Bethe,[‡] in 1929. This theory deals with the splitting of the atomic multiplets by the electric field arising from regularly distributed charges and electric dipoles of the surroundings, and can be shown to be equivalent to the ligand field theory to some approximation. Therefore, some people, mainly solid state physicists use the term, crystalline field theory, in place of the ligand field theory. Strictly speaking, the ligand field theory is more general than the crystalline field theory; this point will be discussed in detail in Chapter X.

Although the origin of the ligand field theory is very old, it is only in recent years that the importance of the theory has been fully recognized. This was achieved by successful applications of the theory to the interpretation of such optical spectra as those mentioned above as well as to the interpretation of microwave absorption spectra and paramagnetic susceptibilities in paramagnetic crystals.

Historically, it was Finkelstein and Van Vleck[§] who first applied the crystalline field theory to interpreting optical spectra of paramagnetic crystals involving *d*-electrons. They studied the absorption lines of a chromium alum crystal located at 6700 Å, which correspond to the *R* lines of ruby, and concluded that the excited states responsible to these lines are Stark split components, Kramers doublets, of the ^2G term of a

‡ H. Bethe, *Ann. Physik* **3**, 133 (1929).

§ R. Finkelstein and J. H. Van Vleck, *J. Chem. Phys.* **8**, 790 (1940).

free Cr^{3+} ion. This conclusion is in agreement with the result of recent detailed studies by the use of the ligand field theory.

In 1951, Hartmann, Schlaefer, and Ilse‡ published a series of papers in which they discussed the origin of broad absorption bands observed in inorganic metal complexes. By using the crystalline field theory, they concluded that these absorption bands were due to the transitions between the crystalline-field split components of the lowest term of the central metal ion, and showed that the number of broad absorption peaks near the visible region could be explained by their theory, at least when the central metal ion has no more than five *d*-electrons.

Since 1954 much work has been done on the optical spectra of *d*-electron systems. Through this work, the ligand field theory has become very successful in explaining both the absorption lines and bands observed in many kinds of metal complexes and insulators involving *d*-electrons. The most successful example of the application of the ligand field theory is the analysis of the optical spectrum of ruby (see Chapter V). It has recently been found that the theory is even applicable to those excited states of ruby whose excitation energies are as high as 45,000 cm^{-1}. It has also been shown that the accuracy of the theory is generally comparable to that of the theory of atomic multiplets.

In the ligand field theory there are two schemes, the strong-field scheme and the weak-field scheme, which will be explained in detail in Chapter X. In this book we exclusively use the strong-field scheme for the following reasons: (1) The strong-field scheme, taking no account of the configuration interaction, yields a good first-order approximation for the problems of *d*-electrons in crystals. (2) There is a strong similarity of the concepts between the strong-field scheme and the theory of atomic multiplets. The latter smoothly goes to the former if one-electron atomic orbitals are replaced by one-electron molecular orbitals. From Chapter I through Chapter IV the derivation of terms and the calculation of term energies in the d^N-electron systems in a cubic field are fully discussed on simple theoretical bases. The results obtained in these chapters are compared with experiments in Chapter V in order to show to what extent the theory is successful.

From Chapter I through Chapter V, no account is taken of the spin-orbit interaction and low-symmetry fields arising from a small distortion of a cubic system. These interactions split terms in a cubic field, resulting in fine structures of the multiplets. Fine structures of multiplets in crystals, including additional splittings induced by external pertur-

‡ H. Hartmann and H. L. Schlaefer, *Z. Phys.* **197**, 115 (1951). F. E. Ilse and H. Hartmann, *Z. Phys.* **197**, 239 (1951). H. Hartmann and H. L. Schlaefer, *Z. Naturforsch.* **6a**, 751, 760 (1951).

bations, such as an external magnetic field and an electric field, attract the interest of laser engineers as well as that of solid-state spectroscopists. Chapter VI through Chapter VIII are devoted to the theory of fine structures of the multiplets, Zeeman effects, and linear Stark effects in crystals.

Up to this point in the book, the nuclear framework in crystals is assumed to be rigid. However, in actual problems, it vibrates even at 0°K, and the interaction of electrons with the nuclear vibration brings important effects on the multiplets. In particular, in degenerate electronic states, the interaction induces static and dynamic Jahn–Teller effects. Fundamental problems relating to such an interaction shall be dealt with in Chapter IX.

Finally, Chapter X is devoted to the elucidation of the physical picture of the ligand field theory. This starts with theoretical efforts to explain from the first principle the values of physical parameters appearing in the ligand field theory, such as a cubic-field splitting parameter, Coulomb interaction parameters, spin-orbit interaction parameter, and so forth. It is shown that the traditional picture of the crystalline field theory introduced by Bethe is no longer applicable. Instead, covalency or virtual electron transfer between a metal ion and ligands, which is neglected in Bethe's model, plays an important role in producing ligand fields, and so forth. We might be able to mention that, although applications of the ligand field theory have been very successful, the physical model of the theory has not been firmly established. For expanding the area to which the ligand field theory can be applied, it is necessary to establish a rigid physical basis of the theory.

For reading the present text, the following books would be helpful as general references:

Ballhausen, C. J., "Introduction to Ligand Field Theory." McGraw-Hill, New York, 1962. Recommended as a reference book for beginners.

Condon, E. U. and Shortley, G. H., "The Theory of Atomic Spectra." Cambridge Univ. Press, London and New York, 1957.

Di Bartolo, B., "Optical Interactions in Solids." Wiley, New York, 1968.

Griffith, J. S., "The Theory of Transition-Metal Ions." Cambridge Univ. Press, London and New York, 1964. Recommended as a reference book for detailed studies.

Hamermesh, M., "Group Theory and Its Application to Physical Problems." Addison-Wesley, Reading, Massachusetts, 1962.

Jørgensen, C. K., "Absorption Spectra and Chemical Bonding in Complexes." Pergamon, Oxford, 1962.

McClure, D. S., "Electronic Spectra of Molecules and Ions in Crystals, Part II, Solid State Physics" (F. Seitz and D. Turnbull, eds.), Vol. 9, p. 399. Academic Press, New York, 1959.

Orgel, L. E., "An Introduction to Transition-Metal Chemistry." Methuen (Wiley), London 1960.

Watanabe, H., "Operator Methods in Ligand Field Theory." Prentice-Hall, Englewood Cliffs, New Jersey, 1966.

Wigner, E. P., "Group Theory and Its Application to the Quantum Mechanics of Atomic Spectra." Academic Press, New York, 1959.

Chapter I SINGLE *d*-ELECTRON IN A LIGAND FIELD

1.1 Single *d*-Electron in a Cubic Field

In this section we consider a hydrogen atom surrounded by six point-charges, $-Ze$, as indicated in Fig. 1.1. Here $Z > 0$ for the negative charge and $Z < 0$ for the positive charge. The distance between the hydrogen atom and each point-charge is a, so that the system has cubic symmetry. In this case, in addition to the Coulomb field due to

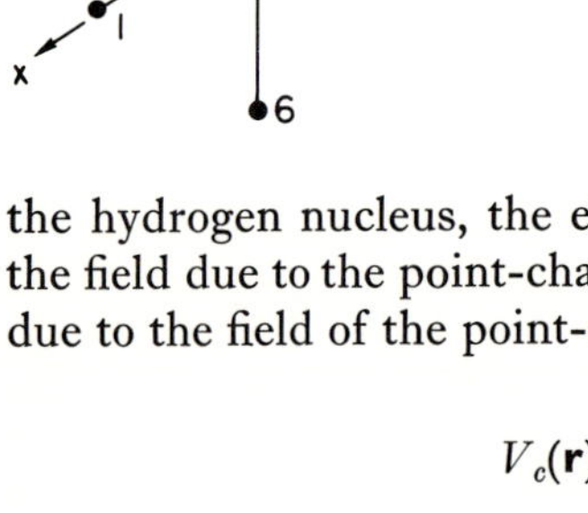

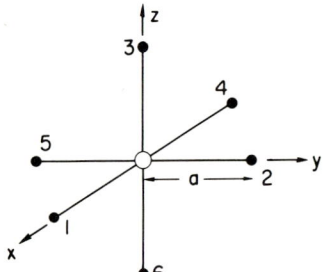

FIG. 1.1. Hydrogen atom surrounded by six point-charges; O_h symmetry. ●, $-Ze$ point charge. ○, hydrogen atom.

the hydrogen nucleus, the electron of the hydrogen atom is exposed to the field due to the point-charges. The potential energy V_c of the electron due to the field of the point-charges is given as

$$V_c(\mathbf{r}) = \sum_{i=1}^{6} Ze^2/| \mathbf{R}_i - \mathbf{r} |, \tag{1.1}$$

where **r** is the electron coordinate and $\mathbf{R}_i$ is the position vector of the ith point-charge. Then, the Schrödinger equation for the electron in this system is

$$[-(\hbar^2/2m)\,\Delta + U(r) + V_c(\mathbf{r})]\,\varphi(\mathbf{r}) = \epsilon\varphi(\mathbf{r}), \tag{1.2}$$

where $U(r)$ is the potential energy due to the field of the hydrogen nucleus, $\varphi(\mathbf{r})$ the wavefunction, and ϵ the energy eigenvalue.

In what follows, the perturbation method will be used to solve (1.2) by assuming V_c to be a small perturbation on a free hydrogen atom. For this purpose, it is convenient to expand V_c in terms of Legendre polynomials as follows:

$$V_c(\mathbf{r}) = Ze^2 \sum_{i=1}^{6} \sum_{k=0}^{\infty} (r_<{}^k/r_>^{k+1})\, P_k(\cos\omega_i), \tag{1.3}$$

in which $r_<$ is the lesser and $r_>$ the greater of a and r, and ω_i is the angle between vectors $\mathbf{R}_i$ and $\mathbf{r}$. When a is much larger than the radius of the hydrogen atom, we may replace, to a good approximation, $r_>$ and $r_<$ in (1.3) by a and r, respectively, and obtain

$$V_c(\mathbf{r}) = Ze^2 \sum_{i=1}^{6} \sum_{k=0}^{\infty} a^{-1}(r/a)^k P_k(\cos\omega_i). \tag{1.4}$$

Furthermore, the addition theorem for spherical harmonics shows

$$P_k(\cos\omega_i) = [4\pi/(2k+1)] \sum_{m=-k}^{k} Y_{km}(\theta\varphi) Y_{km}^*(\theta_i\varphi_i), \tag{1.5}$$

where $Y_{km}\,(\theta\varphi)$'s are the spherical harmonics, and (r, θ, φ) and (a, θ_i, φ_i) are the polar coordinates of **r** and $\mathbf{R}_i$, respectively. Here, $Y_{km}^*(\theta_i\varphi_i)$ is the complex conjugate of $Y_{km}(\theta_i\varphi_i)$ and is equal to $(-1)^m Y_{k-m}(\theta_i\varphi_i)$. From (1.5) and (1.4), V_c is given as a function of the electron coordinate **r** as follows:

$$V_c(\mathbf{r}) = \sum_{k=0}^{\infty} \sum_{m=-k}^{k} r^k q_{km} C_m^{(k)}(\theta\varphi), \tag{1.6}$$

where

$$q_{km} = \left(\frac{4\pi}{2k+1}\right)^{1/2} \frac{Ze^2}{a^{k+1}} \sum_{i=1}^{6} Y_{km}^*(\theta_i\varphi_i), \tag{1.7}$$

and

$$C_m^{(k)}(\theta\varphi) = \left(\frac{4\pi}{2k+1}\right)^{1/2} Y_{km}(\theta\varphi). \tag{1.8}$$

Since $(\theta_1\varphi_1)$, $(\theta_2\varphi_2)$,... are known to be

$$\left(\frac{\pi}{2}\ 0\right),\quad \left(\frac{\pi}{2}\ \frac{\pi}{2}\right),\ldots,$$

respectively, the q_{km}'s are given as

$$q_{k0} = \left(\frac{2}{2k+1}\right)^{1/2} \frac{Ze^2}{a^{k+1}} \left[\Theta_{k0}(0) + 4\Theta_{k0}\left(\frac{\pi}{2}\right) + \Theta_{k0}(\pi)\right], \tag{1.9a}$$

$$q_{km} = \left(\frac{2}{2k+1}\right)^{1/2} \frac{Ze^2}{a^{k+1}} \Theta_{km}\left(\frac{\pi}{2}\right)\left[1 + \exp\left(i\frac{m\pi}{2}\right) + \exp(im\pi) + \exp\left(i\frac{3m\pi}{2}\right)\right] \quad (m : \text{even} \neq 0), \tag{1.9b}$$

$$q_{km} = 0 \quad (m : \text{odd}), \tag{1.9c}$$

in which Θ_{km} is defined by

$$Y_{km}(\theta\varphi) = (2\pi)^{-1/2}\Theta_{km}(\theta)e^{im\varphi}. \tag{1.10}$$

The explicit forms of Θ_{km}'s are given in Table 1.1. By inserting the explicit forms of Θ_{km} into (1.9 a–c), the explicit form of $V_c(\mathbf{r})$ is obtained from (1.9 a–c) and (1.6) as follows:

$$V_c(\mathbf{r}) = \frac{6Ze^2}{a} + \frac{7Ze^2}{2a^5} r^4 \left\{C_0^{(4)}(\theta\varphi) + \left(\frac{5}{14}\right)^{1/2} [C_4^{(4)}(\theta\varphi) + C_{-4}^{(4)}(\theta\varphi)]\right\} + \frac{3Ze^2}{4a^7} r^6 \left\{C_0^{(6)}(\theta\varphi) - \left(\frac{7}{2}\right)^{1/2} [C_4^{(6)}(\theta\varphi) + C_{-4}^{(6)}(\theta\varphi)]\right\} + \cdots. \tag{1.11}$$

The first term in (1.11) represents the potential energy of the electron located at the position of the hydrogen nucleus and elevates all the energy levels of the hydrogen atom by the same amount, $6Ze^2/a$. The other terms split some of the degenerate energy levels as shown later. The field giving rise to the potential energy whose angular dependence is given as (1.11) is called a cubic field. This angular dependence is due to the geometrical arrangement of the point charges as shown in Fig. 1.1.

Before discussing the details of the perturbation calculation, let us calculate the matrix elements of V_c which are necessary in the following arguments and estimate the order of magnitudes of their values. Since the first term of V_c which is independent of the electron coordinate appears in all the diagonal matrix elements, to make the following arguments simple we leave out the first term of V_c and shift the origin

TABLE 1.1

EXPLICIT FORMS OF $\Theta_{lm}(\theta)$

$$\Theta_{lm}(\theta) = (-1)^m \left(\frac{(2l+1)(l-m)!}{2(l+m)!}\right)^{1/2} \sin^m\theta \frac{d^m}{(d\cos\theta)^m} P_l(\cos\theta),$$

$$\Theta_{l-m}(\theta) = (-1)^m \Theta_{lm}(\theta), \qquad m \geqslant 0$$

Θ_{00}	$1/\sqrt{2}$
Θ_{10}	$(\sqrt{3}/\sqrt{2}) \cos\theta$
$\Theta_{1\pm1}$	$\mp(\sqrt{3}/2) \sin\theta$
Θ_{20}	$(\sqrt{5}/2\sqrt{2})(2\cos^2\theta - \sin^2\theta)$
$\Theta_{2\pm1}$	$\mp(\sqrt{15}/2) \cos\theta \sin\theta$
$\Theta_{2\pm2}$	$(\sqrt{15}/4) \sin^2\theta$
Θ_{30}	$(\sqrt{7}/2\sqrt{2})(2\cos^3\theta - 3\cos\theta\sin^2\theta)$
$\Theta_{3\pm1}$	$\mp(\sqrt{21}/4\sqrt{2}) \sin\theta(4\cos^2\theta - \sin^2\theta)$
$\Theta_{3\pm2}$	$(\sqrt{105}/4) \cos\theta \sin^2\theta$
$\Theta_{3\pm3}$	$\mp(\sqrt{35}/4\sqrt{2}) \sin^3\theta$
Θ_{40}	$(3/8\sqrt{2})(8\cos^4\theta - 24\cos^2\theta\sin^2\theta + 3\sin^4\theta)$
$\Theta_{4\pm1}$	$\mp(3\sqrt{5}/4\sqrt{2}) \cos\theta\sin\theta(4\cos^2\theta - 3\sin^2\theta)$
$\Theta_{4\pm2}$	$(3\sqrt{5}/8) \sin^2\theta(6\cos^2\theta - \sin^2\theta)$
$\Theta_{4\pm3}$	$\mp(3\sqrt{35}/4\sqrt{2}) \cos\theta \sin^3\theta$
$\Theta_{4\pm4}$	$(3\sqrt{35}/16) \sin^4\theta$

of the energy by $6Ze^2/a$. Therefore, in what follows, we shall deal with V_c^0 given by

$$V_c^0 = V_c - (6Ze^2/a). \tag{1.12}$$

Denote the hydrogen wavefunction as

$$\varphi_{nlm}(\mathbf{r}) = R_{nl}(r) Y_{lm}(\theta\varphi). \tag{1.13}$$

The matrix element of V_c^0 between the states with quantum numbers (nlm) and $(n'l'm')$ is given by

$$\langle\varphi_{nlm}| V_c^0 | \varphi_{n'l'm'}\rangle = \int d\mathbf{r}\, \varphi_{nlm}^*(\mathbf{r})\, V_c^0(\mathbf{r})\, \varphi_{n'l'm'}(\mathbf{r}). \tag{1.14}$$

For calculating (1.14), it is necessary to evaluate the integrals of the following type:

$$c^k(l'm', l''m'') = \int d\varphi\, d\theta \sin\theta\, Y^*_{l'm'}(\theta\varphi)\, C^{(k)}_m(\theta\varphi)\, Y_{l''m''}(\theta\varphi). \tag{1.15}$$

Integration over φ directly indicates that (1.15) is nonvanishing only when

$$m = m' - m''. \tag{1.16}$$

Then $c^k(lm, l'm')$ can be calculated by using the explicit forms of Θ_{lm}'s; their values are tabulated in Table 1.2. It should be noted that $c^k(lm, l'm')$ is nonvanishing only when

$$k + l + l' = \text{even}, \tag{1.17}$$

and

$$|l - l'| \leqslant k \leqslant l + l'. \tag{1.18}$$

Relation (1.18) tells us that the matrix elements of V_c^0 between the p-states are vanishing, resulting in the absence of the cubic field splitting of the p-levels. Equation (1.18) also shows that, in calculating the matrix elements of V_c^0 between the d-states, the terms proportional to r^k $(k > 4)$ in V_c give a vanishing contribution.

Now let us calculate the matrix elements in (1.14) with $n = n'$ and $l = l' = 2$. The calculation is straightforward and the nonvanishing matrix elements are given as follows:

$$\langle\varphi_{nd\pm2} \mid V_c^0 \mid \varphi_{nd\pm2}\rangle = Dq, \tag{1.19a}$$

$$\langle\varphi_{nd\pm1} \mid V_c^0 \mid \varphi_{nd\pm1}\rangle = -4Dq, \tag{1.19b}$$

$$\langle\varphi_{nd0} \mid V_c^0 \mid \varphi_{nd0}\rangle = 6Dq, \tag{1.19c}$$

$$\langle\varphi_{nd\pm2} \mid V_c^0 \mid \varphi_{nd\mp2}\rangle = 5Dq, \tag{1.19d}$$

where

$$D = 35Ze^2/4a^5, \tag{1.20}$$

$$q = (2/105)\langle r^4\rangle_{nd}, \tag{1.21}$$

and

$$\langle r^m\rangle_{nd} = \int dr\, r^{2+m} \mid R_{nd}(r)|^2. \tag{1.22}$$

It should be remarked that D in (1.20) depends upon the point-charges, and that q in (1.21) reflects the properties of the electron of the central atom. The physical meaning of D may easily be understood if V_c^0 is reexpressed in the form

$$V_c^0 = D(x^4 + y^4 + z^4 - \tfrac{3}{5}r^4) + \cdots. \tag{1.23}$$

TABLE 1.2

NUMERICAL VALUES[a] OF $c^k(lm, l'm') = (-1)^{m-m'}c^k(l'm', lm)$ FOR $l \leqslant 3, l' \leqslant 3$

l	l'	m	m'	$k =$ 1		3		5	
$l + l' =$ odd									
s	p	0	±1	$-1/\sqrt{3}$					
		0	0	$+1/\sqrt{3}$					
s	f	0	±3			-1			
		0	±2			$+1$	$\times 1/\sqrt{7}$		
		0	±1			-1			
		0	0			$+1$			
p	d	±1	±2	$-\sqrt{6}$		$+\sqrt{3}$			
		±1	±1	$+\sqrt{3}$		-3			
		±1	0	-1		$+3\sqrt{2}$			
		0	±2	0	$\times 1/\sqrt{15}$	$+\sqrt{15}$	$\times 1/7\sqrt{5}$		
		0	±1	$-\sqrt{3}$		$-2\sqrt{6}$			
		0	0	$+2$		$+3\sqrt{3}$			
		±1	∓2	0		$+3\sqrt{5}$			
		±1	∓1	0		$-\sqrt{30}$			
d	f	±2	±3	$-\sqrt{15}$		$+\sqrt{10}$		-1	
		±2	±2	$+\sqrt{5}$		$-2\sqrt{5}$		$+\sqrt{5}$	
		±2	±1	-1		$+2\sqrt{6}$		$-\sqrt{15}$	
		±2	0	0		$-2\sqrt{5}$		$+\sqrt{35}$	
		±1	±3	0		$+5$		$-\sqrt{7}$	
		±1	±2	$-\sqrt{10}$		$-\sqrt{15}$		$+2\sqrt{6}$	
		±1	±1	$+2\sqrt{2}$		$+\sqrt{2}$		$-5\sqrt{2}$	
		±1	0	$-\sqrt{3}$		$+\sqrt{2}$		$+4\sqrt{5}$	
		0	±3	0	$\times 1/\sqrt{35}$	$+5$	$\times 1/3\sqrt{35}$	$-2\sqrt{7}$	$\times 1/6\sqrt{254}$
		0	±2	0		0		$+3\sqrt{7}$	
		0	±1	$-\sqrt{6}$		-3		$-3\sqrt{10}$	
		0	0	$+3$		$+4$		$+10$	
		±2	∓3	0		0		$-\sqrt{210}$	
		±2	∓2	0		0		$+3\sqrt{14}$	
		±2	∓1	0		$+\sqrt{10}$		$-\sqrt{70}$	
		±1	∓3	0		0		$-2\sqrt{21}$	
		±1	∓2	0		$+5$		$+4\sqrt{7}$	
		±1	∓1	0		$-\sqrt{15}$		$-\sqrt{105}$	

[a] Unlisted ones are zero.

TABLE 1.2 (*continued*)

l	l'	m	m'	$k=0$	2		4	
$l+l'$ = even								
s	s	0	0	+1				
s	d	0	± 2		+1	$\times 1/\sqrt{5}$		
		0	± 1		-1			
		0	0		+1			
p	p	± 1	± 1	+1	-1	$\times 1/5$		
		± 1	0	0	$+\sqrt{3}$			
		0	0	+1	+2			
		± 1	∓ 1	0	$-\sqrt{6}$			
p	f	± 1	± 3		$+3\sqrt{5}$	$\times 1/5\sqrt{7}$	-1	$\times 1/3\sqrt{21}$
		± 1	± 2		$-\sqrt{30}$		$+\sqrt{3}$	
		± 1	± 1		$+3\sqrt{2}$		$-\sqrt{6}$	
		± 1	0		-3		$+\sqrt{10}$	
		0	± 3		0		$-\sqrt{7}$	
		0	± 2		$+\sqrt{15}$		$+2\sqrt{3}$	
		0	± 1		$-2\sqrt{6}$		$-\sqrt{15}$	
		0	0		$+3\sqrt{3}$		+4	
		± 1	∓ 3		0		$-2\sqrt{7}$	
		± 1	∓ 2		0		$+\sqrt{21}$	
		± 1	∓ 1		$+\sqrt{3}$		$-\sqrt{15}$	
d	d	± 2	± 2	+1	-2	$\times 1/7$	+1	$\times 1/21$
		± 2	± 1	0	$+\sqrt{6}$		$-\sqrt{5}$	
		± 2	0	0	-2		$+\sqrt{15}$	
		± 1	± 1	+1	+1		-4	
		± 1	0	0	+1		$+\sqrt{30}$	
		0	0	+1	+2		+6	
		± 2	∓ 2	0	0		$+\sqrt{70}$	
		± 2	∓ 1	0	0		$-\sqrt{35}$	
		± 1	∓ 1	0	$-\sqrt{6}$		$-2\sqrt{10}$	

TABLE 1.2 (*continued*)

l	l'	m	m'	$k = 0$	2	4	6
$l + l'$ = even							
f	f	± 3	± 3	$+1$	-5	$+3$	-5
		± 3	± 2	0	$+5$	$-\sqrt{30}$	$+5\sqrt{7}$
		± 3	± 1	0	$-\sqrt{10}$	$+3\sqrt{6}$	$-10\sqrt{7}$
		± 3	0	0	0	$-3\sqrt{7}$	$+10\sqrt{21}$
		± 2	± 2	$+1$	0	-7	$+30$
		± 2	± 1	0	$+\sqrt{15}$	$+4\sqrt{2}$	$-5\sqrt{105}$
		± 2	0	0	$-2\sqrt{5}$	$-\sqrt{3}$	$+20\sqrt{14}$
		± 1	± 1	$+1$	$+3$	$+1$	-75
		± 1	0	0	$+\sqrt{2}$	$+\sqrt{15}$	$+25\sqrt{14}$
		0	0	$+1$	$+4$	$+6$	$+100$
		± 3	∓ 3	0	0	0	$-10\sqrt{231}$
		± 3	∓ 2	0	0	0	$+5\sqrt{462}$
		± 3	∓ 1	0	0	$+\sqrt{42}$	$-5\sqrt{210}$
		± 2	∓ 2	0	0	$+\sqrt{70}$	$+30\sqrt{14}$
		± 2	∓ 1	0	0	$-\sqrt{14}$	$-15\sqrt{42}$
		± 1	∓ 1	0	$-2\sqrt{6}$	$-2\sqrt{10}$	$-10\sqrt{105}$
					$\times 1/15$	$\times 1/33$	$\times 1/429$

The values of the matrix elements in (1.19) are of the order of magnitude of $Ze^2\langle r^4\rangle_{nd}/a^5$, which is $\sim 10^{-3}$ atomic unit ($\sim$0.027 eV) when $Z = 1$, $\langle r^4\rangle_{nd} = 1$ au, and $a = 4$ au. This value is much smaller than the energy separation, $\sim$0.6 eV, between the levels with $n = 3$ and 4 of the hydrogen atom. This justifies the perturbation calculation within the $3d$-states neglecting the nondiagonal elements between the $n = 3$ and 4 states. It should, however, be noted that this argument on the values of the matrix elements and the energy separation of the energy levels of the central atom are valid only in our model problem. In the real problems of $3d$-electrons in insulating crystals, the quantities corresponding to the matrix elements of V_c^0 and the energy separation between the $n = 3$ and 4 states are much larger than those discussed here.

Problem 1.1. Derive (1.23). ◇

Now, according to the perturbation theory, the perturbed energies of the $3d$-level of the hydrogen atom due to the presence of the point-charges are given by solving the secular equation,

$$\begin{vmatrix} \epsilon_3{}^0 + Dq - \epsilon & 0 & 0 & 0 & 5Dq \\ 0 & \epsilon_3{}^0 - 4Dq - \epsilon & 0 & 0 & 0 \\ 0 & 0 & \epsilon_3{}^0 + 6Dq - \epsilon & 0 & 0 \\ 0 & 0 & 0 & \epsilon_3{}^0 - 4Dq - \epsilon & 0 \\ 5Dq & 0 & 0 & 0 & \epsilon_3{}^0 + Dq - \epsilon \end{vmatrix} = 0, \tag{1.24}$$

where

$$\epsilon_3{}^0 = \epsilon_3 + (6Ze^2/a), \tag{1.25}$$

and ϵ_3 is the energy of the $3d$-state of the hydrogen atom. The bases of the secular matrix for (1.24) are arranged in the order, $m = 2, 1, 0, -1, -2$. As easily seen, (1.24) splits into three one-dimensional and one two-dimensional determinantal equations and the energy eigenvalues are obtained as

$$\epsilon^{(1)} = \epsilon_3{}^0 + 6Dq, \tag{1.25a}$$

$$\epsilon^{(2)} = \epsilon_3{}^0 - 4Dq, \tag{1.25b}$$

where $\epsilon^{(1)}$ and $\epsilon^{(2)}$ are, respectively, doubly and triply degenerate. It is customary to call the states with energies $\epsilon^{(1)}$ and $\epsilon^{(2)}$, the e_g and t_{2g} states, respectively, and to denote $\epsilon^{(1)}$ $\epsilon^{(2)}$ as $\epsilon(e_g)$ and $\epsilon(t_{2g})$, respectively. The reason why these notations are used will be explained later. The result given in (1.25) shows that, by the effect of the point-charges, the $3d$-level of the hydrogen atom, which has fivefold degeneracy, is split into doubly

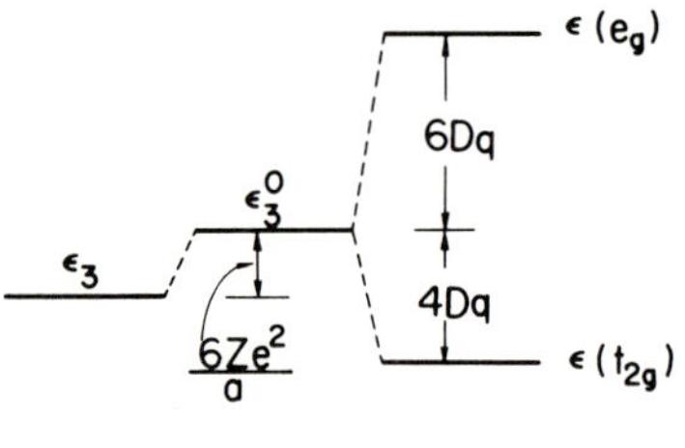

FIG. 1.2. Splitting of the $3d$-level of a hydrogen atom in a cubic field.

and triply degenerate levels as indicated in Fig. 1.2. The magnitude of the splitting is given as

$$\epsilon(e_g) - \epsilon(t_{2g}) = 10Dq. \tag{1.26}$$

That the center-of-mass of the energy levels does not change by the effect of $V_c{}^0$ can be seen from (1.25) as

$$\tfrac{1}{5}[2\epsilon(e_g) + 3\epsilon(t_{2g})] = \epsilon_3{}^0. \tag{1.27}$$

Diagonalizing the secular matrix, we can also obtain the wavefunctions associated with the split levels as follows:

$$\varphi_{3d0}\,, \qquad (\varphi_{3d2} + \varphi_{3d-2})/\sqrt{2}, \qquad \text{for the } e_g\text{-state},$$

$$\varphi_{3d1}\,, \qquad \varphi_{3d-1}\,, \qquad (\varphi_{3d2} - \varphi_{3d-2})/\sqrt{2}, \qquad \text{for the } t_{2g}\text{-state}.$$

The functions for the t_{2g}-state are not real, but can be made real by applying a unitary transformation within the subspace belonging to the t_{2g}-state: Any unitary transformation within a degenerate subspace does not change any physical situation. Let the unitary transformation be defined as

$$\varphi_\gamma = \sum_i \varphi_i U_{i\gamma}\,, \tag{1.28}$$

where φ_i's are

$$\varphi_1 = \varphi_{3d1}\,, \qquad \varphi_2 = \varphi_{3d-1}\,, \qquad \varphi_3 = (\varphi_{3d2} - \varphi_{3d-2})/\sqrt{2}. \tag{1.29}$$

By using the unitary transformation matrix,

$$\mathbf{U} = \begin{bmatrix} \dfrac{i}{\sqrt{2}} & -\dfrac{1}{\sqrt{2}} & 0 \\ \dfrac{i}{\sqrt{2}} & \dfrac{1}{\sqrt{2}} & 0 \\ 0 & 0 & -i \end{bmatrix}, \tag{1.30}$$

real wavefunctions for the t_{2g}-state are obtained as follows:

$$\begin{aligned} \varphi_\xi = (i/\sqrt{2})(\varphi_{3d1} + \varphi_{3d-1}) &= (15/4\pi)^{1/2} \sin\theta \cos\theta \sin\varphi\, R_{3d}(r) \\ &= (15/4\pi)^{1/2} (yz/r^2)\, R_{3d}(r), \end{aligned} \tag{1.31a}$$

$$\begin{aligned} \varphi_\eta = -(1/\sqrt{2})(\varphi_{3d1} - \varphi_{3d-1}) &= (15/4\pi)^{1/2} \sin\theta \cos\theta \cos\varphi\, R_{3d}(r) \\ &= (15/4\pi)^{1/2} (zx/r^2)\, R_{3d}(r), \end{aligned} \tag{1.31b}$$

$$\begin{aligned} \varphi_\zeta = -(i/\sqrt{2})(\varphi_{3d2} - \varphi_{3d-2}) &= (15/4\pi)^{1/2} \sin^2\theta \cos\varphi \sin\varphi\, R_{3d}(r) \\ &= (15/4\pi)^{1/2} (xy/r^2)\, R_{3d}(r). \end{aligned} \tag{1.31c}$$

Similarly, the wavefunctions obtained in (1.28) for the e_g-state may be written as follows:

$$\begin{aligned} \varphi_u = \varphi_{3d0} &= (5/16\pi)^{1/2} (3\cos^2\theta - 1)\, R_{3d}(r) \\ &= (5/16\pi)^{1/2} [(3z^2 - r^2)/r^2]\, R_{3d}(r), \end{aligned} \tag{1.32a}$$

$$\begin{aligned} \varphi_v = (\varphi_{3d2} + \varphi_{3d-2})/\sqrt{2} &= (15/16\pi)^{1/2} \sin^2\theta \cos 2\varphi\, R_{3d}(r) \\ &= (15/16\pi)^{1/2} [(x^2 - y^2)/r^2]\, R_{3d}(r). \end{aligned} \tag{1.32b}$$

For intuitive discussions, it is useful to know the behaviors of these wavefunctions in real space. The most convenient way of illustrating the wavefunctions is to fix the value of r and plot the angular dependence of the wavefunctions representing their magnitude at angle ($\theta\varphi$) by a radial length. This method is particularly useful to know the angular behaviors of the wavefunctions. The wavefunctions obtained in (1.31) and (1.32) are illustrated in Fig. 1.3 by using this method. From Fig. 1.3

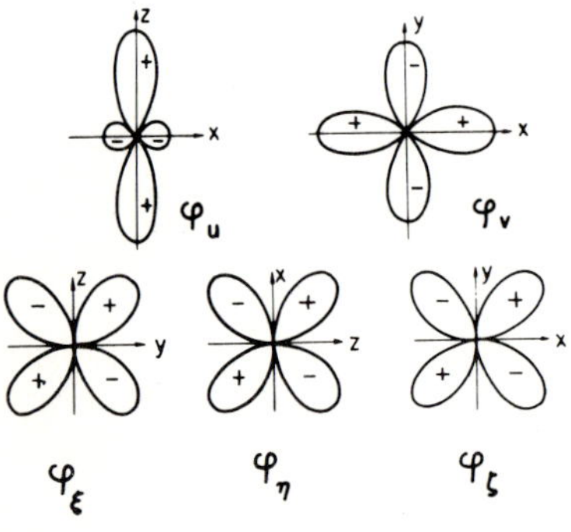

FIG. 1.3. Angular behaviors of wavefunctions, φ_ξ, φ_η, φ_ζ, φ_u, and φ_v.

it is clear that φ_ξ, φ_η, φ_ζ have the same energy, whereas it is not why φ_u and φ_v do. However, if one makes linear combinations of φ_u and φ_v as

$$\varphi_v^{(x)} = -\frac{\sqrt{3}}{2}\varphi_u - \frac{1}{2}\varphi_v \propto y^2 - z^2, \tag{1.33a}$$

$$\varphi_v^{(y)} = \frac{\sqrt{3}}{2}\varphi_u - \frac{1}{2}\varphi_v \propto z^2 - x^2, \tag{1.33b}$$

it is clear from Fig. 1.4 that φ_v, $\varphi_v^{(x)}$, and $\varphi_v^{(y)}$ have the same energy. Note that φ_v and $\varphi_v^{(x)}$ are linearly independent. This tells us that the φ_u and φ_v are the functions of a degenerate state. Since φ_u and φ_v extend toward

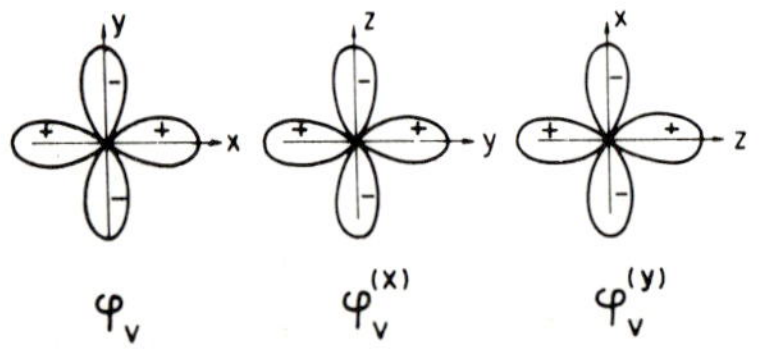

FIG. 1.4. Angular behaviors of wavefunctions, φ_v, $\varphi_v^{(x)}$, and $\varphi_v^{(y)}$.

the point-charges while φ_ξ, φ_η, and φ_ζ avoid them, it can easily be understood that the energy of the e_g-state is higher than that of the t_{2g}-state if the point-charges are negative ($Z > 0$).

In this section we have used notations t_{2g} and e_g for the states in a cubic field. The meaning of these notations will be explained in the next section in the light of the group theory. Although the wavefunctions of the t_{2g}- and e_g-states have been given in terms of the d-functions of the hydrogen atom, a more general interpretation of these wavefunctions will also be given in the next section.

Problem 1.2. Show that D in (1.19) is given by

$$D = 175de/4a^6$$

when six point-dipoles with dipole moment d pointing outward (the positive poles are outside) are placed at the places of the point-charges. ◇

Problem 1.3. By using the perturbation method, calculate the splitting of the $4f$-level of the hydrogen atom in the cubic field due to the point-charges. Also obtain the wavefunctions associated with the split levels. ◇

1.2 Group Theoretical Preliminaries

In this section it will be shown that, considering the symmetry of the system alone, we can predict the qualitative nature of the splitting of the energy level and the angular behavior of the wavefunctions, which were discussed in the previous section by the use of a particular model.

1.2.1 Symmetry Operations in *O*-Group

Let us again consider the system of a hydrogen atom surrounded by six point-charges as shown in Fig. 1.1. Suppose that this system is rotated around the x, y, and z axes by angles $2\pi/4$, $4\pi/4$, and $6\pi/4$. The rotated system is identical to the original one, as all the point-charges are identical. We call these rotations *symmetry operations*, and denote, for example, the rotations around the z-axis by $2\pi/4$, $4\pi/4$, and $6\pi/4$ as $C_4(z)$, $C_4^2(z)$, and $C_4^3(z)$, respectively. Here, the direction of the rotation is defined as the same as that of a right-hand screw which is progressing toward the positive direction of the rotation axis. Note that for $C_4(\bar{z})$ the positive direction of rotation axis is the negative direction of the z-axis. It is clear that successive operation $C_4(z)\, C_4(z)$ leads to $C_4^2(z)$. The x, y, and z axes are called fourfold symmetry (or rotation) axes. There are nine symmetry operations of this type. Furthermore, the following rotational operations are also symmetry operations:

C_3, C_3^2: rotations around the [111], [$\bar{1}$11], [1$\bar{1}$1], and [11$\bar{1}$] axes by angles $2\pi/3$ and $4\pi/3$. We denote, for example, the rotation around the [11$\bar{1}$] axis by $2\pi/3$ and $4\pi/3$ as $C_3(xy\bar{z})$ and $C_3^2(xy\bar{z})$, respectively. There are eight symmetry operations of this type.

C_2 : rotations around the [110], [1$\bar{1}$0], [101], [10$\bar{1}$], [011], and [01$\bar{1}$] axes by angle π. We denote, for example, the rotation around the [01$\bar{1}$] axis by π as $C_2(y\bar{z})$. There are six symmetry operations of this type.

After all, by including the identity operation E which does not move the system, twenty-four symmetry operations have been found in our system. However, it should be remarked that there are other kinds of symmetry operations which leave our system invariant. They are those involving inversion with respect to the center of the system. For the moment we will ignore these symmetry operations, which will be taken into account in a later part of this section.

By symmetry operation R, a point P whose coordinate is $\mathbf{r}(xyz)$ is transformed to P' whose coordinate is $\mathbf{r}'(x'y'z')$. This transformation is simply expressed as

$$\mathbf{r}' = R\mathbf{r}. \tag{1.34}$$

For example, for $R = C_4(z)$ it is seen from Fig. 1.1 that $x' = -y$, $y' = x$, and $z' = z$. From the result of the $C_4(z)$ transformation of point P, rotation $C_4(z)$ may be represented by

$$C_4(z) = \begin{bmatrix} x \rightarrow & -y \\ y \rightarrow & x \\ z \rightarrow & z \end{bmatrix}. \tag{1.35}$$

One of other methods of representing the rotational operations of our system, which favors our geometrical intuition more than (1.35), is to indicate how the six point-charges are transformed by the symmetry operations. For example, by rotation $C_4(z)$ point-charge 1 goes to the position of point-charge 2, 2 to that of 4, 4 to that of 5, 5 to that of 1, 3 to that of 3, and 6 to that of 6. Denoting such transformation simply as $\begin{pmatrix} 1 & 2 & 3 & 4 & 5 & 6 \\ 2 & 4 & 3 & 5 & 1 & 6 \end{pmatrix}$, which indicates that the point-charge with the upper number goes to the position of the point-charge with the number indicated just below, we may also represent rotation $C_4(z)$ as

$$C_4(z) = \begin{pmatrix} 1 & 2 & 3 & 4 & 5 & 6 \\ 2 & 4 & 3 & 5 & 1 & 6 \end{pmatrix}. \tag{1.36}$$

We shall use this method of representing the symmetry operations for a little while.

By using the representations as given in (1.36), one can confirm that an aggregate of our twenty-four symmetry operations satisfies the following four conditions:

(1) A product of two arbitrary operations within the aggregate is also an operation within the aggregate. The product R_2R_1 means the successive operation of R_2 after operation R_1. For example,

$$C_4(y)C_4(z) = \begin{pmatrix} 1 & 2 & 3 & 4 & 5 & 6 \\ 6 & 2 & 1 & 3 & 5 & 4 \end{pmatrix}\begin{pmatrix} 1 & 2 & 3 & 4 & 5 & 6 \\ 2 & 4 & 3 & 5 & 1 & 6 \end{pmatrix}$$

$$= \begin{pmatrix} 1 & 2 & 3 & 4 & 5 & 6 \\ 2 & 3 & 1 & 5 & 6 & 4 \end{pmatrix} = C_3(xyz). \tag{1.37}$$

It should be noted that the operations do not commute with each other as shown by an example,

$$C_4(z)C_4(y) = \begin{pmatrix} 1 & 2 & 3 & 4 & 5 & 6 \\ 2 & 4 & 3 & 5 & 1 & 6 \end{pmatrix}\begin{pmatrix} 1 & 2 & 3 & 4 & 5 & 6 \\ 6 & 2 & 1 & 3 & 5 & 4 \end{pmatrix}$$

$$= \begin{pmatrix} 1 & 2 & 3 & 4 & 5 & 6 \\ 6 & 4 & 2 & 3 & 1 & 5 \end{pmatrix} = C_3(\bar{x}yz)$$

$$\neq C_4(y)C_4(z). \tag{1.38}$$

(2) Let R_1, R_2, and R_3 be operations within the aggregate. Then, they satisfy $(R_1R_2)\,R_3 = R_1(R_2R_3)$. For example, when $R_1 = C_4(y)$, $R_2 = C_4(z)$, and $R_3 = C_4(z)$, using (1.37), we have

$$[C_4(y)C_4(z)]C_4(z) = C_3(xyz)C_4(z)$$

$$= \begin{pmatrix} 1 & 2 & 3 & 4 & 5 & 6 \\ 2 & 3 & 1 & 5 & 6 & 4 \end{pmatrix}\begin{pmatrix} 1 & 2 & 3 & 4 & 5 & 6 \\ 2 & 4 & 3 & 5 & 1 & 6 \end{pmatrix}$$

$$= \begin{pmatrix} 1 & 2 & 3 & 4 & 5 & 6 \\ 3 & 5 & 1 & 6 & 2 & 4 \end{pmatrix} = C_2(zx), \tag{1.39}$$

and

$$C_4(y)[C_4(z)C_4(z)] = C_4(y)C_4{}^2(z)$$

$$= \begin{pmatrix} 1 & 2 & 3 & 4 & 5 & 6 \\ 6 & 2 & 1 & 3 & 5 & 4 \end{pmatrix}\begin{pmatrix} 1 & 2 & 3 & 4 & 5 & 6 \\ 4 & 5 & 3 & 1 & 2 & 6 \end{pmatrix}$$

$$= \begin{pmatrix} 1 & 2 & 3 & 4 & 5 & 6 \\ 3 & 5 & 1 & 6 & 2 & 4 \end{pmatrix} = C_2(zx). \tag{1.40}$$

(3) There is an identity operation E which satisfies $RE = ER = R$ for any operation R within the aggregate.

(4) There is an inverse operation R^{-1} within the aggregate which satisfies $RR^{-1} = R^{-1}R = E$ for each operation R within the aggregate. For example, $C_4(z)^{-1} = C_4^3(z)$.

The aggregate of operations which satisfies the above-mentioned four conditions is called a *group*, and the operations are called the *elements* of the group. If the number of elements are finite as the present example, the group is called a *finite group*. Furthermore, if all the symmetry operations belonging to a finite group keep a lattice point unmoved, the group is called a *point-group*. Thirty-two point-groups have been known to exist. The group to which our twenty-four symmetry operations belong is one of the point-groups called O-group, octahedral group.

The twenty-four symmetry operations of the O-group can further be classified into five *classes*, in each of which symmetry operations are the rotations around equivalent axes by the same angle. Here, the equivalent axes are those which are transformed to each other by the symmetry operations of the O-group: for example, the x, y, and z axes are equivalent. The classification is shown in the tabulation. Operation $C_4^3(x)$

Classes	Symmetry operations
$\hat{E}$	E
$\hat{C}_4$	$C_4(x)$, $C_4(y)$, $C_4(z)$, $C_4^3(x)$, $C_4^3(y)$, $C_4^3(z)$
$\hat{C}_4^2$	$C_4^2(x)$, $C_4^2(y)$, $C_4^2(z)$
$\hat{C}_3$	$C_3(xyz)$, $C_3(\bar{x}yz)$, $C_3(x\bar{y}z)$, $C_3(xy\bar{z})$, $C_3^2(xyz)$, $C_3^2(\bar{x}yz)$, $C_3^2(x\bar{y}z)$, $C_3^2(xy\bar{z})$
$\hat{C}_2$	$C_2(xy)$, $C_2(yz)$, $C_2(zx)$, $C_2(\bar{x}y)$, $C_2(\bar{y}z)$, $C_2(\bar{z}x)$.

is identical to $C_4(\bar{x})$, and the $\bar{x}$-axis is equivalent to the x-axis. Therefore, $C_4^3(x)$ and $C_4(x)$ belong to the same class. A similar argument can be applied to $C_3(xyz)$ and $C_3^2(xyz)$.

In order to define the class in a more general way, let us consider two rotational operations $C(i)$ and $C(j)$ around the equivalent i and j axes, respectively, by the same angle, and assume that the axis i is transformed into j by symmetry operation R in the same group. Then, the following relation may be proved:

$$RC(i)R^{-1} = C(j). \tag{1.41}$$

For example, the x-axis is transformed to the y-axis by $C_4(z)$ and

$$C_4(z)C_4(x)C_4(z)^{-1}$$

$$= \begin{pmatrix} 1 & 2 & 3 & 4 & 5 & 6 \\ 2 & 4 & 3 & 5 & 1 & 6 \end{pmatrix} \begin{pmatrix} 1 & 2 & 3 & 4 & 5 & 6 \\ 1 & 3 & 5 & 4 & 6 & 2 \end{pmatrix} \begin{pmatrix} 1 & 2 & 3 & 4 & 5 & 6 \\ 5 & 1 & 3 & 2 & 4 & 6 \end{pmatrix}$$

$$= \begin{pmatrix} 1 & 2 & 3 & 4 & 5 & 6 \\ 6 & 2 & 1 & 3 & 5 & 4 \end{pmatrix} = C_4(y). \tag{1.42}$$

Equation (1.41) defines the transformation of operations: we say that operation $C(i)$ is transformed to $C(j)$ by R. With this terminology, the definition of the class may be stated as follows: The class is an aggregate of the elements of a group which are transformed to each other by appropriate elements of the group.

Finally it is worth pointing out that all the elements of a group may be generated from a smaller number of elements of the group. In our O-group, one can show that successive operations of two elements $C_4(z)$ and $C_4(y)$ generate all of the twenty-four elements. The generation of $C_3(xyz)$, $C_3(\bar{x}yz)$, and $C_2(zx)$ from $C_4(z)$ and $C_4(y)$ has been already shown in (1.37), (1.38), and (1.39), respectively. The elements whose successive operations generate all the elements of a group are called *generating elements*. In general, there are many ways of choosing the generating elements.

1.2.2 Irreducible Representations

Let us first consider the transformation of function $f(\mathbf{r})$ by symmetry operation R. The function may be considered to represent an electron distribution. Therefore, the transformation of a function may be interpreted as that of an electron distribution. For example, for $f(\mathbf{r}) = x$ the electron distribution may be illustrated as in Fig. 1.5. When

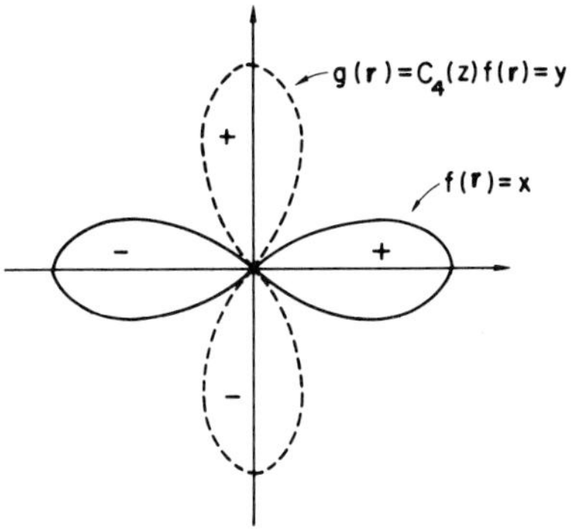

Fig. 1.5. Transformation of function $f(\mathbf{r}) = x$.

operation $C_4(z)$ is applied, the electron distribution is transformed to that indicated by the broken curve which is represented by $g(\mathbf{r}) = y$. Therefore, we express this transformation as $C_4(z)\, x = y$. Similarly $C_4(z)\, y = -x$ and $C_4(z)\, z = z$. With respect to the transformation of functions x, y, and z, therefore, rotation $C_4(z)$ may be represented as

$$C_4(z) = \begin{bmatrix} x \rightarrow & y \\ y \rightarrow & -x \\ z \rightarrow & z \end{bmatrix}_{\text{funct}} . \tag{1.43}$$

The subscript "funct" indicates that the symmetry operation in this case operates on the functions x, y, and z instead of the coordinates. If we regard the transformation in (1.43) as the point transformation such as (1.35), the representation is just that of $C_4(z)^{-1}$. This relationship is not accidental, and the reason is explained as follows: Let the electron distribution corresponding to $f(\mathbf{r})$ be indicated by the solid contour and the transformed one by the broken contour as indicated in Fig. 1.6.

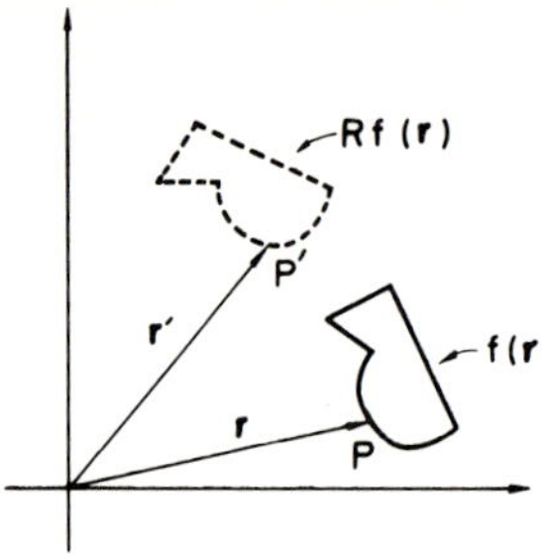

FIG. 1.6. Transformation of function $f(\mathbf{r})$.

Point P on the solid contour is transformed to point P' on the broken one by the transformation. From Fig. 1.6, it is immediately seen that

$$Rf(\mathbf{r}') = f(\mathbf{r}). \tag{1.44}$$

Since $\mathbf{r}' = R\mathbf{r}$, (1.44) is reexpressed as

$$Rf(\mathbf{r}) = f(R^{-1}\mathbf{r}), \tag{1.45}$$

which explains the relation between the representations (1.43) and (1.35).

As seen in (1.31a), φ_ξ is proportional to yz which may be considered to be the product of functions y and z. Therefore,

$$\begin{aligned} C_4(z)\varphi_\xi &= -(15/4\pi)^{1/2} zx R_{3d}(r)/r^2 \\ &= -\varphi_\eta \,. \end{aligned} \tag{1.46a}$$

Similarly,

$$\begin{aligned} C_4(z)\varphi_\eta &= \varphi_\xi\,, \\ C_4(z)\varphi_\zeta &= -\varphi_\zeta\,, \\ C_4(z)\varphi_u &= \varphi_u\,, \\ C_4(z)\varphi_v &= -\varphi_v\,. \end{aligned} \tag{1.46b}$$

Considering vector $\boldsymbol{\varphi}(\varphi_\xi, \varphi_\eta, \varphi_\zeta, \varphi_u, \varphi_v)$ in a five-dimensional space, we may express these transformation in the following compact form:

$$C_4(z)\boldsymbol{\varphi} = [\varphi_\xi\varphi_\eta\varphi_\zeta\varphi_u\varphi_v] \cdot \mathbf{D}(C_4(z))\,, \tag{1.47}$$

where

$$\mathbf{D}(C_4(z)) = \left[\begin{array}{ccc|cc} 0 & 1 & 0 & & \\ -1 & 0 & 0 & & 0 \\ 0 & 0 & -1 & & \\ \hline & 0 & & 1 & 0 \\ & & & 0 & -1 \end{array}\right]. \tag{1.48}$$

With respect to functions x, y, and z, rotation $C_4(y)$ is represented as

$$C_4(y) = \begin{bmatrix} x \rightarrow -z \\ y \rightarrow \ \ y \\ z \rightarrow \ \ x \end{bmatrix}_{\text{funct}}. \tag{1.49}$$

Therefore,

$$C_4(y)\varphi_\xi = \varphi_\zeta\,, \tag{1.50a}$$

$$C_4(y)\varphi_\eta = -\varphi_\eta\,, \tag{1.50b}$$

$$C_4(y)\varphi_\zeta = -\varphi_\xi\,, \tag{1.50c}$$

$$C_4(y)\varphi_u = -\frac{1}{2}\varphi_u + \frac{\sqrt{3}}{2}\varphi_v\,, \tag{1.50d}$$

$$C_4(y)\varphi_v = \frac{\sqrt{3}}{2}\varphi_u + \frac{1}{2}\varphi_v\,, \tag{1.50e}$$

which gives

$$\mathbf{D}(C_4(y)) = \left[\begin{array}{ccc|cc} 0 & 0 & -1 & & \\ 0 & -1 & 0 & & 0 \\ 1 & 0 & 0 & & \\ \hline & & & -\frac{1}{2} & \frac{\sqrt{3}}{2} \\ & 0 & & & \\ & & & \frac{\sqrt{3}}{2} & \frac{1}{2} \end{array}\right]. \tag{1.51}$$

It is straightforward to derive the relation,

$$\mathbf{D}(C_4(y)C_4(z)) = \mathbf{D}(C_3(xyz))$$

$$= \left[\begin{array}{ccc:cc} 0 & 0 & 1 & & \\ 1 & 0 & 0 & & 0 \\ 0 & 1 & 0 & & \\ \hdashline & & & -\frac{1}{2} & -\frac{\sqrt{3}}{2} \\ & 0 & & \frac{\sqrt{3}}{2} & -\frac{1}{2} \end{array}\right]$$

$$= \mathbf{D}(C_4(y)) \cdot \mathbf{D}(C_4(z)). \tag{1.52}$$

In general, for any elements R_1 and R_2 of the O-group, we can show

$$\mathbf{D}(R_1R_2) = \mathbf{D}(R_1) \cdot \mathbf{D}(R_2). \tag{1.53}$$

In this case the aggregate of matrices $\mathbf{D}(R)$'s is called *representation D* of the O-group. The representation has a concrete form, namely matrices in the present case, in contrast to the rather abstract nature of the group. Equations (1.35), (1.36), and (1.43) are also some representations of the O-group. In representation D, φ_ξ, φ_η, φ_ζ, φ_u, and φ_v are called the *bases* of the representation.

It is noticed in (1.48) and (1.51) that $\mathbf{D}(C_4(z))$ and $\mathbf{D}(C_4(y))$ have a characteristic form,

$$\left[\begin{array}{ccc:cc} \times & \times & \times & & \\ \times & \times & \times & & 0 \\ \times & \times & \times & & \\ \hdashline & & & \times & \times \\ & 0 & & & \\ & & & \times & \times \end{array}\right], \tag{1.54}$$

where the matrix elements indicated by $\times$ are not zero in general. More generally, $\mathbf{D}(R)$ for any element R of the O-group can be shown to have the form (1.54), as $C_4(z)$ and $C_4(y)$ are the generating elements of the O-group and $\mathbf{D}(R)$ for any R may be obtained from (1.48) and (1.51) by using (1.53). In this case, representation D may be reduced to two representations of smaller dimensions,

$$D = D^{(T_2)} + D^{(E)}, \tag{1.55}$$

by taking φ_ξ, φ_η, and φ_ζ as the bases of $D^{(T_2)}$ and φ_u and φ_v as those of $D^{(E)}$. These two sets of the bases are never admixed by the symmetry operations of the O-group. It is obvious that both $D^{(T_2)}$ and $D^{(E)}$ are the representations of the O-group. The representation which may be reduced to the representations of smaller dimensions is called a *reducible representation.*

However, it should be noted that, even if representation matrices with some bases do not have forms like (1.54), the representation could be a reducible one. In order to see this, we have to introduce the concept of *equivalent representations.* Let $\mathbf{T}$ be an arbitrary matrix with the same dimension as that of representation D of a group. We assume $\det T \neq 0$. Then, we can show that the aggregate of matrices $\mathbf{D}'(R)$ given for all the elements R of the group by

$$\mathbf{D}'(R) = \mathbf{T}\mathbf{D}(R)\mathbf{T}^{-1} \tag{1.56}$$

is also a representation of the group, as

$$\begin{aligned}\mathbf{D}'(R_1)\mathbf{D}'(R_2) &= \mathbf{T}\mathbf{D}(R_1)\mathbf{T}^{-1}\mathbf{T}\mathbf{D}(R_2)\mathbf{T}^{-1}\\ &= \mathbf{T}\mathbf{D}(R_1)\mathbf{D}(R_2)\mathbf{T}^{-1}\\ &= \mathbf{T}\mathbf{D}(R_1R_2)\mathbf{T}^{-1}\\ &= \mathbf{D}'(R_1R_2).\end{aligned} \tag{1.57}$$

Representations D and D' are called equivalent representations. Transformation (1.56) is called a *similarity transformation.* If a representation matrix can be brought into a form like (1.54) by a similarity transformation, the representation is a reducible one. The representations which are not reducible are called *irreducible representations.* Representations $D^{(T_2)}$ and $D^{(E)}$ can be shown to be irreducible representations.

Problem 1.4. Construct the five-dimensional representation of the O-group with bases φ_{3dm} ($m = 2, 1, 0, -1, -2$) and reduce it. ◇

In this subsection, we have shown that the t_{2g} and e_g wavefunctions are the bases of irreducible representations $D^{(T_2)}$ and $D^{(E)}$, respectively. In general, the wavefunctions of a state can be the bases of an irreducible representation of the group whose symmetry operations leave the system invariant. In order to show this, let us consider the Schrödinger equation for a system

$$\mathscr{H}\varphi_k = \epsilon_k\varphi_k\,. \tag{1.58}$$

If the system is invariant to symmetry operations R of group G, the Hamiltonian operator is invariant when it is transformed by R;

$$R\mathscr{H}R^{-1} = \mathscr{H}. \tag{1.59}$$

Operating R on (1.58) from the left, we obtain

$$\begin{aligned} R\mathscr{H}\varphi_k &= R\mathscr{H}R^{-1}R\varphi_k \\ &= \mathscr{H}R\varphi_k = \epsilon_k R\varphi_k . \end{aligned} \tag{1.60}$$

Equation (1.60) means that, if φ_k is an eigenfunction with energy eigenvalue ϵ_k, then $R\varphi_k$ is also the eigenfunction with the same eigenvalue ϵ_k. When state k is degenerate in g-fold, $R\varphi_{kj}$ should, in general, be given by a linear combination of the wavefunctions φ_{ki} $(i = 1, 2,..., g)$ of the degenerate state:

$$R\varphi_{kj} = \sum_{i=1}^{g} \varphi_{ki} D_{ij}^{(k)}(R), \tag{1.61}$$

where the $D_{ij}^{(k)}(R)$'s are numerical coefficients. Equation (1.61) immediately means that matrix $\mathbf{D}^{(k)}(R)$, appearing in (1.61), is the representation for R with bases φ_{ki} $(i = 1, 2,..., g)$. Besides the case of accidental degeneracy in state k, there is no reason to restrict the summation in (1.6) to particular φ_{ki}'s among the g wavefunctions of the degenerate state, so that $D^{(k)}$ is an irreducible representation of group G.

Problem 1.5. Derive the irreducible representations for $C_4(z)$ and $C_4(y)$ of the O-group with bases φ_x, φ_y, and φ_z of the p-state. ◇

Problem 1.6. Confirm the irreducible representations given in the tabulation. ◇

Bases	E	$C_4(z)$	$C_4^2(z)$	$C_3(xyz)$	$C_2(xy)$
φ_u, φ_v	$\begin{bmatrix}1 & 0\\0 & 1\end{bmatrix}$	$\begin{bmatrix}1 & 0\\0 & -1\end{bmatrix}$	$\begin{bmatrix}1 & 0\\0 & 1\end{bmatrix}$	$\begin{bmatrix}-\frac{1}{2} & -\frac{\sqrt{3}}{2}\\\frac{\sqrt{3}}{2} & -\frac{1}{2}\end{bmatrix}$	$\begin{bmatrix}1 & 0\\0 & -1\end{bmatrix}$
$\varphi_\xi, \varphi_\eta, \varphi_\zeta$	$\begin{bmatrix}1 & 0 & 0\\0 & 1 & 0\\0 & 0 & 1\end{bmatrix}$	$\begin{bmatrix}0 & 1 & 0\\-1 & 0 & 0\\0 & 0 & -1\end{bmatrix}$	$\begin{bmatrix}-1 & 0 & 0\\0 & -1 & 0\\0 & 0 & 1\end{bmatrix}$	$\begin{bmatrix}0 & 0 & 1\\1 & 0 & 0\\0 & 1 & 0\end{bmatrix}$	$\begin{bmatrix}0 & -1 & 0\\-1 & 0 & 0\\0 & 0 & 1\end{bmatrix}$
$\varphi_x, \varphi_y, \varphi_z$	$\begin{bmatrix}1 & 0 & 0\\0 & 1 & 0\\0 & 0 & 1\end{bmatrix}$	$\begin{bmatrix}0 & -1 & 0\\1 & 0 & 0\\0 & 0 & 1\end{bmatrix}$	$\begin{bmatrix}-1 & 0 & 0\\0 & -1 & 0\\0 & 0 & 1\end{bmatrix}$	$\begin{bmatrix}0 & 0 & 1\\1 & 0 & 0\\0 & 1 & 0\end{bmatrix}$	$\begin{bmatrix}0 & 1 & 0\\1 & 0 & 0\\0 & 0 & -1\end{bmatrix}$

1.2.3 Characters of Representations

In the previous subsection, it was shown that the wavefuntions of an eigenstate were the bases of an irreducible representation of the symmetry group whose symmetry operations leave the system invariant. Conversely speaking, the eigenstate may be characterized by the irreducible representations whose bases are the wavefunction of the state. Then, questions arise, concerning how many kinds of irreducible representations we can have for a group and how each kind of irreducible representations is characterized. The latter question arises because there are many equivalent irreducible representations. We have to seek a quantity which is invariant to a similarity transformation (1.56). This quantity can be shown to be the diagonal sum (trace) of the representation matrix, as

$$\begin{aligned}\sum_{m} D'_{mm}(R) &= \sum_{mnl} T_{ml} D_{ln}(R)(T^{-1})_{nm} \\ &= \sum_{nl} D_{ln}(R) \sum_{m} (T^{-1})_{nm} T_{ml} \\ &= \sum_{nl} D_{ln}(R)\, \delta_{nl} = \sum_{n} D_{nn}(R). \end{aligned} \tag{1.62}$$

Hereafter, the diagonal sum of the kth irreducible representation $\mathbf{D}^{(k)}(R)$ will be written as

$$\chi^{(k)}(R) = \sum_{m} D^{(k)}_{mm}(R), \tag{1.63}$$

and will be called the *character* of the kth irreducible representation $D^{(k)}$ for operation R. Since symmetry operations belonging to the same class are related to each other by transformation (1.41), the character of the irreducible representation is the same for all the symmetry operations in the same class. Therefore, the character would better be considered to be given for a class rather than for each element.

Now returning to the first question, let us consider the number of the inequivalent irreducible representations appearing in a group. For this purpose, we first give the orthogonality relation for the matrix elements of irreducible representations $D^{(\alpha)}$ and $D^{(\beta)}$;

$$\sum_{R} D^{(\alpha)}_{pq}(R)\, D^{(\beta)}_{rs}(R^{-1}) = \frac{h}{n_\alpha}\, \delta_{\alpha\beta}\, \delta_{ps}\, \delta_{qr}\,, \tag{1.64}$$

which may be proved by use of Shur's lemmas. The proof of (1.64) will be left to any standard textbook of the group theory. In (1.64) h is the number of elements in the group (called *order* of the group), and n_α is the dimension of irreducible representation $D^{(\alpha)}$ (called *degree* of the *k*th irreducible representation). Putting $p = q$ and $r = s$ and summing over p and r in (1.64), we obtain

$$\sum_R \chi^{(\alpha)}(R)\,\chi^{(\beta)}(R^{-1}) = h\,\delta_{\alpha\beta}\,. \tag{1.65a}$$

Since for finite groups every representation is equivalent to a unitary representation, in our problem of the finite group

$$D_{rs}^{(\beta)}(R^{-1}) = D_{sr}^{(\beta)}(R)^*. \tag{1.66}$$

Therefore, in this case (1.65a) may be written as

$$\sum_R \chi^{(\alpha)}(R)\,\chi^{(\beta)}(R)^* = h\,\delta_{\alpha\beta}\,. \tag{1.65b}$$

Equations (1.65a, b) are called the *orthogonality relation of the first kind for characters*. By denoting the class to which R belongs as $\hat{R}$ and remembering that $\chi^{(k)}(R)$'s for all R in class $\hat{R}$ take the same value $\chi^{(k)}(\hat{R})$, then (1.65b) may be reexpressed as

$$\sum_{\hat{R}} h_{\hat{R}}\chi^{(\alpha)}(\hat{R})\,\chi^{(\beta)}(\hat{R})^* = h\,\delta_{\alpha\beta}\,, \tag{1.67}$$

where $h_{\hat{R}}$ is the number of elements in class $\hat{R}$. Equation (1.67) tells us that vectors $\boldsymbol{\chi}^{(k)}$'s in the m-dimensional space whose components are $(h_{\hat{R}_1})^{1/2}\chi^{(k)}(\hat{R}_1)$, $(h_{\hat{R}_2})^{1/2}\chi^{(k)}(\hat{R}_2),\ldots, (h_{\hat{R}_m})^{1/2}\chi^{(k)}(\hat{R}_m)$ are mutually orthogonal. Here m is the number of classes in the group of interest. Since the number of mutually orthogonal vectors can not exceed the dimension of the vector space, it may be concluded that the number of vectors $\boldsymbol{\chi}^{(k)}$'s, in other words, the number of nonequivalent irreducible representations q is equal to or smaller than m:

$$q \leqslant m. \tag{1.68}$$

In order to show another inequality which together with (1.68) finally determines q, we have to give the *orthogonality relation of the second kind for characters*, which is expressed as follows:

$$\sum_k \chi^{(k)}(\hat{R}_1)\,\chi^{(k)}(\hat{R}_2^{-1}) = \frac{h}{h_{\hat{R}_1}}\,\delta(\hat{R}_1\hat{R}_2) \tag{1.69a}$$

or for unitary representations

$$\sum_k \chi^{(k)}(\hat{R}_1)\,\chi^{(k)}(\hat{R}_2)^* = \frac{h}{h_{\hat{R}_1}}\,\delta(\hat{R}_1\hat{R}_2). \tag{1.69b}$$

Readers will find the derivation of (1.69) in any textbook of the group theory (for example, see Hamermesh, p. 110). Equation (1.69) shows that the q-dimensional vectors $\boldsymbol{\chi}(\hat{R})$'s whose components are $\chi^{(k_1)}(\hat{R})$, $\chi^{(k_2)}(\hat{R}),\ldots, \chi^{(k_q)}(\hat{R})$ are mutually orthogonal. Since the number of mutually orthogonal vectors cannot exceed the dimension of the space, we have an inequality,

$$m \leqslant q. \tag{1.70}$$

Equations (1.68) and (1.70) lead us to the conclusion that the number of inequivalent irreducible representations is equal to the number of classes. In our O-group, therefore, five inequivalent irreducible representations are predicted.

Equation (1.69) also gives us a useful relation which determines dimension n_k of irreducible representation $D^{(k)}$. Assuming $\hat{R}_1 = \hat{R}_2 = \hat{E}$ in (1.69) and noting $\chi^{(k)}(\hat{E}) = n_k$, we obtain

$$\sum_k n_k{}^2 = h \qquad (n_k : \text{nonvanishing integers}). \tag{1.71}$$

In the O-group, $h = 24$ and k runs from 1 through 5. Therefore, n_k's are uniquely determined as follows:

$$1^2 + 1^2 + 2^2 + 3^2 + 3^2 = 24. \tag{1.72}$$

Now let us obtain all the values of $\chi^{(k)}(\hat{R})$ in the O-group. One-dimensional irreducible representations are given by numbers, among which the simplest one has the value unity for all the elements of the group. This representation certainly satisfies the requirement for the representation of the group as given in (1.53) and is clearly irreducible. The irreducible representation like this is called an *identity representation*, and always exists for any group. Furthermore, in Problem 1.6 we have already obtained the irreducible representations with the sets of bases (φ_u, φ_v), $(\varphi_\xi, \varphi_\eta, \varphi_\zeta)$, and $(\varphi_x, \varphi_y, \varphi_z)$ for an element in each class, which will be labeled as E, T_2, and T_1 irreducible representations, respectively. That T_2 and T_1 are inequivalent may be seen from the difference in the characters of these representations for classes $\hat{C}_4$ and $\hat{C}_2$. Therefore, the characters of two- and two three-dimensional inequivalent irreducible representations are in our hands. The characters of the remaining one-dimensional irreducible representation can be determined by using the orthogonality relations (1.67) and (1.69b). The character table of the O-group thus completed is given in Table 1.3.

Character tables of thirty-two point-groups have already been obtained; they are given in Appendix I.

TABLE 1.3

CHARACTER TABLE OF THE O-GROUP

Irreducible representations			Characters				
Mulliken	Bethe	BSW [a]	$\hat{E}$	[b]$6\hat{C}_4$	$3\hat{C}_4^2$	$8\hat{C}_3$	$6\hat{C}_2$
A_1	Γ_1	Γ_1	1	1	1	1	1
A_2	Γ_2	Γ_2	1	—1	1	1	—1
E	Γ_3	Γ_{12}	2	0	2	—1	0
T_1	Γ_4	Γ_{15}	3	1	—1	0	—1
T_2	Γ_5	Γ_{25}	3	—1	—1	0	1

[a] Notations by L. P. Bouckaert, R. Smoluchowski, and E. P. Wigner.

[b] The number of elements in a class.

1.2.4 SPLITTING OF ENERGY LEVELS

As an example of how to use the character tables, we shall discuss the splitting of energy levels when the symmetry of a system is lowered. In Section 1.1, using a simple model, we calculated the splitting of the $3d$-level of a hydrogen atom when it was placed in a cubic field. In this subsection, however, we will show that the qualitative nature of the splitting may easily be derived from a more general point of view by the use of characters.

In a free hydrogen atom an electron is exposed to a spherically symmetric potential field of the nucleus, so that the electron Hamiltonian is invariant to the rotations around the nucleus by arbitrary angles. An aggregate of the rotations around the axis of arbitrary directions by arbitrary angles clearly forms a group, which is called a *continuous rotation group*. In this group the number of elements is infinite and uncountable. Since the rotations by the same angle belong to the same class irrespective of the direction of the rotation axes, the number of classes in this group is also infinite and uncountable.

According to the result in Section 1.2.2, the hydrogen wavefunctions of the nl-state,

$$\varphi_{nlm}(\mathbf{r}) = R_{nl}(r)\, Y_{lm}(\theta\varphi) \qquad (m = l, l-1, \ldots, -l)$$

should be the bases of $(2l+1)$-dimensional irreducible representation $D^{(l)}$ of the continuous rotation group. In order to obtain the characters in this group, let us consider the irreducible representation for rotation R_α

around the z-axis by angle α. The choice of this particular rotation axis is sufficient for our purpose, as the characters do not depend upon the direction of the rotational axis. From (1.61) representation matrix $\mathbf{D}^{(l)}$ is given by

$$R_\alpha \varphi_{nlm} = \sum_{m'=l}^{-l} \varphi_{nlm'} D^{(l)}_{m'm} \,. \tag{1.73}$$

By using the relation,

$$R_\alpha Y_{lm}(\theta\varphi) = e^{-im\alpha} Y_{lm}(\theta\varphi), \tag{1.74}$$

we obtain $\mathbf{D}^{(l)}(\alpha)$ from (1.73) as follows:

$$\mathbf{D}^{(l)}(\alpha) = \begin{bmatrix} e^{-il\alpha} & & & & \\ & e^{-i(l-1)\alpha} & & & 0 \\ & & \cdot & & \\ & & & \cdot & \\ & 0 & & & e^{il\alpha} \end{bmatrix} \tag{1.75}$$

from which character $\chi^{(l)}(\alpha)$ is calculated as

$$\chi^{(l)}(\alpha) = \sum_{m=l}^{-l} e^{-im\alpha} = \frac{\sin(l+\frac{1}{2})\alpha}{\sin\frac{1}{2}\alpha} \,. \tag{1.76}$$

When a hydrogen atom is placed in the system of cubic symmetry, the electron Hamiltonian is no longer invariant under rotations around arbitrary rotational axes by arbitrary angles, but is only invariant under the rotations around specific rotational axes by specific angles, i.e., under the symmetry operations in the O-group which form only a small part of the symmetry operations in the continuous rotation group (in this case the O-group is called a *subgroup* of the continuous rotation group). In what follows, it will be shown that in the O-group representation $D^{(l)}$ is reducible, in general.

Let us assume that $D^{(l)}$ may be reduced to irreducible representations $D^{(k)}$'s of the O-group as follows:

$$D^{(l)} = \sum_k c(kl)\, D^{(k)}, \tag{1.77}$$

where $c(kl)$ indicates the number of the same $D^{(k)}$ that appear. By using characters, (1.77) may be expressed as

$$\chi^{(l)}(\hat{R}) = \sum_k c(kl)\, \chi^{(k)}(\hat{R}), \tag{1.78}$$

or, by using (1.67) further,

$$c(kl) = \frac{1}{h} \sum_{\hat{R}} h_{\hat{R}} \chi^{(l)}(\hat{R}) \chi^{(k)}(\hat{R})^*. \tag{1.79}$$

In (1.79) $\chi^{(l)}(\hat{R})$ is calculated by inserting for α in (1.76) a particular value characteristic to the rotation angle of R in class $\hat{R}$. Then $\chi^{(l)}(\hat{R})$ for each class of the O-group is obtained as follows:

$$\chi^{(l)}(\hat{E}) = \chi^{(l)}(\alpha = 0) = 2l + 1, \tag{1.80a}$$

$$\chi^{(l)}(\hat{C}_4) = \chi^{(l)}\left(\alpha = \frac{\pi}{2}\right) = \begin{cases} (-1)^{\frac{1}{2}l}, & \text{for } l = 2m \\ (-1)^{\frac{1}{2}(l-1)}, & \text{for } l = 2m+1 \end{cases} \quad (m\text{: integers}) \tag{1.80b}$$

$$\chi^{(l)}(\hat{C}_4^{\,2}) = \chi^{(l)}(\hat{C}_2) = \chi^{(l)}(\alpha = \pi) = (-1)^l, \tag{1.80c}$$

$$\chi^{(l)}(\hat{C}_3) = \chi^{(l)}\left(\alpha = \frac{2}{3}\pi\right) = \begin{cases} 1, & \text{for } l = 3m \\ 0, & \text{for } l = 3m+1 \\ -1, & \text{for } l = 3m+2 \end{cases} \quad (m\text{: integers}). \tag{1.80d}$$

Since $\chi^{(k)}(\hat{R})$'s for all $\hat{R}$ have already been obtained in Table 1.1, $c(k\ l{=}2)$ can now be calculated from (1.79) and (1.80) as follows:

$$\begin{aligned} c(A_1 2) &= 0, \qquad c(A_2 2) = 0, \\ c(E\,2) &= 1, \qquad c(T_1 2) = 0, \\ c(T_2 2) &= 1, \end{aligned} \tag{1.81}$$

which means

$$D^{(2)} = D^{(E)} + D^{(T_2)}. \tag{1.82}$$

Bearing in mind that the states of the system are characterized by the irreducible representations of the group to which the symmetry of the system belongs, we notice from (1.82) that the d-level of a hydrogen atom splits into a doubly degenerate level and a triply degenerate level in a cubic field in agreement with the result of the previous calculation by the use of a specific model. However, by the symmetry arguments alone, as given in this subsection, the magnitude of the energy separation between the split levels, e_g and t_{2g}, cannot be predicted.

In this way (1.79) predicts the splittings of the energy levels with various l in a cubic field. The result is given in Table 1.4.

In concluding this subsection let us consider one more example of the splitting of the e_g and t_{2g} levels in a tetragonal field. A simple system having the field of this symmetry is obtained by changing in Fig. 1.1 the

TABLE 1.4

l		Splitting in a cubic field
0	S	A_1
1	P	T_1
2	D	$E + T_2$
3	F	$A_2 + T_1 + T_2$
4	G	$A_1 + E + T_1 + T_2$
5	H	$E + 2T_1 + T_2$
6	I	$A_1 + A_2 + E + T_1 + 2T_2$

distances between the central atom and the point-charges 3 and 6 to $b(b \neq a)$ while keeping the other point-charges unmoved. This system

FIG. 1.7. Hydrogen atom surrounded by six point-charges; D_{4h} symmetry.

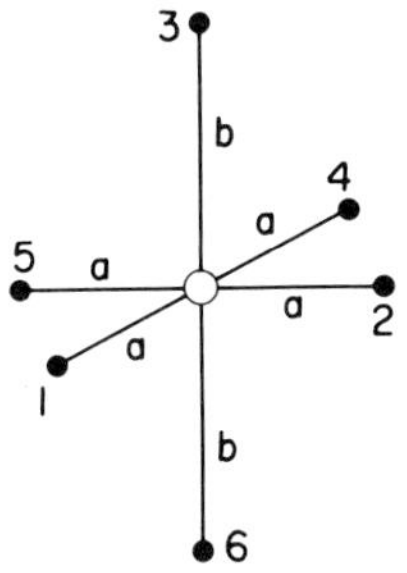

is illustrated in Fig. 1.7. This system is invariant under the symmetry operations shown in the accompanying tabulation.

Class	Symmetry operations
$\hat{E}$	E
$2\hat{C}_4$	$C_4(z)$, $C_4{}^3(z)$
$\hat{C}_4{}^2$	$C_4{}^2(z)$
$2\hat{C}_2$	$C_2(xy)$, $C_2(x\bar{y})$
$2\hat{C}_2{}'$	$C_2(x)$, $C_2(y)$.

The group having these eight symmetry operations is called D_4-group. It is evident that the D_4-group is a subgroup of the O-group. There are five classes in the D_4-group. Therefore, (1.71) in the present case can be written as

$$1^2 + 1^2 + 1^2 + 1^2 + 2^2 = 8, \tag{1.83}$$

which indicates that we have four one-dimensional and one two-dimensional irreducible representations in the D_4-group. The character table of this group has been obtained as shown in Table 1.5.

TABLE 1.5

CHARACTER TABLE OF THE D_4-GROUP

Bases	Irred. repres.	$\hat{E}$	$2\hat{C}_4$	$\hat{C}_4^2$	$2\hat{C}_2$	$2\hat{C}_2'$
φ_u	A_1	1	1	1	1	1
φ_z	A_2	1	1	1	—1	—1
φ_v	B_1	1	—1	1	—1	1
φ_ζ	B_2	1	—1	1	1	—1
(φ_x, φ_y) $(\varphi_\xi, \varphi_\eta)$	E	2	0	—2	0	0

Using Table 1.3 and Table 1.5, we obtain

$$D^{(E)} = D^{(A_1)} + D^{(B_1)},$$
$$D^{(T_2)} = D^{(B_2)} + D^{(E)}, \tag{1.84}$$

from a relation similar to (1.79). Therefore, by symmetry arguments only, one can predict that the e_g-level splits into two nondegenerate levels, and the t_{2g}-level into a nondegenerate level and doubly degenerate levels. This situation is visualized in Fig. 1.8. The reason why suffix g is attached to the irreducible representations of the D_4-group as well as for those for the O-group will be explained in the next subsection.

Problem 1.7. Let the distances between the hydrogen atom and point-charges 1, 2, and 3 in Fig. 1.1 be a, and those between the hydrogen atom and point-charges 4, 5, and 6 be b ($a \neq b$). Examine the splittings of the e_g and t_{2g} levels in this system. ◇

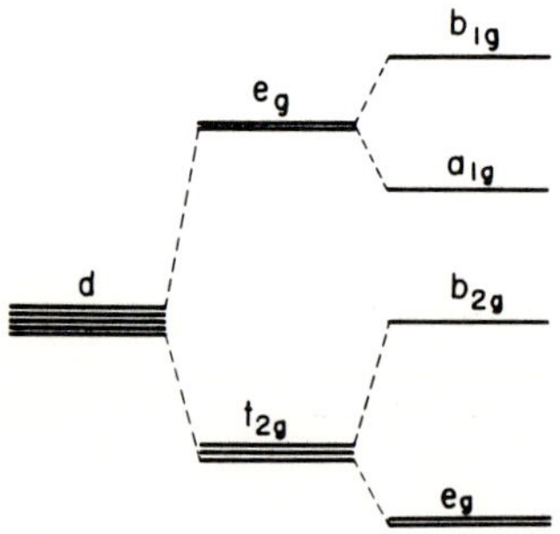

FIG. 1.8. Splitting of the *d*-level in a field of D_{4h} symmetry.

1.2.5 Inversion Symmetry

So far, for simplicity, we have confined ourselves to simple rotations. However, as mentioned in Section 1.2.1, the system illustrated in Fig. 1.1 is invariant under the other kinds of symmetry operations involving inversion I with respect to the center of the system. Inversion I may be represented as

$$I = \begin{pmatrix} 1 & 2 & 3 & 4 & 5 & 6 \\ 4 & 5 & 6 & 1 & 2 & 3 \end{pmatrix}, \tag{1.85}$$

so that clearly

$$I^2 = E. \tag{1.86}$$

Thus, two symmetry operations I and E form a group called C_i-group. It can further be shown that I commutes with all the operations of the O-group. For example, from (1.36) and (1.85)

$$\begin{aligned} IC_4(z) &= \begin{pmatrix} 1 & 2 & 3 & 4 & 5 & 6 \\ 4 & 5 & 6 & 1 & 2 & 3 \end{pmatrix}\begin{pmatrix} 1 & 2 & 3 & 4 & 5 & 6 \\ 2 & 4 & 3 & 5 & 1 & 6 \end{pmatrix} \\ &= \begin{pmatrix} 1 & 2 & 3 & 4 & 5 & 6 \\ 5 & 1 & 6 & 2 & 4 & 3 \end{pmatrix} \end{aligned} \tag{1.87a}$$

and

$$C_4(z)I = \begin{pmatrix} 1 & 2 & 3 & 4 & 5 & 6 \\ 5 & 1 & 6 & 2 & 4 & 3 \end{pmatrix}, \tag{1.87b}$$

which show

$$IC_4(z) = C_4(z)I\,. \tag{1.88}$$

In general, if two groups G_1 and G_2 have no common element besides the identity element and any element g_{1i} of G_1 commutes with any element g_{2j} of G_2, all the products $g_{1i}g_{2j}$ can be shown to form a group. This group G is called the *direct product* of groups G_1 and G_2, and is expressed as

$$G = G_1 \times G_2\,. \tag{1.89}$$

Problem 1.8. Prove that all the products $g_{1i}g_{2j}$ form a group. ◇

For the direct product one can show that, if $g_{1\mu}$ and $g_{1\nu}$ are the elements of the same class in G_1 connected by the transformation

$$g_{1\mu} = g_{1i}\,g_{1\nu}\,g_{1i}^{-1}, \tag{1.90}$$

and $g_{2\mu'}$ and $g_{2\nu'}$ are the elements of the same class in G_2 related as

$$g_{2\mu'} = g_{2j}\,g_{2\nu'}\,g_{2j}^{-1}, \tag{1.91}$$

$g_{1\mu}g_{2\mu'}$ and $g_{1\nu}g_{2\nu'}$ are the elements of the same class in G because of the relation

$$(g_{1i}\,g_{2j})\,g_{1\nu}\,g_{2\nu'}(g_{1i}\,g_{2j})^{-1} = g_{1\mu}\,g_{2\mu'}\,. \tag{1.92}$$

Therefore, the number of classes in G is the product $m_1 m_2$ of the number of classes in G_1, m_1, and that in G_2, m_2.

The matrix whose elements are given by

$$D^{(k_1k_2)}_{\mu\mu',\nu\nu'}(g_{1i}\,g_{2j}) = D^{(k_1)}_{\mu\nu}(g_{1i}) \times D^{(k_2)}_{\mu'\nu'}(g_{2j}) \tag{1.93}$$

is obviously a representation of G and has dimension $q = q_1 \times q_2$ if the dimensions of $D^{(k_1)}$ and $D^{(k_2)}$ are q_1 and q_2, respectively. And $D^{(k_1)}_{\mu\nu}(g_{1i})$ and $D^{(k_2)}_{\mu'\nu'}(g_{2j})$ are the matrix elements of the k_1 and k_2 irreducible representations of G_1 and G_2, respectively. Furthermore, $D^{(k_1k_2)}$ is irreducible, as the number of inequivalent irreducible representations of G exceeds the number of classes in G, $m_1 m_2$, if $D^{(k_1k_2)}$ is reducible.

Now, returning to our problem, since the O- and C_i-groups have no common element besides E and any element of the O-group commutes with that of the C_i-group, we can make a direct product,

$$O_h = O \times C_i\,. \tag{1.94}$$

Then, the O_h-group involves forty-eight elements and ten classes. By using the character table for the C_i-group which is easily obtained as shown in Table 1.6, the character table of the O_h-group is obtained as shown in Table 1.7.

In Table 1.6, the base of the irreducible representation g belongs to *even parity*, or the parity of this base is even, and the parity of the u base is *odd*. The parity of the bases of A_{1g}, A_{2g}, E_g, T_{1g}, and T_{2g} is even and that of A_{1u}, A_{2u}, E_u, T_{1u}, and T_{2u} is odd. The parity of the d-wavefunction is clearly even. Therefore, the bases (φ_u, φ_v) and $(\varphi_\xi, \varphi_\eta, \varphi_\zeta)$ are the bases of the E_g and T_{2g} irreducible representations of the O_h-group. This is the reason why we have used notation e_g and t_{2g} for the split components of the d-level. The p wavefunctions belong to the odd parity, so that $(\varphi_x, \varphi_y, \varphi_z)$ are the bases of T_{1u} in the O_h-group. The

TABLE 1.6

CHARACTER TABLE OF THE C_i-GROUP

Irred. repres.	$\hat{E}$	$\hat{I}$
g	1	1
u	1	—1

TABLE 1.7

CHARACTER TABLE OF THE O_h-GROUP[a]

Irred. repres.	$\hat{E}$	$6\hat{C}_4$	$3\hat{C}_4^2$	$8\hat{C}_3$	$6\hat{C}_2$	I	$6\widehat{IC}_4$	$3\widehat{IC}_4^2$	$8\widehat{IC}_3$	$6\widehat{IC}_2$
A_{1g}										
A_{2g}										
E_g			T					T		
T_{1g}										
T_{2g}										
A_{1u}										
A_{2u}										
E_u			T					$-T$		
T_{1u}										
T_{2u}										

[a] Here T is the character table given in Table 1.3, and $-T$ means that all the signs in Table 1.3 should be inverted.

system illustrated in Fig. 1.7 is also invariant under the inversion, and we can make a direct product,

$$D_{4h} = D_4 \times C_i \, . \tag{1.95}$$

We have attached suffix g to the irreducible representations of the D_4-group in illustrating Fig. 1.8, as their parity is even.

One of the important results obtained in this section is that the wavefunctions of the e_g and t_{2g} states are the bases of the irreducible representations E_g and T_{2g} of the O_h-group, respectively. This conclusion has been obtained without using any approximation. However, the result obtained in Section 1.1, that these wavefunctions are constructed from only the d-functions, is based on an approximate treatment, i.e., the perturbation method. Therefore, it may be mentioned that the e_g and t_{2g} wavefunctions could be some admixture of various atomic functions, for example, those with $l = 2, 4, 6$, etc., as long as they are the bases of the irreducible representations E_g and T_{2g}. In what follows we will use the notations of the wavefunctions e_g and t_{2g} merely to indicate that they are the bases of the irreducible representations.

In this section we have not explained all the topics of the group theory that is necessary for later discussions. For example, we have not mentioned double groups, Kronecker products, reflection symmetry, and so on. These topics will be discussed in the following chapters as they become necessary.

Chapter II TWO ELECTRONS IN A CUBIC FIELD

2.1 Formulation of the Two-Electron Problem

2.1.1 The Hamiltonian and Slater Determinant

In this chapter we shall be concerned with the problem of obtaining the eigenstates and the energy eigenvalues of the two-electron systems, in which two electrons are accomodated in the t_{2g} and e_g shells in various ways. Here the t_{2g} and e_g shells mean the aggregates of the degenerate one-electron orbitals denoted by t_{2g} and e_g, respectively, in the previous chapter. There are six and four ways of accommodating an electron in the t_{2g} and e_g shells, respectively.

The reason why these two shells are particularly dealt with is that they are in many cases the outermost shells in the systems of the iron group elements in a cubic field in which we are interested. In these systems the inner shells of the iron group ions are completely filled and neighboring atoms, ions, or molecules which are called *ligands* also have closed-shell configurations with energies lower than those of the t_{2g} and e_g shells. We call the electrons in both the inner shells of the central metal ion and the closed shells of the ligands simply the inner-shell electrons. In the following treatments we assume a picture of the electrons in the t_{2g} and e_g shells (valence electrons) moving and interacting with each other in a field coming from both the inner-shell electrons and the nuclei of the metal ion and the ligands. The potential energy of a valence electron due to this field is denoted by $V(\mathbf{r})$, without expressing the detailed form of $V(\mathbf{r})$. One point we know exactly about $V(\mathbf{r})$ is that it has cubic symmetry.

With this picture, the electron Hamiltonian of the system is given by

$$\mathscr{H} = f_1 + f_2 + g_{12}, \tag{2.1}$$

where

$$f_i = -\tfrac{1}{2}\Delta_i + V(\mathbf{r}_i) \qquad (i = 1, 2) \tag{2.2}$$

is a one-electron operator acting on electron i, and

$$g_{12} = 1/r_{12} \tag{2.3}$$

is a two-electron operator representing the Coulomb interaction between electrons 1 and 2. In these expressions the atomic units are employed. With the Hamiltonian in (2.1) the Schrödinger equation is given as

$$\mathscr{H}\Psi(\mathbf{r}_1\boldsymbol{\sigma}_1, \mathbf{r}_2\boldsymbol{\sigma}_2) = E\Psi(\mathbf{r}_1\boldsymbol{\sigma}_1, \mathbf{r}_2\boldsymbol{\sigma}_2), \tag{2.4}$$

where $\boldsymbol{\sigma}_i$ $(i = 1, 2)$ is the spin-coordinate of electron i and takes two values $\frac{1}{2}$ and $-\frac{1}{2}$. In order to solve this Schrödinger equation, we use the perturbation method in which g_{12} is assumed to be a small perturbation on the noninteracting electron system. This method is more than a mere approximation. It provides us not only with a good insight into the essence of the many-electron problem but also with an exact information of the solution concerning symmetry properties. In the perturbation treatment we split the Hamiltonian as

$$\mathscr{H} = \mathscr{H}_0 + \mathscr{H}_1, \tag{2.5}$$

where

$$\mathscr{H}_0 = f_1 + f_2 \qquad \text{and} \qquad \mathscr{H}_1 = g_{12}.$$

Then, we first solve the equation

$$\mathscr{H}_0\Psi_0(\mathbf{r}_1\boldsymbol{\sigma}_1, \mathbf{r}_2\boldsymbol{\sigma}_2) = E_0\Psi_0(\mathbf{r}_1\boldsymbol{\sigma}_1, \mathbf{r}_2\boldsymbol{\sigma}_2). \tag{2.6}$$

The solution of (2.6) is easily obtained if the solution of the following equation for a single electron is known:

$$f\phi_k(\mathbf{r}\boldsymbol{\sigma}) = \epsilon_k\phi_k(\mathbf{r}\boldsymbol{\sigma}), \qquad \epsilon_k = \langle\phi_k \,|\, f \,|\, \phi_k\rangle. \tag{2.7}$$

Since operator f involves no spin-coordinate, $\phi_k(\mathbf{r}\boldsymbol{\sigma})$ is given by the product of orbital function $\varphi_\mu(\mathbf{r})$ and spin function $\alpha(\boldsymbol{\sigma})$ or $\beta(\boldsymbol{\sigma})$, and is called a *spin-orbital*. Here $\varphi_\mu(\mathbf{r})$ is the eigenfunction of operator f which involves electron space-coordinate only, and spin functions $\alpha(\boldsymbol{\sigma})$ and $\beta(\boldsymbol{\sigma})$ are defined as follows:

$$\alpha(\sigma = \tfrac{1}{2}) = 1, \qquad \alpha(\sigma = -\tfrac{1}{2}) = 0, \tag{2.8a}$$

$$\beta(\sigma = \tfrac{1}{2}) = 0, \qquad \beta(\sigma = -\tfrac{1}{2}) = 1. \tag{2.8b}$$

By using the solution of (2.7), the eigenfunction and eigenvalue of (2.6) are given as

$$\Psi_{0kl}(\mathbf{r}_1\boldsymbol{\sigma}_1, \mathbf{r}_2\boldsymbol{\sigma}_2) = \phi_k(\mathbf{r}_1\boldsymbol{\sigma}_1)\,\phi_l(\mathbf{r}_2\boldsymbol{\sigma}_2), \tag{2.9a}$$

$$E_{0kl} = \epsilon_k + \epsilon_l\,. \tag{2.9b}$$

According to the Pauli principle, wavefunctions for electrons should be antisymmetric with respect to the exchange of electrons. By using the mathematical properties of determinants, the antisymmetric wavefunction of (2.9) can be written as

$$\Psi^A_{0kl}(\mathbf{r}_1\boldsymbol{\sigma}_1, \mathbf{r}_2\boldsymbol{\sigma}_2) = \frac{1}{\sqrt{2}}\begin{vmatrix} \phi_k(\mathbf{r}_1\boldsymbol{\sigma}_1) & \phi_l(\mathbf{r}_1\boldsymbol{\sigma}_1) \\ \phi_k(\mathbf{r}_2\boldsymbol{\sigma}_2) & \phi_l(\mathbf{r}_2\boldsymbol{\sigma}_2) \end{vmatrix}$$

$$\equiv |\,\phi_k\phi_l\,|. \tag{2.10}$$

This determinant is called *Slater determinant.* Factor $1/\sqrt{2}$ is normalization factor. Hereafter, we leave out superscript A of the wavefunction for simplicity and often use the abbreviation given in the last expression of (2.10). This abbreviation includes the normalization factor. Of course, one finds

$$|\,\phi_k\phi_l\,| = -\,|\,\phi_l\phi_k\,|. \tag{2.11}$$

2.1.2 Terms

Let us first consider the states obtained by accommodating two electrons in the t_{2g} shell. It has already been known that the orbital-functions associated with this shell are $\varphi_\xi(\mathbf{r})$, $\varphi_\eta(\mathbf{r})$, and $\varphi_\zeta(\mathbf{r})$, which will simply be written as $\xi(\mathbf{r})$, $\eta(\mathbf{r})$, and $\zeta(\mathbf{r})$, respectively. Since two spin-functions $\alpha(\boldsymbol{\sigma})$ and $\beta(\boldsymbol{\sigma})$ are available, we have six spin-orbitals available in this shell; $\xi(\mathbf{r})\,\alpha(\boldsymbol{\sigma})$, $\xi(\mathbf{r})\,\beta(\boldsymbol{\sigma})$, $\eta(\mathbf{r})\,\alpha(\boldsymbol{\sigma})$,... . Therefore, there are six ways of placing the first electron in the shell. However, according to Pauli principle, the second electron cannot be placed in the spin-orbital where the first one is already accommodated, so that the number of ways of accommodating two electrons is given by ${}_6C_2 = 6!/4!2! = 15$. Abbreviating spin-orbitals, for example, $\xi\alpha$ and $\xi\beta$ as ξ and $\bar{\xi}$, respectively, the Slater determinants corresponding to these fifteen states are given as

$$|\,\xi\eta\,|,\quad |\,\bar{\xi}\bar{\eta}\,|,\quad |\,\eta\zeta\,|,\quad |\,\bar{\eta}\bar{\zeta}\,|,\quad |\,\zeta\xi\,|,\quad |\,\bar{\zeta}\bar{\xi}\,|, \tag{2.12a}$$

$$|\,\xi\bar{\eta}\,|,\quad |\,\bar{\xi}\eta\,|,\quad |\,\eta\bar{\zeta}\,|,\quad |\,\bar{\eta}\zeta\,|,\quad |\,\zeta\bar{\xi}\,|,\quad |\,\bar{\zeta}\xi\,|, \tag{2.12b}$$

$$|\,\xi\bar{\xi}\,|,\quad |\,\eta\bar{\eta}\,|,\quad |\,\zeta\bar{\zeta}\,|. \tag{2.12c}$$

These states are visualized in Fig. 2.1.

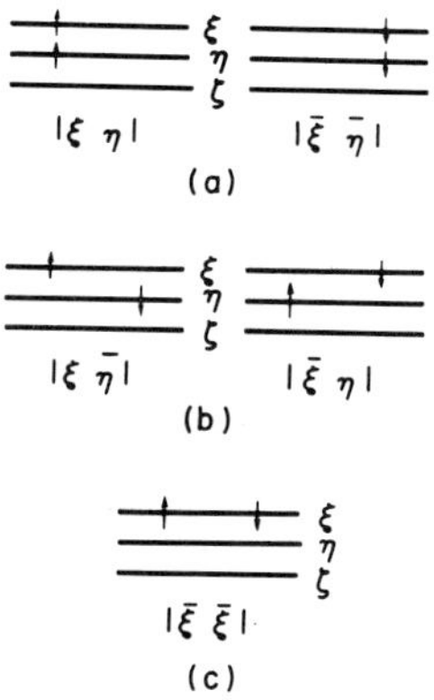

FIG. 2.1. Visualized Slater determinants.

Since the ξ, η, and ζ orbitals are degenerate, one has

$$\langle \xi \,|\, f \,|\, \xi \rangle = \langle \eta \,|\, f \,|\, \eta \rangle = \langle \zeta \,|\, f \,|\, \zeta \rangle. \tag{2.13}$$

By using (2.13) one can show that, as long as the Coulomb interaction between the two electrons is neglected, the fifteen states in (2.12a, b, c) have the same energy. Therefore, in this case we have one energy level with fifteenfold degeneracy. In what follows, this high degeneracy will be shown to be partially removed by the effect of the Coulomb interaction.

According to the perturbation theory, the perturbed energy levels and the wavefunctions associated with them are obtained by diagonalizing the 15 × 15 matrix of $\mathscr{H}_1$ whose base functions are those Slater determinants given in (2.12). This procedure is quite similar to that used in Section 1.1 for calculating the splitting of the d-level and the wavefunctions, and it is greatly simplified with the aid of the group theory as discussed in Section 1.2.

Let us first note that the total Hamiltonian $\mathscr{H}_0 + \mathscr{H}_1$ is invariant under the transformation by R of the O_h-group. According to the group theory, the eigenfunctions associated with a certain energy level of this system are the bases of an irreducible representation, say Γ, of the O_h-group, and the energy level may be labeled as Γ. The nondiagonal matrix elements of the Hamiltonian between the states with different Γ's are all vanishing. Therefore, if we find the linear combinations of the Slater determinants given in (2.12) which are the bases of irreducible representations, the matrix with these bases would already be partially diagonal. The remaining task is to diagonalize the matrix of a smaller dimension which is given between the states with the same Γ.

So far we have considered the orbital part only. Now we will consider the spin part. Since the Hamiltonian involves no spin operator, it commutes with $\mathbf{S}$, consequently $\mathbf{S}^2$, where $\mathbf{S}$ is the resultant spin angular momentum $\mathbf{S} = \mathbf{s}_1 + \mathbf{s}_2$. This means that $\mathbf{S}^2$ is a constant of motion and has a definite value $S(S+1)$ in the eigenstate.

As a whole the energy levels are characterized by $S\Gamma$ and have $(2S+1) \times (\Gamma)$-fold degeneracy. Here, (Γ) represents the dimension of irreducible representation Γ. Degeneracy $(2S+1)$ occurs because the eigenvalues M_s of S_z in the state with $\mathbf{S}^2 = S(S+1)$ are $S, S-1, \ldots, -S+1, -S$ and the energy eigenvalues are independent of these values of M_s. The energy levels characterized by $S\Gamma$ are called *terms* and are denoted by ${}^{2S+1}\Gamma$. The $(2S+1)(\Gamma)$ wavefunctions of the ${}^{2S+1}\Gamma$ term are expressed as $\Psi(\alpha S\Gamma M\gamma)$, in which M is the abbreviation of M_s, γ denotes one of the bases of irreducible representation Γ, for example $\gamma = \xi, \eta$, and ζ for $\Gamma = T_{2g}$, and α is some quantum number which distinguishes the states with the same $S\Gamma$.

Once one finds a suitable linear combination of Slater determinants to express $\Psi(\alpha S\Gamma M\gamma)$, the matrix of $\mathscr{H}_1$ calculated with these wavefunctions is partially diagonalized: All the matrix elements between the states with different $S\Gamma$ are zero. Therefore, in the next section we will discuss the method of constructing $\Psi(\alpha S\Gamma M\gamma)$ from Slater determinants, which satisfies the following requirements: First, for an arbitrary operation R of the O_h-group, it must be transformed in the same way as the bases of irreducible representation $D^{(\Gamma)}$:

$$R\Psi(\alpha S\Gamma M\gamma) = \sum_{\gamma'} \Psi(\alpha S\Gamma M\gamma') \, D^{(\Gamma)}_{\gamma'\gamma}(R). \tag{2.14}$$

Second, it must satisfy the equations

$$\mathbf{S}^2\Psi(\alpha S\Gamma M\gamma) = S(S+1)\,\Psi(\alpha S\Gamma M\gamma), \tag{2.15}$$

$$S_z\Psi(\alpha S\Gamma M\gamma) = M\Psi(\alpha S\Gamma M\gamma). \tag{2.16}$$

Finally, it must be antisymmetric with respect to electron exchanges.

2.2 Two-Electron Wavefunctions

As preliminaries for constructing $\Psi(\alpha S\Gamma M\gamma)$ from Slater determinants, the first few subsections will be devoted to deriving the bases of an irreducible representation by linear combinations of the products of orbital functions, and also to obtaining the eigenfunction of $\mathbf{S}^2$ and S_z by linear combinations of the products of spin functions.

2.2.1 Product Representations

Let $\varphi(\Gamma_1\gamma_1)$ and $\varphi(\Gamma_2\gamma_2)$ be the bases of irreducible representations Γ_1 and Γ_2, respectively. By operation R of a group, they are transformed to

$$R\varphi(\Gamma_1\gamma_1') = \sum_{\gamma_1} \varphi(\Gamma_1\gamma_1)\, D^{(\Gamma_1)}_{\gamma_1\gamma_1'}(R),$$

$$R\varphi(\Gamma_2\gamma_2') = \sum_{\gamma_2} \varphi(\Gamma_2\gamma_2)\, D^{(\Gamma_2)}_{\gamma_2\gamma_2'}(R). \tag{2.17}$$

Then, how are the $(\Gamma_1) \times (\Gamma_2)$ products of $\varphi(\Gamma_1\gamma_1)$ and $\varphi(\Gamma_2\gamma_2)$ transformed by operation R? In this case variables in $\varphi(\Gamma_1\gamma_1)$ and $\varphi(\Gamma_2\gamma_2)$ may be different from each other, but operation R is applied to both variables at the same time. From (2.17) it is clear that

$$\begin{aligned} R[\varphi(\Gamma_1\gamma_1')\,\varphi(\Gamma_2\gamma_2')] &= R\varphi(\Gamma_1\gamma_1')\,R\varphi(\Gamma_2\gamma_2') \\ &= \sum_{\gamma_1\gamma_2} \varphi(\Gamma_1\gamma_1)\,\varphi(\Gamma_2\gamma_2)\, D^{(\Gamma_1)}_{\gamma_1\gamma_1'}(R)\, D^{(\Gamma_2)}_{\gamma_2\gamma_2'}(R). \end{aligned} \tag{2.18a}$$

For simplicity, we will use the following abbreviations:

$$\psi(\nu) = \varphi(\Gamma_1\gamma_1)\,\varphi(\Gamma_2\gamma_2), \tag{2.19a}$$

$$D_{\nu\nu'}(R) = D^{(\Gamma_1)}_{\gamma_1\gamma_1'}(R)\, D^{(\Gamma_2)}_{\gamma_2\gamma_2'}(R), \tag{2.19b}$$

where ν represents various combinations of γ_1 and γ_2. By using (2.19), (2.18a) is simplified to

$$R\psi(\nu') = \sum_{\nu} \psi(\nu)\, D_{\nu\nu'}(R), \tag{2.18b}$$

which indicates that $D_{\nu\nu'}(R)$ are matrix elements of a representation of the group. Actually, for the elements satisfying $QR = T$, one can show that

$$\sum_{\mu} D_{\nu\mu}(Q)\, D_{\mu\nu'}(R) = \sum_{\delta_1\delta_2} D^{(\Gamma_1)}_{\gamma_1\delta_1}(Q)\, D^{(\Gamma_2)}_{\gamma_2\delta_2}(Q)\, D^{(\Gamma_1)}_{\delta_1\gamma_1'}(R)\, D^{(\Gamma_2)}_{\delta_2\gamma_2'}(R)$$

$$= D^{(\Gamma_1)}_{\gamma_1\gamma_1'}(QR)\, D^{(\Gamma_2)}_{\gamma_2\gamma_2'}(QR)$$

$$= D^{(\Gamma_1)}_{\gamma_1\gamma_1'}(T)\, D^{(\Gamma_2)}_{\gamma_2\gamma_2'}(T)$$

$$= D_{\nu\nu'}(T). \tag{2.20}$$

This representation is called the *product representation* or *Kronecker product* of Γ_1 and Γ_2 and is often expressed simply as $\Gamma_1 \times \Gamma_2$. The dimension of this representation is given by $(\Gamma_1) \times (\Gamma_2)$. It is important to note that, in general, the product representation is reducible. This may be seen in the following example: Let both Γ_1 and Γ_2 be T_{1u} of the O_h-group and their bases be (x_1, y_1, z_1) and (x_2, y_2, z_2), respectively. One may think that these bases are the coordinates of electrons 1 and 2. From nine bases of the product representation, x_1x_2, x_1y_2,..., z_1z_2, one can make a linear combination,

$$(1/\sqrt{3})(x_1x_2 + y_1y_2 + z_1z_2) = (1/\sqrt{3})(\mathbf{r}_1 \cdot \mathbf{r}_2).$$

Since this is a scalar, it is not changed by any operation of the O_h-group. Therefore, this linear combination is the base of the A_{1g} representation. One can also make linear combinations, $(1/\sqrt{2})(y_1z_2 - z_1y_2)$, $(1/\sqrt{2})(z_1x_2 - x_1z_2)$, $(1/\sqrt{2})(x_1y_2 - y_1x_2)$. Since these are the components of vector $(1/\sqrt{2})(\mathbf{r}_1 \times \mathbf{r}_2)$, they are obviously the bases of the T_{1g} representation. The remaining linear combinations,

$$[(1/\sqrt{2})(y_1z_2 + z_1y_2),\quad (1/\sqrt{2})(z_1x_2 + x_1z_2),\quad (1/\sqrt{2})(x_1y_2 + y_1x_2)]$$

and

$$[(1/\sqrt{6})(3z_1z_2 - \mathbf{r}_1 \cdot \mathbf{r}_2),\qquad (1/\sqrt{2})(x_1x_2 - y_1y_2)]$$

may be seen as the bases of T_{2g} and E_g, respectively, by comparing them with the t_{2g} and e_g wavefunctions given in (1.31) and (1.32). Of course, one may confirm this by applying the symmetry operations of the O_h-group and calculating the characters. After all, we have seen that the product representation $T_{1u} \times T_{1u}$ can be reduced to four irreducible representations: A_{1g}, E_g, T_{1g}, and T_{2g}.

The reduction prodcedure of product representation $T_{1u} \times T_{1u}$

may be summarized as follows: We apply unitary transformation $\mathbf{U}$ to nine bases $\psi(\nu)$'s of the product representation, and obtain

$$\psi(\Gamma\gamma) = \sum \psi(\nu)\, U_{\nu,\Gamma\gamma} \tag{2.21}$$

which are classified into the sets of the bases of irreducible representations $\Gamma(\Gamma = A_{1g}, E_g, T_{1g}, T_{2g})$. When $\psi(\Gamma\gamma)$'s are used as the new bases of the product representation, the matrix of the product representation is given by

$$\mathbf{U}^{-1}\mathbf{D}(R)\mathbf{U} = \left[\begin{array}{c|c|c|c} \mathbf{D}^{(A_{1g})}(R) & 0 & 0 & 0 \\ \hline 0 & \mathbf{D}^{(E_g)}(R) & 0 & 0 \\ \hline 0 & 0 & \mathbf{D}^{(T_{1g})}(R) & 0 \\ \hline 0 & 0 & 0 & \mathbf{D}^{(T_{2g})}(R) \end{array}\right], \tag{2.22}$$

where four matrices of the irreducible representations are located on the diagonal. In our example, the unitary transformation was found intuitively.

For merely finding which irreducible representations are obtained by reducing a product representation, it is unnecessary to know the unitary transformation: Instead, we use the relation for characters,

$$\begin{aligned} \chi(R) = \sum_{\nu} D_{\nu\nu}(R) &= \sum_{\gamma_1\gamma_2} D^{(\Gamma_1)}_{\gamma_1\gamma_1}(R)\, D^{(\Gamma_2)}_{\gamma_2\gamma_2}(R) \\ &= \chi^{(\Gamma_1)}(R)\, \chi^{(\Gamma_2)}(R). \end{aligned} \tag{2.23}$$

In our example of $T_{1u} \times T_{1u}$, the characters for the product representation are given from Table 1.7 as shown in the tabulation. By using

	$\hat{I}$ $\hat{E}$	$6\hat{I}\hat{C}_4$ $6\hat{C}_4$	$3\hat{I}\hat{C}_4^{\,2}$ $3\hat{C}_4^{\,2}$	$8\hat{I}\hat{C}_3$ $8\hat{C}_3$	$6\hat{I}\hat{C}_2$ $6\hat{C}_2$
$\chi(\hat{R})$	9	1	1	0	1

(1.79) in which $\chi^{(l)}(\hat{R})$ are replaced by these $\chi(\hat{R})$, one can show that irreducible representations A_{1g}, E_g, T_{1g}, and T_{2g} appear when $T_{1u} \times T_{1u}$ is reduced. This reduction is simply expressed as

$$T_{1u} \times T_{1u} = A_{1g} + E_g + T_{1g} + T_{2g}\,. \tag{2.24}$$

Table 2.1 shows how all the product representations of the O-group are reduced. For the O_h-group, suffices g and u should be attached to the irreducible representations in Table 2.1 according to the rules

$$g \times g = g, \qquad g \times u = u, \qquad u \times u = g. \tag{2.25}$$

TABLE 2.1

$\Gamma_1 \times \Gamma_2 = \Sigma \Gamma_i$ FOR THE O-GROUP

$\Gamma_2 \backslash \Gamma_1$	A_1	A_2	E	T_1	T_2
A_1	A_1	A_2	E	T_1	T_2
A_2		A_1	E	T_2	T_1
E			$A_1 + A_2 + E$	$T_1 + T_2$	$T_1 + T_2$
T_1				$A_1 + E + T_1 + T_2$	$A_2 + E + T_1 + T_2$
T_2					$A_1 + E + T_1 + T_2$

Problem 2.1. Derive Table 2.1. ◇

2.2.2 Clebsch–Gordan Coefficients

To obtain the wavefunctions and the energy matrix in the two-electron system, we have to find the unitary transformation in (2.21). Matrix elements $U_{\nu,\Gamma\gamma}$'s are usually denoted by

$$U_{\nu,\Gamma\gamma} = \langle \Gamma_1\gamma_1\Gamma_2\gamma_2 \mid \Gamma\gamma \rangle, \tag{2.26}$$

and are called *Clebsch–Gordan coefficients.* Since $\mathbf{U}$ is unitary, Clebsch–Gordan (C–G) coefficients must satisfy the following relations:

$$\begin{aligned}(U^{-1})_{\Gamma\gamma,\nu} &= \langle \Gamma\gamma \mid \Gamma_1\gamma_1\Gamma_2\gamma_2 \rangle \\ &= \langle \Gamma_1\gamma_1\Gamma_2\gamma_2 \mid \Gamma\gamma \rangle^*, \end{aligned} \tag{2.27}$$

$$\sum_{\gamma_1\gamma_2} \langle \Gamma\gamma \mid \Gamma_1\gamma_1\Gamma_2\gamma_2 \rangle \langle \Gamma_1\gamma_1\Gamma_2\gamma_2 \mid \Gamma'\gamma' \rangle = \delta(\Gamma\Gamma')\, \delta(\gamma\gamma'), \tag{2.28}$$

$$\sum_{\Gamma\gamma} \langle \Gamma_1\gamma_1\Gamma_2\gamma_2 \mid \Gamma\gamma \rangle \langle \Gamma\gamma \mid \Gamma_1\gamma_1'\Gamma_2\gamma_2' \rangle = \delta(\gamma_1\gamma_1')\, \delta(\gamma_2\gamma_2'). \tag{2.29}$$

In the example of $T_{1u} \times T_{1u}$ a unitary transformation was introduced in order to satisfy (2.22), which in terms of C–G coefficients is expressed for $\Gamma_1 \times \Gamma_2$ as

$$\sum_{\substack{\gamma_1\gamma_1 \\ \gamma_1'\gamma_2'}} \langle \Gamma\gamma \mid \Gamma_1\gamma_1\Gamma_2\gamma_2 \rangle\, D^{(\Gamma_1)}_{\gamma_1\gamma_1'}(R)\, D^{(\Gamma_2)}_{\gamma_2\gamma_2'}(R) \langle \Gamma_1\gamma_1'\Gamma_2\gamma_2' \mid \Gamma'\gamma' \rangle = D^{(\Gamma)}_{\gamma\gamma'}(R)\, \delta(\Gamma\Gamma'). \tag{2.30}$$

Furthermore, by using (2.29), (2.30) can be reexpressed as

$$\sum_{\gamma_1'\gamma_2'} D^{(\Gamma_1)}_{\gamma_1\gamma_1'}(R)\, D^{(\Gamma_2)}_{\gamma_2\gamma_2'}(R) \langle \Gamma_1\gamma_1'\Gamma_2\gamma_2' \mid \Gamma\gamma' \rangle = \sum_{\gamma} \langle \Gamma_1\gamma_1\Gamma_2\gamma_2 \mid \Gamma\gamma \rangle\, D^{(\Gamma)}_{\gamma\gamma'}(R). \tag{2.31}$$

This equation may be compared with (2.21) which in terms of C–G coefficients is given as

$$\sum_{\gamma_1\gamma_2} \varphi(\Gamma_1\gamma_1)\, \varphi(\Gamma_2\gamma_2) \langle \Gamma_1\gamma_1\Gamma_2\gamma_2 \mid \Gamma\gamma \rangle = \psi(\Gamma\gamma). \tag{2.32}$$

Since matrix elements of the irreducible representations appearing in (2.31) are known from symmetry arguments, coupled equations (2.31) with various operations R determine C–G coefficients. In this case it is sufficient to set the equations for only generating elements. The reason is explained in Problem 2.2.

Problem 2.2. Show that, if (2.31) is satisfied for R_1 and R_2, it is also satisfied for $R = R_1R_2$. ◇

As an example of calculating C–G coefficients from (2.31) let us consider the case of $\Gamma_1 = \Gamma_2 = \Gamma = E$ of the O-group. We use generating elements $C_4(z)$ and $C_4(x)$ whose representation matrices $D^{(E)}$ are

$$D^{(E)}(C_4(z)) = \begin{matrix} u \\ v \end{matrix} \begin{bmatrix} 1 & 0 \\ 0 & -1 \end{bmatrix}, \tag{2.33a}$$

$$D^{(E)}(C_4(x)) = \begin{matrix} u \\ \\ v \end{matrix} \begin{bmatrix} -\dfrac{1}{2} & -\dfrac{\sqrt{3}}{2} \\ -\dfrac{\sqrt{3}}{2} & \dfrac{1}{2} \end{bmatrix}. \tag{2.33b}$$

For $R = C_4(z)$, (2.31) shows that nonvanishing C–G coefficients are only $\langle EuEu \mid Eu \rangle$, $\langle EvEv \mid Ev \rangle$, $\langle EuEv \mid Ev \rangle$, and $\langle EvEu \mid Ev \rangle$: for instance, for $R = C_4(z)$, $\gamma_1 = \gamma_2 = u$, and $\gamma' = v$, (2.31) gives

$$\langle EuEu \mid Ev \rangle = -\langle EuEu \mid Ev \rangle. \tag{2.34}$$

For $R = C_4(x)$ and $\gamma_1 = \gamma_2 = \gamma' = u$, by using the result just mentioned, (2.31) is written as

$$\left(-\frac{1}{2}\right)\left(-\frac{1}{2}\right) \langle EuEu \mid Eu \rangle + \left(-\frac{\sqrt{3}}{2}\right)\left(-\frac{\sqrt{3}}{2}\right) \langle EvEv \mid Eu \rangle$$
$$= \langle EuEu \mid Eu \rangle \left(-\frac{1}{2}\right), \tag{2.35}$$

from which the following relation is obtained:

$$\langle EuEu \mid Eu\rangle = -\langle EvEv \mid Eu\rangle. \tag{2.36}$$

For $R = C_4(x)$, $\gamma_1 = u$, and $\gamma_2 = \gamma' = v$, (2.31) is

$$\begin{aligned}\left(-\frac{1}{2}\right)\left(\frac{1}{2}\right)\langle EuEv \mid Ev\rangle + \left(-\frac{\sqrt{3}}{2}\right)\left(-\frac{\sqrt{3}}{2}\right)\langle EvEu \mid Ev\rangle \\ = \langle EuEv \mid Ev\rangle\left(\frac{1}{2}\right),\end{aligned} \tag{2.37}$$

from which

$$\langle EuEv \mid Ev\rangle = \langle EvEu \mid Ev\rangle. \tag{2.38}$$

Furthermore, for $R = C_4(x)$, $\gamma_1 = \gamma_2 = u$, and $\gamma' = v$, (2.31) gives

$$\langle EuEv \mid Ev\rangle + \langle EvEu \mid Ev\rangle = -2\langle EuEu \mid Eu\rangle. \tag{2.39}$$

Results in (2.36), (2.38), and (2.39) are summarized as follows:

$$\begin{aligned}-\langle EuEu \mid Eu\rangle &= \langle EvEv \mid Eu\rangle \\ &= \langle EuEv \mid Ev\rangle = \langle EvEu \mid Ev\rangle.\end{aligned} \tag{2.40}$$

From (2.40) and the requirements for a unitary matrix, C–G coefficients are obtained as shown in Appendix II, where all the C–G coefficients for the O-group are listed. In the Appendix the bases of T_1 are denoted by α, β, and γ. As seen in this appendix, when $\Gamma_1 = \Gamma_2$, we have the relations

$$\langle \Gamma_1\gamma_1\Gamma_1\gamma_2 \mid \Gamma\gamma\rangle = \epsilon(\Gamma_1\Gamma_1\Gamma)\langle \Gamma_1\gamma_2\Gamma_1\gamma_1 \mid \Gamma\gamma\rangle, \tag{2.41}$$

where

$$\begin{aligned}\epsilon(\Gamma_1\Gamma_1\Gamma) = -1 \quad &\text{for} \quad \Gamma_1 = E \quad \text{and} \quad \Gamma = A_2, \\ &\phantom{\text{for}} \quad \Gamma_1 = T_1 \quad \text{and} \quad \Gamma = T_1, \quad \text{and} \\ &\phantom{\text{for}} \quad \Gamma_1 = T_2 \quad \text{and} \quad \Gamma = T_1, \\ \epsilon(\Gamma_1\Gamma_1\Gamma) = 1 \quad &\text{otherwise.}\end{aligned}$$

When $\Gamma_1 \neq \Gamma_2$, we use the phase conventions

$$\langle \Gamma_2\gamma_2\Gamma_1\gamma_1 \mid \Gamma\gamma\rangle = \langle \Gamma_1\gamma_1\Gamma_2\gamma_2 \mid \Gamma\gamma\rangle, \tag{2.42}$$

so that only half the C–G coefficients for $\Gamma_1 \neq \Gamma_2$ are listed in the appendix. For the O_h-group it is sufficient to attach suffices g and u to Γ according to the rules given in (2.25). By using these C–G coefficients the base functions of irreducible representations can be constructed from linear combinations of the products of two base functions of irreducible representations as shown in (2.32).

In concluding this subsection, it is worth pointing out that sometimes one may calculate C–G coefficients in a simple fashion without using (2.31). The previously mentioned arguments on reducing $T_{1u} \times T_{1u}$ is one of the cases. One of the other cases is seen in the derivation (see Problem 2.3) of C–G coefficients for $T_1 \times T_2$ from those for $T_1 \times T_1$ by using the fact that the bases of T_2 are given by the products of the bases of A_2 and T_1 as $e_2\alpha$, $e_2\beta$, and $e_2\gamma$.

Problem 2.3. Derive C–G coefficients for $T_1 \times T_2$ and $T_2 \times T_2$ from those for $T_1 \times T_1$. ◇

Problem 2.4. Using (2.31) calculate C–G coefficients for $A_2 \times E$ and $E \times T_1$. ◇

2.2.3 Wigner Coefficients

In this subsection, we simply give the well-known formula for constructing the eigenfunctions of $\mathbf{S}^2$ and S_z in terms of the products of two spin functions $\theta(s_1m_1)$ and $\theta(s_2m_2)$. Here, $\mathbf{S} = \mathbf{s}_1 + \mathbf{s}_2$ and $\theta(s_1m_1)$ and $\theta(s_2m_2)$ are the eigenfunctions of $\mathbf{s}_1^2$, s_{1z}, and $\mathbf{s}_2^2$, s_{2z}, respectively.

The eigenfunctions $\Theta(SM)$ of $\mathbf{S}^2$ with $S = s_1 + s_2$, $s_1 + s_2 - 1, \ldots,$ $|s_1 - s_2|$, and S_z with its eigenvalue M are given in terms of the product of two spin functions as follows:

$$\Theta(SM) = \sum_{m_1m_2} \theta(s_1m_1)\,\theta(s_2m_2)\langle s_1m_1s_2m_2 \mid SM\rangle, \tag{2.43}$$

in which coefficients of the linear combination, $\langle s_1m_1s_2m_2 \mid SM\rangle$, are called *Wigner coefficients.*‡ The transformation given in (2.43) is unitary so that Wigner coefficients satisfy the following relations:

$$\langle SM \mid s_1m_1s_2m_2\rangle = \langle s_1m_1s_2m_2 \mid SM\rangle^*, \tag{2.44}$$

$$\sum_{m_1m_2} \langle SM \mid s_1m_1s_2m_2\rangle\langle s_1m_1s_2m_2 \mid S'M'\rangle = \delta(SS')\,\delta(MM'), \tag{2.45}$$

$$\sum_{SM} \langle s_1m_1s_2m_2 \mid SM\rangle\langle SM \mid s_1m_1's_2m_2'\rangle = \delta(m_1m_1')\,\delta(m_2m_2'). \tag{2.46}$$

Wigner coefficients are nonvanishing only when $M = m_1 + m_2$ and $|s_1 - s_2| \leqslant S \leqslant s_1 + s_2$ are satisfied. A general formula for calculating

‡ Sometimes these coefficients are also called Clebsch–Gordan coefficients or vector-coupling coefficients.

Wigner coefficients with arbitrary values of parameters involved is given in Appendix III. Further, simplified formulas for calculating the Wigner coefficients with $s_2 = \frac{1}{2}, 1, \frac{3}{2}, 2$ are also given in Appendix III. From the general formula, we see the relation

$$\langle s_2 m_2 s_1 m_1 \mid SM \rangle = (-1)^{s_1+s_2-S} \langle s_1 m_1 s_2 m_2 \mid SM \rangle. \tag{2.47}$$

In the present two-electron problem, $s_1 = s_2 = \frac{1}{2}$ and the $\Theta(SM)$'s are given as

$$\begin{aligned} \Theta(S{=}1, M{=}1) &= \alpha(\boldsymbol{\sigma}_1)\,\alpha(\boldsymbol{\sigma}_2), \\ \Theta(S{=}1, M{=}0) &= \frac{1}{\sqrt{2}}\,[\alpha(\boldsymbol{\sigma}_1)\,\beta(\boldsymbol{\sigma}_2) + \beta(\boldsymbol{\sigma}_1)\,\alpha(\boldsymbol{\sigma}_2)], \\ \Theta(S{=}1, M{=}-1) &= \beta(\boldsymbol{\sigma}_1)\,\beta(\boldsymbol{\sigma}_2), \end{aligned} \tag{2.48a}$$

$$\Theta(S{=}0, M{=}0) = \frac{1}{\sqrt{2}}\,[\alpha(\boldsymbol{\sigma}_1)\,\beta(\boldsymbol{\sigma}_2) - \beta(\boldsymbol{\sigma}_1)\,\alpha(\boldsymbol{\sigma}_2)], \tag{2.48b}$$

where $\alpha(\boldsymbol{\sigma}) = \theta(\frac{1}{2}\,\frac{1}{2})$ and $\beta(\sigma) = \theta(\frac{1}{2} - \frac{1}{2})$ according to the definition in (2.8).

2.2.4 Wavefunctions

Now we are in a position to construct the two-electron wavefunctions by using the results obtained in the previous subsections. There are two methods of constructing many-electron wavefunctions. One of them is to start from the products of one-electron spin-orbitals and make their linear combination to be the base of irreducible representation Γ and the eigenfunction of $\mathbf{S}^2$ and S_z. However, since the linear combination is not always antisymmetric with respect to the electron exchange, we make it antisymmetric afterward in order to obtain the two-electron wavefunction. This method will be used in the next chapter to obtain many-electron wavefunctions. Another method is to start from Slater determinants which are already antisymmetric. Since they are neither bases of irreducible representations nor the eigenfunctions of $\mathbf{S}^2$, we have to find a suitable linear combination of Slater determinants to make it satisfy requirements of (2.14), (2.15), and (2.16). We use this method in this subsection.

In order to use the latter method, it is necessary to know how Slater determinants behave when the rotation and spin operators are operated on them. Let us denote one-electron spin-orbital $\varphi(t_2\gamma)\,\theta(\frac{1}{2}m)$ in the t_{2g} shell as $\phi(t_2 m\gamma)$. For simplicity, hereafter subscript g of t_{2g} and e_g

will be left out when no confusion occurs. By using properties of determinants, it is easy to show that, if we have

$$R\phi(t_2m\gamma') = \sum_{\gamma} \phi(t_2m\gamma)\, D^{(T_2)}_{\gamma\gamma'}(R), \tag{2.49}$$

we obtain

$$R|\,\phi(t_2m_1\gamma_1')\,\phi(t_2m_2\gamma_2')| = \sum_{\gamma_1\gamma_2} |\,\phi(t_2m_1\gamma_1)\,\phi(t_2m_2\gamma_2)|\, D^{(T_2)}_{\gamma_1\gamma_1'}(R)\, D^{(T_2)}_{\gamma_2\gamma_2'}(R), \tag{2.50}$$

and also that, if we have

$$\mathbf{s}\phi(t_2m'\gamma) = \sum_{m} \phi(t_2m\gamma)\langle m|\,\mathbf{s}\,|\,m'\rangle, \tag{2.51}$$

we obtain

$$\begin{aligned}
&\mathbf{S}|\,\phi(t_2m_1'\gamma_1)\,\phi(t_2m_2'\gamma_2)| \\
&= \mathbf{s}_1|\,\phi(t_2m_1'\gamma_1)\,\phi(t_2m_2'\gamma_2)| + \mathbf{s}_2|\,\phi(t_2m_1'\gamma_1)\,\phi(t_2m_2'\gamma_2)| \\
&= \frac{1}{\sqrt{2}}\begin{vmatrix} \mathbf{s}\phi_{t_2m_1'\gamma_1}(1) & \mathbf{s}\phi_{t_2m_2'\gamma_2}(1) \\ \phi_{t_2m_1'\gamma_1}(2) & \phi_{t_2m_2'\gamma_2}(2) \end{vmatrix} + \frac{1}{\sqrt{2}}\begin{vmatrix} \phi_{t_2m_1'\gamma_1}(1) & \phi_{t_2m_2'\gamma_2}(1) \\ \mathbf{s}\phi_{t_2m_1'\gamma_1}(2) & \mathbf{s}\phi_{t_2m_2'\gamma_2}(2) \end{vmatrix} \\
&= \frac{1}{\sqrt{2}}\begin{vmatrix} \mathbf{s}\phi_{t_2m_1'\gamma_1}(1) & \phi_{t_2m_2'\gamma_2}(1) \\ \mathbf{s}\phi_{t_2m_1'\gamma_1}(2) & \phi_{t_2m_2'\gamma_2}(2) \end{vmatrix} + \frac{1}{\sqrt{2}}\begin{vmatrix} \phi_{t_2m_1'\gamma_1}(1) & \mathbf{s}\phi_{t_2m_2'\gamma_2}(1) \\ \phi_{t_2m_1'\gamma_1}(2) & \mathbf{s}\phi_{t_2m_2'\gamma_2}(2) \end{vmatrix} \\
&= \sum_{m_1} |\,\phi(t_2m_1\gamma_1)\,\phi(t_2m_2'\gamma_2)|\langle m_1|\,\mathbf{s}\,|\,m_1'\rangle \\
&\quad + \sum_{m_2} |\,\phi(t_2m_1'\gamma_1)\,\phi(t_2m_2\gamma_2)|\langle m_2|\,\mathbf{s}\,|\,m_2'\rangle.
\end{aligned} \tag{2.52}$$

Equations (2.50) and (2.52) show that Slater determinant $|\,\phi(t_2m_1\gamma_1)\,\phi(t_2m_2\gamma_2)|$ behaves just like the simple product of one-electron spin-orbitals $\phi(t_2m_1\gamma_1)\,\phi(t_2m_2\gamma_2)$ when the rotation and spin operators are operated. This fact leads us to the conclusion that the linear combination of Slater determinants,

$$\sum_{\substack{m_1m_2\\ \gamma_1\gamma_2}} |\,\phi(t_2m_1\gamma_1)\,\phi(t_2m_2\gamma_2)|\langle \tfrac{1}{2}m_1\tfrac{1}{2}m_2\,|\,SM\rangle\langle T_2\gamma_1T_2\gamma_2\,|\,\Gamma\gamma\rangle, \tag{2.53}$$

is base γ of irreducible representation Γ of the O_h-group and at the same time the eigenfunction of $\mathbf{S}^2$ and S_z with eigenvalues $S(S+1)$ and M, respectively.

So far we have used a similarity between the Slater determinant and the simple product of one-electron spin-orbitals. However, there are two important differences between them. The first difference is that because of the property in (2.11) the linear combination of Slater determinants as (2.53) vanishes identically for certain sets of $S\Gamma$ while the corresponding linear combination of the simple product of one-electron spin-orbitals is always nonvanishing. The possible sets of $S\Gamma$ in our problem of two t_2 electrons are all the possible combinations of $S = 0, 1$, and $\Gamma = A_1, E, T_1, T_2$. For example, when $S = 0$ and $\Gamma = T_1$, (2.53) with $\gamma = \gamma$ is

$$|\xi\bar{\eta}|\Big(\frac{1}{\sqrt{2}}\Big)\Big(\frac{1}{\sqrt{2}}\Big) + |\bar{\xi}\eta|\Big(-\frac{1}{\sqrt{2}}\Big)\Big(\frac{1}{\sqrt{2}}\Big) + |\eta\bar{\xi}|\Big(\frac{1}{\sqrt{2}}\Big)\Big(-\frac{1}{\sqrt{2}}\Big)$$

$$+ |\bar{\eta}\xi|\Big(-\frac{1}{\sqrt{2}}\Big)\Big(-\frac{1}{\sqrt{2}}\Big) = 0. \tag{2.54}$$

This result can be shown to be independent of γ. In this way, among eight possible sets of $S\Gamma$, the wavefunctions of ${}^3A_1$, ${}^3E$, ${}^1T_1$, and ${}^3T_2$ can be shown to be identically zero. Therefore, only four terms ${}^1A_1$, ${}^1E$, ${}^3T_1$, and ${}^1T_2$ are allowed. The number of states in the allowed terms is $1 + 2 + 9 + 3 = 15$, which agrees with the number of possible states predicted by using Pauli principle, ${}_6C_2 = 15$.

The second difference between the linear combination of Slater determinants and that of simple products of one-electron spin-orbitals is that the former linear combination is not normalized, in general, even for the allowed terms while the latter is always normalized. This difference also comes from the property of Slater determinants given in (2.11). For example, for ${}^1A_1$ (2.53) is given as

$$|\xi\bar{\xi}|\Big(\frac{1}{\sqrt{2}}\Big)\Big(\frac{1}{\sqrt{3}}\Big) + |\bar{\xi}\xi|\Big(\frac{-1}{\sqrt{2}}\Big)\Big(\frac{1}{\sqrt{3}}\Big) + |\eta\bar{\eta}|\Big(\frac{1}{\sqrt{2}}\Big)\Big(\frac{1}{\sqrt{3}}\Big)$$

$$+ |\bar{\eta}\eta|\Big(-\frac{1}{\sqrt{2}}\Big)\Big(\frac{1}{\sqrt{3}}\Big) + |\zeta\bar{\zeta}|\Big(\frac{1}{\sqrt{2}}\Big)\Big(\frac{1}{\sqrt{3}}\Big) + |\bar{\zeta}\zeta|\Big(-\frac{1}{\sqrt{2}}\Big)\Big(\frac{1}{\sqrt{3}}\Big)$$

$$= \Big(\frac{2}{3}\Big)^{1/2}[|\xi\bar{\xi}| + |\eta\bar{\eta}| + |\zeta\bar{\zeta}|], \tag{2.55}$$

which is not normalized.

Now it is clear that two-electron wavefunctions associated with terms ${}^{2S+1}\Gamma$ are obtained by normalizing (2.53). In the present problem they are denoted by $\Psi(t_2{}^2\, S\Gamma\, M\gamma)$, and are listed in Table 2.2.

Problem 2.5. Derive $\Psi(t_2{}^2\, {}^3T_1\, M\alpha)$ from (2.53), and confirm the footnotes of Table 2.2. ◇

TABLE 2.2[a]

WAVEFUNCTIONS $\Psi(t_2{}^2S\Gamma M\gamma)$

$$\Psi(t_2{}^2\,{}^1A_1) = [|\,\xi\bar{\xi}\,| + |\,\eta\bar{\eta}\,| + |\,\zeta\bar{\zeta}\,|]/\sqrt{3}$$

$$\Psi(t_2{}^2\,{}^1Eu) = [-|\,\xi\bar{\xi}\,| - |\,\eta\bar{\eta}\,| + 2\,|\,\zeta\bar{\zeta}\,|]/\sqrt{6}$$

$$\Psi(t_2{}^2\,{}^1Ev) = [|\,\xi\bar{\xi}\,| - |\,\eta\bar{\eta}\,|]/\sqrt{2}$$

$$\Psi(t_2{}^2\,{}^3T_1\,M{=}1\,\gamma) = |\,\xi\eta\,|$$

$$\Psi(t_2{}^2\,{}^3T_1\,M{=}0\,\gamma)^{\,b} = [|\,\xi\bar{\eta}\,| - |\,\eta\bar{\xi}\,|]/\sqrt{2}$$

$$\Psi(t_2{}^2\,{}^3T_1\,M{=}-1\,\gamma)^{\,b} = |\,\bar{\xi}\bar{\eta}\,|$$

$$\Psi(t_2{}^2\,{}^1T_2\zeta) = [|\,\xi\bar{\eta}\,| + |\,\eta\bar{\xi}\,|]/\sqrt{2}$$

[a] The unlisted components of the wavefunctions are obtained from the listed one by an appropriate cyclic permutation of ξ, η, and ζ.

[b] Here, $\Psi(t_2{}^2\,{}^3T_1\,M{=}0\,\gamma)$ and $\Psi(t_2{}^2\,{}^3T_1\,M{=}-1\,\gamma)$ may be derived from $\Psi(t_2{}^2\,{}^3T_1\,M{=}1\,\gamma)$ by successive operations of S_- on $\Psi(t_2{}^2\,{}^3T_1\,M{=}1\,\gamma)$.

Similarly $\Psi(e^2S\Gamma\,M\gamma)$ can be obtained as shown in Table 2.3. In this case allowed terms are ${}^1A_1$, ${}^3A_2$, and ${}^1E$. The number of states in these terms is $1 + 3 + 2 = 6$ which agrees with ${}_4C_2 = 6$.

TABLE 2.3

WAVEFUNCTIONS $\Psi(e^2S\Gamma M\gamma)$

$$\Psi(e^2\,{}^1A_1) = [|\,u\bar{u}\,| + |\,v\bar{v}\,|]/\sqrt{2}$$

$$\Psi(e^2\,{}^3A_2\,M{=}1) = |\,uv\,|$$

$$\Psi(e^2\,{}^3A_2\,M{=}0) = [|\,u\bar{v}\,| - |\,v\bar{u}\,|]/\sqrt{2}$$

$$\Psi(e^2\,{}^3A_2\,M{=}-1) = |\,\bar{u}\bar{v}\,|$$

$$\Psi(e^2\,{}^1E\,u) = [-|\,u\bar{u}\,| + |\,v\bar{v}\,|]/\sqrt{2}$$

$$\Psi(e^2\,{}^1E\,v) = [|\,u\bar{v}\,| + |\,v\bar{u}\,|]/\sqrt{2}$$

Finally, let us consider the wavefunctions of the t_2e configuration. In this case all the Slater determinants appearing in (2.53) are linearly independent, so that all the terms are allowed and (2.53) is normalized. Thus, wavefunctions $\Psi(t_2eS\,M\gamma)$ are given as

$$\Psi(t_2eS\Gamma\,M\gamma) = \sum_{\substack{m_1m_2\\ \gamma_1\gamma_2}} |\,\phi(t_2m_1\gamma_1)\,\phi(em_2\gamma_2)|\,\langle\tfrac{1}{2}m_1\tfrac{1}{2}m_2\,|\,SM\rangle\langle T_2\gamma_1E\gamma_2\,|\,\Gamma\gamma\rangle. \quad (2.56)$$

The terms are given by all the possible combinations of $S = 0$, 1 and $\Gamma = T_1$, T_2 : They are ${}^3T_1$, ${}^3T_2$, ${}^1T_1$, and ${}^1T_2$. The number of states in these terms is $9 + 9 + 3 + 3 = 24$ which agrees with $6 \times 4 = 24$. Wavefunctions $\Psi(t_2eS\Gamma\,M\gamma)$ calculated from (2.56) are given in Table 2.4.

Problem 2.6. Derive $\Psi(t_2e\,{}^1T_1\alpha)$ and $\Psi(t_2e\,{}^1T_2\xi)$ from (2.56) and confirm the footnote of Table 2.4. ◇

TABLE 2.4[a]
WAVEFUNCTIONS $\Psi(t_2eS\Gamma M\gamma)$

$\Psi(t_2e\,{}^3T_1\,M{=}1\,\gamma)$	=	$\mid\zeta v\mid$
$\Psi(t_2e\,{}^3T_1\,M{=}0\,\gamma)$	=	$[\mid\zeta\bar{v}\mid + \mid\bar{\zeta}v\mid]/\sqrt{2}$
$\Psi(t_2e\,{}^3T_1\,M{=}-1\,\gamma)$	=	$\mid\bar{\zeta}\bar{v}\mid$
$\Psi(t_2e\,{}^3T_2\,M{=}1\,\zeta)$	=	$\mid\zeta u\mid$
$\Psi(t_2e\,{}^3T_2\,M{=}0\,\zeta)$	=	$[\mid\zeta\bar{u}\mid + \mid\bar{\zeta}u\mid]/\sqrt{2}$
$\Psi(t_2e\,{}^3T_2\,M{=}-1\,\zeta)$	=	$\mid\bar{\zeta}\bar{u}\mid$
$\Psi(t_2e\,{}^1T_1\,\gamma)$	=	$[\mid\zeta\bar{v}\mid - \mid\bar{\zeta}v\mid]/\sqrt{2}$
$\Psi(t_2e\,{}^1T_2\,\zeta)$	=	$[\mid\zeta\bar{u}\mid - \mid\bar{\zeta}u\mid]/\sqrt{2}$

[a] The unlisted components of the wavefunctions are obtained from the listed ones by appropriate cyclic changes of (ξ, η, ζ), (u_x, u_y, u_z), and (v_x, v_y, v_z). Here, $v_x = -(\sqrt{3}/2)u - \frac{1}{2}v$, $v_y = (\sqrt{3}/2)u - \frac{1}{2}v$, $v_z = v$ as shown in (1.33) by using notations v_x, v_y, v_z for $\varphi_v^{(x)}$, $\varphi_v^{(y)}$, $\varphi_v^{(z)}$, and $u_x = -\frac{1}{2}u + (\sqrt{3}/2)v$, $u_y = -\frac{1}{2}u - (\sqrt{3}/2)v$, and $u_z = u$.

2.3 Term Energies

In the previous section we derived the wavefunctions of the $t_2{}^2$, e^2, and t_2e electron configuration. At the same time we learned what kinds of terms appear in these electron configurations. In the case of the $t_2{}^2$

configuration, a single unperturbed level with fifteen-fold degeneracy splits into four terms by the Coulomb interaction. They are non-degenerate ${}^1A_1$, doubly degenerate ${}^1E$, triply degenerate ${}^1T_2$, and ${}^3T_1$ with 9-fold degeneracy including the spin-degeneracy. Since all these terms are labeled with different sets of $S\Gamma$, the matrix of $\mathscr{H}_1$ is completely diagonalized when it is calculated by the use of the wavefunctions obtained in the previous section. This makes the calculation of term energies very simple.

2.3.1 Matrix Elements between Slater Determinants

In order to calculate the matrix of $\mathscr{H}_1$ by using the two-electron wavefunctions, we derive a formula to calculate the following integral:

$$\sum_{\sigma_1\sigma_2} \int d\tau_1\, d\tau_2 \,|\phi(\lambda_1 m_1 \gamma_1)\,\phi(\lambda_2 m_2 \gamma_2)|^*\, g_{12}\, |\phi(\lambda_1' m_1' \gamma_1')\,\phi(\lambda_2' m_2' \gamma_2')|. \quad (2.57)$$

Here, the λ_i's represent t_2 and/or e. Expanding the first Slater determinant in (2.57) into two terms and exchangeing electron numbers in the second term, we obtain

$$(2.57) = \sqrt{2} \sum_{\sigma_1\sigma_2} \int d\tau_1\, d\tau_2 \phi^*_{\lambda_1 m_1 \gamma_1}(\mathbf{r}_1\boldsymbol{\sigma}_1)\,\phi^*_{\lambda_2 m_2 \gamma_2}(\mathbf{r}_2\boldsymbol{\sigma}_2)$$

$$\times\, g_{12} \frac{1}{\sqrt{2}} \begin{vmatrix} \phi_{\lambda_1' m_1' \gamma_1'}(\mathbf{r}_1\boldsymbol{\sigma}_1) & \phi_{\lambda_2' m_2' \gamma_2'}(\mathbf{r}_1\boldsymbol{\sigma}_1) \\ \phi_{\lambda_1' m_1' \gamma_1'}(\mathbf{r}_2\boldsymbol{\sigma}_2) & \phi_{\lambda_2' m_2' \gamma_2'}(\mathbf{r}_2\boldsymbol{\sigma}_2) \end{vmatrix}, \quad (2.58)$$

where factor $\sqrt{2}$ comes from the normalization factor $1/\sqrt{2}$ of the Slater determinant multiplied by 2. Expanding again the Slater determinant in (2.58), we obtain

$$(2.57) = \langle \lambda_1 m_1 \gamma_1, \lambda_2 m_2 \gamma_2 \,|\, g \,|\, \lambda_1' m_1' \gamma_1', \lambda_2' m_2' \gamma_2' \rangle$$

$$- \langle \lambda_1 m_1 \gamma_1, \lambda_2 m_2 \gamma_2 \,|\, g \,|\, \lambda_2' m_2' \gamma_2', \lambda_1' m_1' \gamma_1' \rangle, \quad (2.59)$$

where

$$\langle \lambda_1 m_1 \gamma_1, \lambda_2 m_2 \gamma_2 \,|\, g \,|\, \lambda_1' m_1' \gamma_1', \lambda_2' m_2' \gamma_2' \rangle$$

$$= \sum_{\sigma_1\sigma_2} \int d\tau_1\, d\tau_2 \phi^*_{\lambda_1 m_1 \gamma_1}(\mathbf{r}_1\boldsymbol{\sigma}_1)\,\phi^*_{\lambda_2 m_2 \gamma_2}(\mathbf{r}_2\boldsymbol{\sigma}_2)\, g_{12}$$

$$\times\, \phi_{\lambda_1' m_1' \gamma'}(\mathbf{r}_1\boldsymbol{\sigma}_1)\,\phi_{\lambda_2' m_2' \gamma_2'}(\mathbf{r}_2\boldsymbol{\sigma}_2). \quad (2.60)$$

Since g_{12} is independent of spin coordinates, (2.60) is expressed in terms of orbital functions $\varphi_{\lambda\gamma}(\mathbf{r})$ in place of spin-orbitals $\phi_{\lambda m\gamma}(\mathbf{r}\boldsymbol{\sigma}) = \varphi_{\lambda\gamma}(\mathbf{r})\,\theta_{\frac{1}{2}m}(\boldsymbol{\sigma})$:

$$\langle \lambda_1 m_1 \gamma_1 , \lambda_2 m_2 \gamma_2 \mid r_{12}^{-1} \mid \lambda_1' m_1' \gamma_1', \lambda_2' m_2' \gamma_2' \rangle$$
$$= \delta(m_1 m_1')\,\delta(m_2 m_2')\langle \lambda_1 \gamma_1 \lambda_2 \gamma_2 \mid r_{12}^{-1} \mid \lambda_1' \gamma_1' \lambda_2' \gamma_2' \rangle, \tag{2.61}$$

where

$$\langle \lambda_1 \gamma_1 \lambda_2 \gamma_2 \mid r_{12}^{-1} \mid \lambda_1' \gamma_1' \lambda_2' \gamma_2' \rangle = \int d\boldsymbol{\tau}_1\, d\boldsymbol{\tau}_2 \varphi^*_{\lambda_1\gamma_1}(\mathbf{r}_1)\, \varphi^*_{\lambda_2\gamma_2}(\mathbf{r}_2)$$
$$\times (1/r_{12})\, \varphi_{\lambda_1'\gamma_1'}(\mathbf{r}_1)\, \varphi_{\lambda_2'\gamma_2'}(\mathbf{r}_2)$$
$$\equiv \langle \lambda_1 \gamma_1 \lambda_2 \gamma_2 \mid\mid \lambda_1' \gamma_1' \lambda_2' \gamma_2' \rangle. \tag{2.62}$$

For simplicity we shall often use the abbreviation in the last expression of (2.62). In particular,

$$J(\lambda_1 \gamma_1 \lambda_2 \gamma_2) \equiv \langle \lambda_1 \gamma_1 \lambda_2 \gamma_2 \mid\mid \lambda_1 \gamma_1 \lambda_2 \gamma_2 \rangle \tag{2.63}$$

is called the *Coulomb integral*, and

$$K(\lambda_1 \gamma_1 \lambda_2 \gamma_2) \equiv \langle \lambda_1 \gamma_1 \lambda_2 \gamma_2 \mid\mid \lambda_2 \gamma_2 \lambda_1 \gamma_1 \rangle \tag{2.64}$$

is called the *exchange integral*. It can be shown that

$$J(\lambda_1 \gamma_1 \lambda_2 \gamma_2) \geqslant K(\lambda_1 \gamma_1 \lambda_2 \gamma_2) \geqslant 0. \tag{2.65}$$

From (2.59) one may obtain the following formula for a special case of (2.57):

$$\sum_{\sigma_1\sigma_2} \int d\boldsymbol{\tau}_1\, d\boldsymbol{\tau}_2 \mid \phi(\lambda_1 m_1 \gamma_1)\, \phi(\lambda_2 m_2 \gamma_2)\mid^* g_{12} \mid \phi(\lambda_1 m_1 \gamma_1)\, \phi(\lambda_2 m_2 \gamma_2)\mid$$
$$= J(\lambda_1 \gamma_1 \lambda_2 \gamma_2) - \delta(m_1 m_2)\, K(\lambda_1 \gamma_1 \lambda_2 \gamma_2). \tag{2.66}$$

2.3.2 The t_2^2 Electron Configuration

By using the formulas obtained in the previous subsection, let us calculate the energies of the terms of the t_2^2 electron configuration. We make the full use of the wavefunctions listed in Table 2.2.

For $^1A_1$, one obtains

$$\langle t_2^2\, {}^1A_1 \mid \mathscr{H}_1 \mid t_2^2\, {}^1A_1 \rangle = \tfrac{1}{3} \sum_{(\xi\eta\zeta)} \sum_{\sigma_1\sigma_2} \int d\boldsymbol{\tau}_1\, d\boldsymbol{\tau}_2 \mid \xi\bar{\xi} \mid^* g_{12} \mid \xi\bar{\xi} \mid$$
$$+ \tfrac{1}{3} \sum_{(\xi\eta\zeta)} \sum_{\sigma_1\sigma_2} \int d\boldsymbol{\tau}_1\, d\boldsymbol{\tau}_2 \mid \xi\bar{\xi} \mid^* g_{12} \mid \eta\bar{\eta} \mid$$
$$+ \tfrac{1}{3} \sum_{(\xi\eta\zeta)} \sum_{\sigma_1\sigma_2} \int d\boldsymbol{\tau}_1\, d\boldsymbol{\tau}_2 \mid \xi\bar{\xi} \mid^* g_{12} \mid \zeta\bar{\zeta} \mid, \tag{2.67}$$

where $\sum_{(\xi\eta\zeta)}$ means the summation over the terms obtained by cyclic permutations of ξ, η, and ζ. Since ξ, η, and ζ are real functions, relations such as

$$\langle \xi\xi \mid\mid \eta\eta \rangle = K(\xi\eta), \qquad \text{etc.} \tag{2.68}$$

hold. Therefore, (2.67) can be rewritten as

$$(2.67) = \tfrac{1}{3} \sum_{(\xi\eta\zeta)} J(\xi\xi) + \tfrac{2}{3} \sum_{(\xi\eta\zeta)} K(\xi\eta). \tag{2.69}$$

Furthermore by using relations,

$$\begin{aligned} J(\xi\xi) &= J(\eta\eta) = J(\zeta\zeta), \\ K(\xi\eta) &= K(\eta\zeta) = K(\zeta\xi), \end{aligned} \tag{2.70}$$

which may be derived from the symmetry properties of the orbital functions, (2.67) is finally given as

$$\langle t_2^2\, {}^1A_1 \mid \mathscr{H}_1 \mid t_2^2\, {}^1A_1 \rangle = J(\zeta\zeta) + 2K(\xi\eta). \tag{2.71}$$

For ${}^1E$ one may obtain the term energy by calculating the matrix of $\mathscr{H}_1$ in the ${}^1Eu$ state as follows:

$$\begin{aligned} \langle t_2^2\, {}^1Eu \mid \mathscr{H}_1 \mid t_2^2\, {}^1Eu \rangle &= \tfrac{1}{6}[J(\xi\xi) + J(\eta\eta) + 4J(\zeta\zeta) \\ &\qquad + 2K(\xi\eta) - 4K(\zeta\xi) - 4K(\eta\zeta)] \\ &= J(\zeta\zeta) - K(\xi\eta). \end{aligned} \tag{2.72}$$

The same result is also obtained by calculating the matrix element in the ${}^1Ev$ state.

Problem 2.7. Confirm that $\langle t_2^2\, {}^1Eu \mid \mathscr{H}_1 \mid t_2^2\, {}^1Ev \rangle = 0$. ◇

For ${}^3T_1$, the simplest way of obtaining the term energy is to calculate the matrix element of $\mathscr{H}_1$ in the ${}^3T_1\ M = 1\ \gamma$ state. However, for exercise, we calculate it here in the ${}^3T_1\ M = 0\ \gamma$ state as follows:

$$\begin{aligned} &\langle t_2^2\, {}^3T_1\ M{=}0\ \gamma \mid \mathscr{H}_1 \mid t_2^2\, {}^3T_1\ M{=}0\ \gamma \rangle \\ &= \tfrac{1}{2}\Big[\sum_{\sigma_1\sigma_2} \int d\boldsymbol{\tau}_1\, d\boldsymbol{\tau}_2 \mid \xi\bar{\eta} \mid^* g_{12} \mid \xi\bar{\eta} \mid + \sum_{\sigma_1\sigma_2} \int d\boldsymbol{\tau}_1\, d\boldsymbol{\tau}_2 \mid \eta\bar{\xi} \mid^* g_{12} \mid \eta\bar{\xi} \mid \\ &\qquad - \sum_{\sigma_1\sigma_2} \int d\boldsymbol{\tau}_1\, d\boldsymbol{\tau}_2 \mid \xi\bar{\eta} \mid^* g_{12} \mid \eta\bar{\xi} \mid - \sum_{\sigma_1\sigma_2} \int d\boldsymbol{\tau}_1\, d\boldsymbol{\tau}_2 \mid \eta\bar{\xi} \mid^* g_{12} \mid \xi\bar{\eta} \mid \Big] \\ &= \tfrac{1}{2}[J(\xi\eta) + J(\xi\eta) - K(\xi\eta) - K(\xi\eta)] \\ &= J(\xi\eta) - K(\xi\eta). \end{aligned} \tag{2.73}$$

Comparing the wavefunction of ${}^1T_2\zeta$ with that of ${}^3T_1\, M = 0\,\gamma$, we see that the term energy of the ${}^1T_2$ term is obtained by changing the sign of $K(\xi\eta)$ in (2.73):

$$\langle t_2{}^2\,{}^1T_2\zeta \mid \mathscr{H}_1 \mid t_2{}^2\,{}^1T_2\zeta\rangle = J(\xi\eta) + K(\xi\eta). \tag{2.74}$$

In this way the term energies in $t_2{}^2$ are given in terms of three integrals $J(\zeta\zeta)$, $J(\xi\eta)$, and $K(\xi\eta)$, which are mutually independent, i.e., none of them can be expressed in terms of the others. In order to predict the term locations it is necessary to know the values of these three integrals. However, by using qualitative arguments, one may predict the order of some terms as follows. Since $K(\xi\eta) > 0$, the energy of the ${}^1A_1$ term, $E({}^1A_1)$, is higher than that of ${}^1E$, $E({}^1E)$: $E({}^1A_1) > E({}^1E)$. Similarly $E({}^1T_2) > E({}^3T_1)$. By using the relation, $J(\zeta\zeta) = J(\xi\xi) = J(\eta\eta) > J(\xi\eta)$, one may predict that $E({}^1E) > E({}^3T_1)$ and $E({}^1A_1) > E({}^1T_2)$. However, it is impossible to predict the order of ${}^1E$ and ${}^1T_2$ with qualitative arguments. After all the order of the terms is predicted as $E({}^1A_1) > E({}^1E)$, $E({}^1T_2) > E({}^3T_1)$. Later, it will be shown that, if the t_{2g} functions are d-functions, the ${}^1E$ and ${}^1T_2$ terms are accidentally degenerate. Such a situation is visualized in Fig. 2.2.

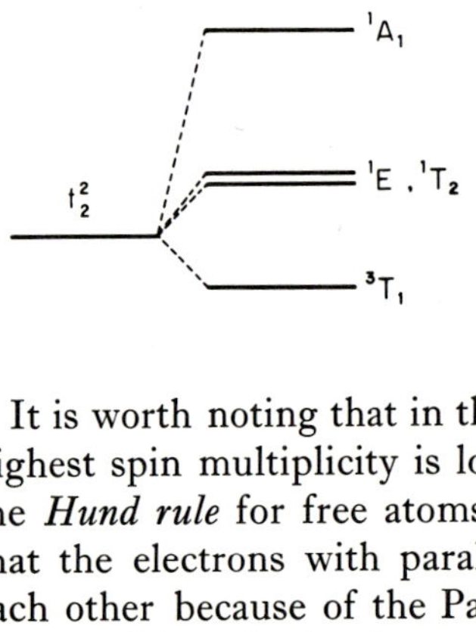

FIG. 2.2. Energy levels arising from the $t_2{}^2$ electron configuration.

It is worth noting that in the present problem the ${}^3T_1$ term having the highest spin multiplicity is lowest in energy. This is in accordance with the *Hund rule* for free atoms and ions. This rule is based on the fact that the electrons with parallel spins are prevented from approaching each other because of the Pauli principle, resulting in the reduction of the repulsive Coulomb energy. This situation is also found in our problem as seen from the fact that

$$\Psi(t_2{}^2\,{}^3T_1\, M{=}1\,\gamma) = (1/\sqrt{2})[\xi(\mathbf{r}_1)\,\eta(\mathbf{r}_2) - \eta(\mathbf{r}_1)\,\xi(\mathbf{r}_2)]\,\alpha(\boldsymbol{\sigma}_1)\,\alpha(\boldsymbol{\sigma}_2)$$

vanishes when $\mathbf{r}_1 = \mathbf{r}_2$.

Problem 2.8. Prove that $J(\xi\xi) = J(\eta\eta) \geqslant J(\xi\eta)$.◇

2.3.3 The e^2 Electron Configuration

By using the wavefunctions listed in Table 2.3, we obtain

$$\langle e^2\,{}^1A_1 \mid \mathscr{H}_1 \mid e^2\,{}^1A_1\rangle = \tfrac{1}{2}[J(uu) + J(vv)] + K(uv), \tag{2.75}$$

$$\langle e^2\,{}^3A_2\, M{=}0 \mid \mathscr{H}_1 \mid e^2\,{}^3A_2\, M{=}0\rangle = J(uv) - K(uv), \tag{2.76}$$

$$\langle e^2\,{}^1Eu \mid \mathscr{H}_1 \mid e^2\,{}^1Eu\rangle = \tfrac{1}{2}[J(uu) + J(vv)] - K(uv). \tag{2.77}$$

On the other hand, one has

$$\langle e^2\,{}^1Ev \mid \mathscr{H}_1 \mid e^2\,{}^1Ev\rangle = J(uv) + K(uv), \tag{2.78}$$

which has to be equal to (2.77). Therefore, the following relation should hold:

$$\tfrac{1}{2}[J(uu) + J(vv)] = J(uv) + 2K(uv). \tag{2.79}$$

By using (2.79), (2.75) can be reexpressed as

$$\langle e^2\,{}^1A_1 \mid \mathscr{H}_1 \mid e^2\,{}^1A_1\rangle = J(uv) + 3K(uv). \tag{2.80}$$

Problem 2.9. Show that $J(uu) = J(vv)$. ◇

Thus, the term energies for the e^2 configuration are given in terms of two integrals, $J(uv)$ and $K(uv)$, which are mutually independent. It is clear that $E({}^1A_1) > E({}^1E) > E({}^3A_2)$ and all the term separations are given by $2K(uv)$. This situation is visualized in Fig. 2.3. Again the term with the highest spin multiplicity is lowest in energy.

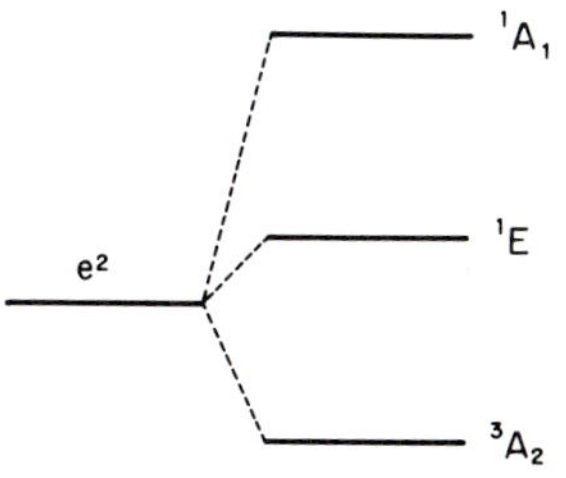

FIG. 2.3. Energy levels arising from the e^2 electron configuration.

2.3.4 The t_2e Electron Configuration

By using the wavefunctions listed in Table 2.4, we obtain

$$\langle t_2e\,{}^3T_1\, M{=}0\,\gamma \mid \mathscr{H}_1 \mid t_2e\,{}^3T_1\, M{=}0\,\gamma\rangle = J(\zeta v) - K(\zeta v), \tag{2.81}$$

$$\langle t_2e\,{}^1T_1\gamma \mid \mathscr{H}_1 \mid t_2e\,{}^1T_1\gamma\rangle = J(\zeta v) + K(\zeta v), \tag{2.82}$$

$$\langle t_2e\,{}^3T_2\, M{=}0\,\zeta \mid \mathscr{H}_1 \mid t_2e\,{}^3T_2\, M{=}0\,\zeta\rangle = J(\zeta u) - K(\zeta u), \tag{2.83}$$

$$\langle t_2e\,{}^1T_2\zeta \mid \mathscr{H}_1 \mid t_2e\,{}^1T_2\zeta\rangle = J(\zeta u) + K(\zeta u). \tag{2.84}$$

Thus, these term energies are given in terms of four integrals, $J(\zeta u)$, $J(\zeta v)$, $K(\zeta u)$, and $K(\zeta v)$ which are mutually independent. It is immediately seen that $E(^1T_1) > E(^3T_1)$ and $E(^1T_2) > E(^3T_2)$. However, it is hard to predict, for example, the order of $^3T_2$ and $^3T_1$ without knowing the magnitudes of the integrals. The order of the terms illustrated in Fig. 2.4 is determined by assuming that the t_2 and e wavefunctions are the d-functions as discussed later.

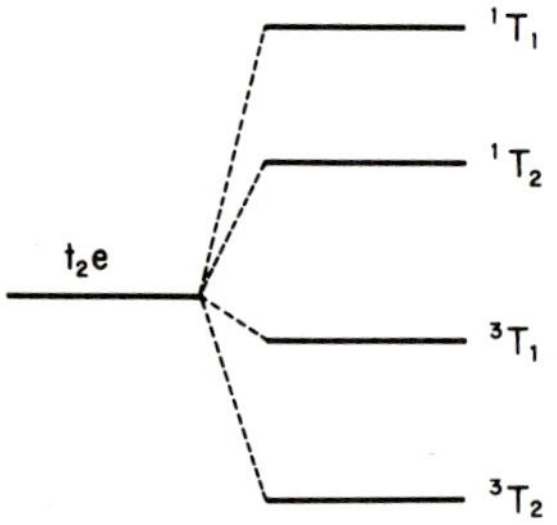

FIG. 2.4. Energy levels arising from the t_2e electron configuration.

2.3.5 Configuration Mixing

So far we have calculated term energies within a single electron configuration: t_2^2, e^2, or t_2e. The states of t_2^2, e^2, and t_2e have, respectively, energies of $2\langle\zeta\,|f|\,\zeta\rangle$, $2\langle v\,|f|\,v\rangle$, and $\langle\zeta\,|f|\,\zeta\rangle + \langle v\,|f|\,v\rangle$ in addition to the Coulomb interaction energies, and, as long as $\langle v\,|f|\,v\rangle - \langle\zeta\,|f|\,\zeta\rangle$ is much larger than the Coulomb interaction energies, it is a good approximation to calculate term energies within a single electron configuration. Here, $\langle v\,|f|\,v\rangle - \langle\zeta\,|f|\,\zeta\rangle$ corresponds to $10Dq$ in (1.26). However, if $\langle v\,|f|\,v\rangle - \langle\zeta\,|f|\,\zeta\rangle$ is not so large, we can no longer neglect the nondiagonal matrix elements of $\mathscr{H}_1$ between the same $S\Gamma$ states of different electron configurations. For example, there are two $^3T_1$ states arising from t_2^2 and t_2e, and these states are admixed due to the Coulomb interaction resulting in shifts of their energies.

Let us first calculate the nondiagonal element of $\mathscr{H}_1$ for $^3T_1$. It is given as

$$\begin{aligned}\langle t_2^2\,{}^3T_1\,M{=}0\,\gamma\,|\,\mathscr{H}_1\,|\,t_2e\,{}^3T_1\,M{=}0\,\gamma\rangle &= \langle\xi\eta\,||\,\zeta v\rangle - \langle\xi\eta\,||\,v\zeta\rangle \\ &= 2\langle\xi\eta\,||\,\zeta v\rangle, \end{aligned} \tag{2.85}$$

in which the relation

$$\langle\xi\eta\,||\,\zeta v\rangle = -\langle\xi\eta\,||\,v\zeta\rangle \tag{2.86}$$

is used. Relation (2.86) can be proved as follows: Since the Coulomb interaction operator $1/r_{12}$ is invariant to any rotation of the O_h-group, one obtains

$$\begin{aligned}\langle \xi\eta \mid\mid \zeta v\rangle &= \langle C_4(z)\xi\, C_4(z)\eta \mid\mid C_4(z)\zeta\, C_4(z)v\rangle \\ &= \langle -\eta\xi \mid\mid -\zeta - v\rangle \\ &= -\langle \xi\eta \mid\mid v\zeta\rangle. \end{aligned} \tag{2.87}$$

There are two ${}^1T_2$ states arising from t_2^2 and t_2e. The nondiagonal element for ${}^1T_2$ is calculated as

$$\begin{aligned}\langle t_2^2\, {}^1T_2\zeta \mid \mathscr{H}_1 \mid t_2e\, {}^1T_2\zeta\rangle &= \langle \xi\eta \mid\mid \zeta u\rangle + \langle \xi\eta \mid\mid u\zeta\rangle \\ &= 2\langle \xi\eta \mid\mid \zeta u\rangle, \end{aligned} \tag{2.88}$$

in which we have used the relation

$$\langle \xi\eta \mid\mid \zeta u\rangle = \langle \xi\eta \mid\mid u\zeta\rangle. \tag{2.89}$$

Relation (2.89) can be proved in just the same way as (2.86) was proved.

Problem 2.10. Derive $\langle \xi\eta \| \zeta v\rangle = \sqrt{3}\,\langle \xi\eta \| \zeta u\rangle$. ◇

For ${}^1A_1$ there are two states arising from t_2^2 and e^2. The nondiagonal element is given as

$$\begin{aligned}\langle t_2^2\, {}^1A_1 \mid \mathscr{H}_1 \mid e^2\, {}^1A_1\rangle &= \frac{1}{\sqrt{6}} \sum_{\substack{(\xi\eta\zeta)\\(uv)}} \langle \xi\xi \mid\mid uu\rangle \\ &= \frac{1}{\sqrt{6}} \sum_{\substack{(\xi\eta\zeta)\\(uv)}} K(\xi u). \end{aligned} \tag{2.90a}$$

Among the six terms in (2.90a), only $K(\zeta u)$ and $K(\zeta v)$ may be taken to be independent:

$$\begin{aligned} K(\xi u) &= K(\eta u) = \tfrac{1}{4}K(\zeta u) + \tfrac{3}{4}K(\zeta v), \\ K(\xi v) &= K(\eta v) = \tfrac{3}{4}K(\zeta u) + \tfrac{1}{4}K(\zeta v). \end{aligned} \tag{2.91}$$

Then, (2.90a) is reexpressed as

$$\langle t_2^2\, {}^1A_1 \mid \mathscr{H}_1 \mid e^2\, {}^1A_1\rangle = (3/2)^{1/2}[K(\zeta u) + K(\zeta v)]. \tag{2.90b}$$

Finally for ${}^1E$, one obtains

$$\langle t_2^2\, {}^1Eu \mid \mathscr{H}_1 \mid e^2\, {}^1Eu\rangle = -(3/2)^{1/2}[K(\zeta u) - K(\zeta v)]. \tag{2.92}$$

Problem 2.11. Derive (2.91). Note that $K(\xi u_x) = K(\eta u_y) = K(\zeta u)$. ◇

Now we have obtained all the necessary nondiagonal matrix elements of $\mathscr{H}_1$. Term energies of ${}^3T_1$, ${}^1T_2$, ${}^1A_1$, and ${}^1E$ are obtained by diagonalizing the two-dimensional matrix of $\mathscr{H}_0 + \mathscr{H}_1$. For instance, term energies of two ${}^3T_1$ states are calculated by solving the secular equation,

$$\begin{matrix} t_2^2 \\ t_2 e \end{matrix} \begin{vmatrix} 2\langle\zeta\,|f|\,\zeta\rangle + J(\xi\eta) - K(\xi\eta) - E & 2\langle\xi\eta\,||\,\zeta v\rangle \\ 2\langle\xi\eta\,||\,\zeta v\rangle & \langle\zeta\,|f|\,\zeta\rangle + \langle v\,|f|\,v\rangle + J(\zeta v) - K(\zeta v) - E \end{vmatrix} = 0. \tag{2.93}$$

The eigenfunctions are given as

$$\begin{aligned} \Psi(a\,{}^3T_1\,M{=}0\,\gamma) &= \cos\theta\Psi(t_2^2\,{}^3T_1\,M{=}0\,\gamma) - \sin\theta\Psi(t_2e\,{}^3T_1\,M{=}0\,\gamma), \\ \Psi(b\,{}^3T_1\,M{=}0\,\gamma) &= \sin\theta\Psi(t_2^2\,{}^3T_1\,M{=}0\,\gamma) + \cos\theta\Psi(t_2e\,{}^3T_1\,M{=}0\,\gamma), \end{aligned} \tag{2.94}$$

where θ is determined by

$$\tan 2\theta = \frac{4\langle\xi\eta\,||\,\zeta v\rangle}{10Dq + J(\zeta v) - K(\zeta v) - J(\xi\eta) + K(\xi\eta)}, \tag{2.95}$$

and

$$10Dq = \langle v\,|f|\,v\rangle - \langle\zeta\,|f|\,\zeta\rangle. \tag{2.96}$$

In this section we have shown that term energies of the system having two electrons in the t_{2g} and e_g shells are given in terms of ten two-electron integrals $J(\zeta\zeta)$, $J(\xi\eta)$, $K(\xi\eta)$; $J(uv)$, $K(uv)$; $J(\zeta u)$, $K(\zeta u)$; $J(\zeta v)$, $K(\zeta v)$; $\langle\xi\eta\,\|\,\zeta u\rangle$, in addition to $10Dq$ representing the splitting of one-electron levels, t_{2g} and e_g. In many cases it is very difficult to know the accurate wavefunctions of t_{2g} and e_g and consequently to evaluate these integrals. Therefore, these integrals are left as parameters to be determined by experiments. However, in doing so, we immediately meet a difficulty: The number of parameters is too big to be determined from the limited experimental information. In the next subsection a reasonable approximation will be made to reduce the number of these parameters.

Problem 2.12. Calculate the term energies and the wevefunctions for ${}^1T_2$ by taking into account the configuration mixing. ◇

2.3.6 Slater Integrals

One of the ways of reducing the number of two-electron integrals introduced in the previous subsections is to assume a simple angular dependence for the t_{2g} and e_g wavefunctions. If d-electrons in crystals are relatively localized around the iron-group metal ions and deformation of atomic d-orbitals is not drastic, it is reasonable to assume as the first approximation that the t_{2g} and e_g wavefunctions have pure d-character and are given in the forms of (1.31) and (1.32). However, this approximation does not necessarily mean that radial part $R_{3d}(r)$ in (1.31) and (1.32) is that of a free atom or ion. In this sense, we will write $R_d(r)$ for $R_{3d}(r)$. This point will be discussed in more detail in the chapters on optical spectra and molecular orbitals. In this subsection we will use this approximation to show how ten parameters are expressed in terms of fewer parameters.

Since the t_{2g} and e_g wavefunctions are now assumed to be linear combination of $\varphi_{dm}(\mathbf{r}) = R_d(r)\, Y_{2m}(\theta\varphi)$ $(m = 2, 1, 0, -1, -2,)$, the two electron integrals derived in the previous subsection are given in terms of those involving $\varphi_{dm}(\mathbf{r})$. For example, one has

$$\begin{aligned} J(\zeta\zeta) &= \langle\zeta\zeta \,|\,|\, \zeta\zeta\rangle \\ &= \int d\boldsymbol{\tau}_1\, d\boldsymbol{\tau}_2 \left(\frac{i}{\sqrt{2}}\right) [\varphi^*_{d2}(\mathbf{r}_1) - \varphi^*_{d-2}(\mathbf{r}_1)] \left(\frac{i}{\sqrt{2}}\right) [\varphi^*_{d2}(\mathbf{r}_2) - \varphi^*_{d-2}(\mathbf{r}_2)] \\ &\quad \times \frac{1}{r_{12}} \left(\frac{-i}{\sqrt{2}}\right) [\varphi_{d2}(\mathbf{r}_1) - \varphi_{d-2}(\mathbf{r}_1)] \left(\frac{-i}{\sqrt{2}}\right) [_{d2}(\mathbf{r}_2) - \varphi_{d-2}(\mathbf{r}_2)] \\ &= \tfrac{1}{4}[\langle 22 \,|\,|\, 22\rangle + \langle -2\,-2 \,|\,|\, -2\,-2\rangle + \langle 2\,-2 \,|\,|\, 2\,-2\rangle \\ &\quad + \langle -22 \,|\,|\, -22\rangle + \langle 2\,-2 \,|\,|\, -22\rangle + \langle -22 \,|\,|\, 2\,-2\rangle], \end{aligned} \tag{2.97}$$

where

$$\langle m_1 m_2 \,|\,|\, m_1' m_2'\rangle = \int d\boldsymbol{\tau}_1\, d\boldsymbol{\tau}_2\, \varphi^*_{dm_1}(\mathbf{r}_1)\, \varphi^*_{dm_2}(\mathbf{r}_2) \frac{1}{r_{12}} \varphi_{dm_1'}(\mathbf{r}_1)\, \varphi_{dm_2'}(\mathbf{r}_2). \tag{2.98}$$

In deriving (2.97) we have used the fact that (2.98) is nonvanishing only when $m_1 + m_2 = m_1' + m_2'$. Expanding $1/r_{12}$ in terms of Legendre polynomials and using (1.5) and (1.8), we obtain

$$\frac{1}{r_{12}} = \sum_k \frac{r_<^k}{r_>^{k+1}} \sum_q (-1)^q C_q^{(k)}(\theta_1\varphi_1)\, C_{-q}^{(k)}(\theta_2\varphi_2), \tag{2.99}$$

where $r_<$ is the lesser and $r_>$ is the greater of r_1 and r_2. By using (2.99) and (1.15), (2.98) can be expressed as

$$\begin{aligned} \langle m_1 m_2 \,|\,|\, m_1' m_2'\rangle &= \sum_{kq} (-1)^q c^k(2m_1, 2m_1')\, c^k(2m_2, 2m_2') \\ &\qquad \times \delta(q + m_1', m_1)\, \delta(-q + m_2', m_2)\, F^k(dd) \\ &= \delta(m_1 + m_2, m_1' + m_2')(-1)^{m_1 - m_1'} \sum_k c^k(2m_1, 2m_1') \\ &\qquad \times c^k(2m_2, 2m_2')\, F^k(dd), \end{aligned} \tag{2.100}$$

where

$$F^k(dd) = \int_0^\infty r_1^2\, dr_1 \int_0^\infty r_2^2\, dr_2\, R_d^2(r_1)\, R_d^2(r_2)\, r_<^k / r_>^{k+1}. \tag{2.101}$$

By using the table of $c^k(lm, l'm')$ in Table 1.2, $\langle m_1 m_2 \| m_1' m_2' \rangle$ appearing in (2.97) are given as

$$\langle 22 \,|\,|\, 22 \rangle = \langle -2\, -2 \,|\,|\, -2\, -2 \rangle = \langle 2\, -2 \,|\,|\, 2\, -2 \rangle = \langle -22 \,|\,|\, -22 \rangle$$
$$= F^0 + \frac{4}{49}F^2 + \frac{1}{441}F^4, \tag{2.102}$$

$$\langle 2\, -2 \,|\,|\, -22 \rangle = \langle -22 \,|\,|\, 2\, -2 \rangle = \frac{70}{441}F^4, \tag{2.103}$$

where the F^k's are the abbreviation of the $F^k(dd)$'s. To simplify the results, the F_k's are often used in place of the F^k's:

$$F_0 = F^0, \qquad F_2 = \frac{1}{49}F^2, \qquad F_4 = \frac{1}{441}F^4. \tag{2.104}$$

Integrals F_k or F^k are called the *Slater integrals* or *Slater–Condon parameters.* Now $J(\zeta\zeta)$ is given in terms of the Slater integrals as

$$J(\zeta\zeta) = F_0 + 4F_2 + 36F_4\,. \tag{2.105}$$

It is also convenient to use the following parameters introduced by Racah:

$$A = F_0 - 49F_4\,, \qquad B = F_2 - 5F_4\,, \qquad C = 35F_4\,, \tag{2.106}$$

which are called the *Racah parameters.* In terms of the Racah parameters,

$$J(\zeta\zeta) = A + 4B + 3C. \tag{2.107}$$

Similarly all the ten two-electron integrals can be expressed by using the three Racah parameters as shown in Table 2.5.

TABLE 2.5

TEN TWO-ELECTRON INTEGRALS IN TERMS OF THE RACAH PARAMETERS

$J(\zeta\zeta) = A + 4B + 3C$	$J(uv) = A - 4B + C$
$J(\xi\eta) = A - 2B + C$	$K(uv) = 4B + C$
$K(\xi\eta) = 3B + C$	
$J(\zeta u) = A - 4B + C$	$J(\zeta v) = A + 4B + C$
$K(\zeta u) = 4B + C$	$K(\zeta v) = C$
$\langle \xi\eta \,|\,|\, \zeta u \rangle = \sqrt{3}B$	

Problem 2.13. By assuming $\langle \zeta \,|\, f \,|\, \zeta \rangle = \langle v \,|\, f \,|\, v \rangle$ and t_{2g} and e_g with d-character, diagonalize two-dimensional energy matrices for $^3T_1$, $^1T_2$, $^1A_1$, and ^1E, and show that the terms in each of the following sets are degenerate:

Set I:	$a\,^3T_1$,	$t_2e\,^3T_2$,	$e^2\,^3A_2$.	
Set II:	$a\,^1E$,	$a\,^1T_2$.		
Set III:	$a\,^1A_1$,	$b\,^1T_2$,	$t_2e\,^1T_1$,	$b\,^1E$.

Here, $E(b\,^{2S+1}\Gamma) > E(a\,^{2S+1}\Gamma)$. ◇

Chapter III MANY ELECTRONS IN A CUBIC FIELD

3.1 Many-Electron Wavefunctions

3.1.1 WAVEFUNCTIONS OF THE t_2^3 CONFIGURATIONS

The method of obtaining the wavefunctions of the t_2^3 electron configuration we are describing here is easily extended to the general cases of n-electrons ($n > 3$) in a shell. Later we will show that, once the wavefunctions of the t_2^n and e^m configurations are known, those of the $t_2^n e^m$ configurations are easily obtained.

We start from a system of one t_2' electron added to the system of the t_2^2 configuration. Here, t_2' means that it has the same symmetry as that of t_2, but it is different from the t_2 orbital: One may imagine the case in which t_2 is the split component of the atomic $3d$-orbital and t_2' that of the $5g$-orbital both of which belong to irreducible representation T_{2g} of the O_h-group. To obtain the wavefunctions of this system, one may use a method similar to that employed in obtaining the wavefunctions of the t_2e configuration. Since there are four terms, ${}^1A_1$, ${}^1E$, ${}^1T_2$, and ${}^3T_1$ of t_2^2, we obtain fifteen terms by adding t_2' to t_2^2 as shown in Table 3.1.

In order to obtain the wavefunctions of these terms, let us first study the function,

$$\Psi'(t_2^2(S_0\Gamma_0)\, t_2'S\Gamma) = \sum_{\substack{M_0 m_3 \\ \gamma_0 \gamma_3}} \Psi(t_2^2 S_0 \Gamma_0 M_0 \gamma_0)\, \phi(t_2' m_3 \gamma_3) \times \langle S_0 M_0 \tfrac{1}{2} m_3 \mid SM \rangle \langle \Gamma_0 \gamma_0 T_2 \gamma_3 \mid \Gamma\gamma \rangle, \quad (3.1)$$

TABLE 3.1

THE ALLOWED TERMS OF $t_2^2(S_0\Gamma_0)t_2'S\Gamma$

$S_0\Gamma_0$	$S\Gamma$
$^1A_1$	$^2T_2$
^1E	$^2T_1$, $^2T_2$
$^1T_2$	$^2A_1$, ^2E, $^2T_1$, $^2T_2$
$^3T_1$	$^2A_2$, ^2E, $^2T_1$, $^2T_2$ $^4A_2$, ^4E, $^4T_1$, $^4T_2$

where $\Psi(t_2^2S_0\Gamma_0M_0\gamma_0)$ is the wavefunction of t_2^2 involving electrons 1 and 2, and $\phi(t_2'm_3\gamma_3)$ the spin-orbital of the added t_2' electron, electron 3. The function given by (3.1) is base γ of irreducible representation Γ of the O_h-group and also the eigenfunction of $\mathbf{S}^2$ and S_z with eigenvalues $S(S+1)$ and M, respectively. Here, $\mathbf{S} = \mathbf{S}_0 + \mathbf{s}_3$ ($s_3 = \frac{1}{2}$). However, this function is not antisymmetric with respect to the exchange of electrons 1 and 3, and electrons 2 and 3, although it is antisymmetric to the exchange of electrons 1 and 2. Therefore, the next task to do is to make function (3.1) totally antisymmetric without destroying its two characters: (1) It is the base γ of irreducible representation Γ; (2) It is the eigenfunction of $\mathbf{S}^2$ and S_z.

In terms of Slater determinants, (3.1) can be expressed as

$$\Psi' = \sum_{k_1k_2k_3} C_{k_1k_2k_3} \,|\,\phi_{k_1}\phi_{k_2}\,|_{(1,2)} \times \phi_{k_3}(3), \tag{3.2}$$

where k_i represents sets of quantum numbers $(t_2m_i\gamma_i)$ for $i = 1, 2$, k_3 represents $(t_2'm_3\gamma_3)$. Subscript (1, 2) of the Slater determinant indicates that it involves electron coordinates $\mathbf{r}_1\boldsymbol{\sigma}_1$ and $\mathbf{r}_2\boldsymbol{\sigma}_2$. The $C_{k_1k_2k_3}$'s are the numerical coefficients determined from (3.1). For making (3.2) totally antisymmetric, it is sufficient to make the following linear combination:

$$\frac{1}{\sqrt{3}} \sum_{k_1k_2k_3} C_{k_1k_2k_3}[\,|\,\phi_{k_1}\phi_{k_2}\,|_{(1,2)} \times \phi_{k_3}(3) - |\,\phi_{k_1}\phi_{k_2}\,|_{(1,3)} \times \phi_{k_3}(2) + |\,\phi_{k_1}\phi_{k_2}\,|_{(2,3)} \times \phi_{k_3}(1)], \tag{3.3}$$

which, according to the property of determinants, can be reexpressed as

$$\Psi(t_2^2(S_0\Gamma_0)\,t_2'S\Gamma\,M\,\gamma) = \sum_{k_1k_2k_3} C_{k_1k_2k_3}\,|\,\phi_{k_1}\phi_{k_2}\phi_{k_3}\,|. \tag{3.4}$$

This function clearly keeps the two characteristics which function Ψ'

had, and in addition to this it is totally antisymmetric. Factor $1/\sqrt{3}$ in (3.3) was introduced to let (3.4) be normalized. Therefore,

$$\Psi(t_2^2(S_0\Gamma_0)\, t_2' S\Gamma\, M\, \gamma)$$

is the wavefunction of the $t_2^2t_2'$ electron configuration.

Now, in order to obtain the wavefunctions of the t_2^3 configuration, we replace t_2' in (3.4) by t_2. Then, the Slater determinants appearing in (3.4) are no longer mutually independent, and (3.4) vanishes identically for some particular sets of $S\Gamma$ in Table 3.1 as was seen in the case of t_2^2. In other words, terms with these sets of $S\Gamma$ are not allowed. In the allowed cases, however, wavefunctions (3.4) with a given $S\Gamma$ become essentially identical to each other even if they are constructed from different sets of $S_0\Gamma_0$. Therefore, the number of the allowed terms of t_2^3 is greatly reduced from fifteen in Table 3.1. Nonvanishing functions obtained by the replacement $t_2' \to t_2$ are usually not normalized, so that we have to normalize them to obtain wavefunctions $\Psi(t_2^3 S\Gamma M\gamma)$.

As a nonvanishing case of (3.4), let us calculate $\Psi(t_2^3\, {}^4A_2\, M = \frac{3}{2})$, which can be constructed from $S_0\Gamma_0 = {}^3T_1$ as seen in Table 3.1. From Table 2.2 we know that

$$\Psi(t_2^2\, {}^3T_1\, M{=}1\, \alpha) = |\, \eta\zeta\, |, \tag{3.5a}$$

$$\Psi(t_2^2\, {}^3T_1\, M{=}1\, \beta) = |\, \zeta\xi\, |, \tag{3.5b}$$

$$\Psi(t_2^2\, {}^3T_1\, M{=}1\, \gamma) = |\, \xi\eta\, |. \tag{3.5c}$$

By using the C–G coefficients for $\langle T_1\gamma_1 T_2\gamma_2 \mid A_2 e_2\rangle$ in Appendix II and the Wigner coefficients $\langle 1M_0 \frac{1}{2} m_3 \mid \frac{3}{2}\frac{3}{2}\rangle = \delta(M_0 1)\, \delta(m_3 \frac{1}{2})$, (3.2) is given as

$$\begin{aligned}\Psi'\left(t_2^2(^3T_1)\, t_2'\, {}^4A_2\, M{=}\frac{3}{2}\right) = -\frac{1}{\sqrt{3}} [&|\, \eta\zeta\, |_{(1,2)} \times \xi'(3)\, \alpha(3) \\ &+ |\, \zeta\xi\, |_{(1,2)} \times \eta'(3)\, \alpha(3) \\ &+ |\, \xi\eta\, |_{(1,2)} \times \zeta'(3)\, \alpha(3)].\end{aligned} \tag{3.6}$$

Antisymmetrizing (3.6), we obtain

$$\Psi\left(t_2^2(^3T_1)\, t_2'\, {}^4A_2\, M{=}\frac{3}{2}\right) = -\frac{1}{\sqrt{3}} [|\, \eta\zeta\xi'\, | + |\, \zeta\xi\eta'\, | + |\, \xi\eta\zeta'\, |], \tag{3.7}$$

which corresponds to (3.4). By performing the replacement, $\xi' \to \xi$, $\eta' \to \eta$, $\zeta' \to \zeta$, and noting that $|\, \xi\eta\zeta\, | = |\, \zeta\xi\eta\, | = |\, \eta\zeta\xi\, |$, the normalized wavefunction is obtained as

$$\Psi(t_2^3\, {}^4A_2\, M{=}3/2) = -|\, \xi\eta\zeta\, |. \tag{3.8}$$

As a vanishing case of (3.4), $\Psi(t_2^3\ {}^4E\ M{=}\frac{3}{2}\,u)$ is calculated from $S_0\Gamma_0 = {}^3T_1$ as follows: (3.2) in this case is

$$\Psi'(t_2^2({}^3T_1)\,t_2'\ {}^4E\ M{=}\frac{3}{2}\,u)$$
$$= -\frac{1}{\sqrt{2}}\,|\,\eta\zeta\,|_{(1,2)} \times \xi'(3)\,\alpha(3) + \frac{1}{\sqrt{2}}\,|\,\zeta\xi\,|_{(1,2)} \times \eta'(3)\,\alpha(3), \qquad (3.9)$$

so that (3.4) is proportional to $|\,\eta\zeta\xi'\,| - |\,\zeta\xi\eta'\,|$ which is vanishing by the replacement, $\xi' \to \xi$, $\eta' \to \eta$. Thus, the ${}^4E$ term of t_2^3 is not allowed to appear.

As an example of obtaining the nonvanishing identical wavefunctions from different sets of $S_0\Gamma_0$, let us calculate $\Psi(t_2^3\ {}^2E\ M{=}\frac{1}{2}\,u)$ from $S_0\Gamma_0 = {}^3T_1$ and ${}^1T_2$. For $S_0\Gamma_0 = {}^3T_1$, (3.2) is given as

$$\Psi'\left(t_2^2({}^3T_1)\,t_2'\ {}^2E\ M{=}\frac{1}{2}\,u\right) = -\frac{1}{2\sqrt{3}}\,[-|\,\bar{\eta}\zeta\,|_{(1,2)} \times \xi'(3)\,\alpha(3)$$
$$- |\,\eta\bar{\zeta}\,|_{(1,2)} \times \xi'(3)\,\alpha(3) + |\,\bar{\zeta}\xi\,|_{(1,2)} \times \eta'(3)\,\alpha(3)$$
$$+ |\,\zeta\bar{\xi}\,|_{(1,2)} \times \eta'(3)\,\alpha(3)] + \frac{1}{\sqrt{3}}\,[-|\,\eta\zeta\,|_{(1,2)}$$
$$\times\ \xi'(3)\,\beta(3) + |\,\zeta\xi\,|_{(1,2)} \times \eta'(3)\,\beta(3)], \qquad (3.10)$$

so that the normalized wavefunction of t_2^3 is

$$\Psi\left(t_2^3\ {}^2E\ M{=}\frac{1}{2}\,u\right) = \frac{1}{\sqrt{2}}\,[|\,\xi\bar{\eta}\zeta\,| - |\,\bar{\xi}\eta\zeta\,|]. \qquad (3.11)$$

On the other hand, for $S_0\Gamma_0 = {}^1T_2$, (3.2) is given as

$$\Psi'\left(t_2^2({}^1T_2)\,t_2'\ {}^2E\ M{=}\frac{1}{2}\,u\right) = \frac{1}{2\sqrt{3}}\,[2|\,\xi\bar{\eta}\,|_{(1,2)} \times \zeta'(3)\,\alpha(3)$$
$$+ 2|\,\eta\bar{\xi}\,|_{(1,2)} \times \zeta'(3)\,\alpha(3) - |\,\zeta\bar{\xi}\,|_{(1,2)} \times \eta'(3)\,\alpha(3)$$
$$- |\,\xi\bar{\zeta}\,|_{(1,2)} \times \eta'(3)\,\alpha(3) - |\,\eta\bar{\zeta}\,|_{(1,2)} \times \xi'(3)\,\alpha(3)$$
$$- |\,\zeta\bar{\eta}\,|_{(1,2)} \times \xi'(3)\,\alpha(3)]. \qquad (3.12)$$

From (3.12) it is clear that the normalized wavefunction thus obtained is identical to (3.11).

In this way nonvanishing wavefunctions are obtained only for terms ${}^4A_2$, ${}^2E$, ${}^2T_1$, and ${}^2T_2$. The wavefunctions for ${}^2E$, ${}^2T_1$, or ${}^2T_2$ constructed from various sets of $S_0\Gamma_0$ are identical to each other. Thus, we know that among fifteen terms of $t_2{}^2t_2{}'$ only four terms are allowed for the $t_2{}^3$ electron configuration. The number of states with $M = \frac{1}{2}$ in these allowed terms is $1 + 2 + 3 + 3 = 9$ which agrees with the number of independent Slater determinants with $M = \frac{1}{2}$ as follows:

$$\begin{array}{lllll} |\bar{\xi}\eta\zeta|, & |\xi\bar{\eta}\zeta|, & |\xi\eta\bar{\zeta}|, & |\xi\eta\bar{\eta}|, & |\xi\zeta\bar{\zeta}|, \\ |\eta\zeta\bar{\zeta}|, & |\eta\xi\bar{\xi}|, & |\zeta\xi\bar{\xi}|, & |\zeta\eta\bar{\eta}|. & \end{array} \tag{3.13}$$

All the wavefunctions of $t_2{}^3$ are given in Table 3.2.

TABLE 3.2

$\Psi(t_{2g}^3 S\Gamma M\gamma)$

$S\Gamma$	M	γ	Ψ
$^4A_2$	$\frac{3}{2}$	e_2	$-\lvert\xi\eta\zeta\rvert$
^2E	$\frac{1}{2}$	u	$\frac{1}{\sqrt{2}}[\lvert\xi\bar{\eta}\zeta\rvert - \lvert\bar{\xi}\eta\zeta\rvert]$
		v	$\frac{1}{\sqrt{6}}[2\lvert\xi\eta\bar{\zeta}\rvert - \lvert\xi\bar{\eta}\zeta\rvert - \lvert\bar{\xi}\eta\zeta\rvert]$
$^2T_1$	$\frac{1}{2}$	α	$\frac{1}{\sqrt{2}}[\lvert\xi\eta\bar{\eta}\rvert - \lvert\xi\zeta\bar{\zeta}\rvert]$
		β	$\frac{1}{\sqrt{2}}[\lvert\eta\zeta\bar{\zeta}\rvert - \lvert\eta\xi\bar{\xi}\rvert]$
		γ	$\frac{1}{\sqrt{2}}[\lvert\zeta\xi\bar{\xi}\rvert - \lvert\zeta\eta\bar{\eta}\rvert]$
$^2T_2$	$\frac{1}{2}$	ξ	$\frac{1}{\sqrt{2}}[\lvert\xi\eta\bar{\eta}\rvert + \lvert\xi\zeta\bar{\zeta}\rvert]$
		η	$\frac{1}{\sqrt{2}}[\lvert\eta\zeta\bar{\zeta}\rvert + \lvert\eta\xi\bar{\xi}\rvert]$
		ζ	$\frac{1}{\sqrt{2}}[\lvert\zeta\xi\bar{\xi}\rvert + \lvert\zeta\eta\bar{\eta}\rvert]$

The method described in this subsection can easily be extended and applied to obtaining the wavefunctions of $t_2{}^n$ $(n > 3)$ and e^m $(m > 2)$. However, we will show in the next chapter that the allowed terms of t_2^{6-n} are just those of $t_2{}^n$, and that the allowed terms of e^{4-m} are just those of e^m. Furthermore, there is a simple correlation between the wavefunctions for the terms of t_2^{6-n} and $t_2{}^n$ and also between those for the terms of e^{4-m} and e^m. Therefore, it is unnecessary to calculate the wavefunctions of $t_2{}^4$, $t_2{}^5$, and e^3 by using the method described here.

Problem 3.1. Derive wavefunctions for ${}^2T_1$ and ${}^2T_2$ of $t_2{}^3$ and confirm that the wavefunctions for ${}^2T_1$ or ${}^2T_2$ derived from different sets of $S_0\Gamma_0$ are identical to each other. ◇

3.1.2 Wavefunctions of $t_2{}^n e^m$

In order to obtain the wavefunctions of the $t_2{}^n e^m$ electron configuration we use the wavefunctions of $t_2{}^n$ and those of e^m, which are assumed to be known already.

As a simple example, let us consider the case of $t_2{}^2e$. Since the allowed terms of $t_2{}^2$ are ${}^1A_1$, ${}^1E$, ${}^1T_2$, and ${}^3T_1$, the allowed terms of $t_2{}^2e$ are those ten listed in Table 3.3. The wavefunctions are obtained from (3.4)

TABLE 3.3

The Allowed Terms of $t_2{}^2(S_0\Gamma_0)eS\Gamma$

$S_0\Gamma_0$	$S\Gamma$
${}^1A_1$	${}^1E$
${}^1E$	${}^2A_1$, ${}^2A_2$, ${}^2E$
${}^1T_2$	${}^2T_1$, ${}^2T_2$
${}^3T_1$	${}^2T_1$, ${}^2T_2$, ${}^4T_1$, ${}^4T_2$

in which t_2' is replaced by e, and k_3' by the sets of quantum numbers $(em_3\gamma_3)$.

Several examples will be shown below. $\Psi(t_2{}^2({}^3T_1)\, e\, {}^4T_2\, M{=}\tfrac{3}{2}\,\zeta)$ is obtained as follows: since one has

$$\Psi'(t_2{}^2({}^3T_1)e\, {}^4T_2\, M{=}\tfrac{1}{2}\,\zeta) = -|\, \xi\eta\, |_{(1,2)} \times v(3)\,\alpha(3), \tag{3.14}$$

Eq. (3.4) gives

$$\Psi(t_2{}^2({}^3T_1)e\, {}^4T_2\, M{=}\tfrac{3}{2}\,\zeta) = -|\, \xi\eta v\, |. \tag{3.15}$$

As an example of obtaining the wavefunctions with the same set of $S\Gamma$ but with different sets of $S_0\Gamma_0$, let us construct

$$\Psi(t_2{}^2({}^1A_1)e\, {}^2E\, M{=}\tfrac{1}{2}\,u) \quad \text{and} \quad \Psi(t_2{}^2({}^1E)e\, {}^2E\, M{=}\tfrac{1}{2}\,u).$$

Since one has

$$\Psi'\left(t_2^2(^1A_1)e\ ^2E\ M{=}\frac{1}{2}\ u\right)$$
$$= \frac{1}{\sqrt{3}}\,[|\ \xi\bar{\xi}\ |_{(1,2)} + |\ \eta\bar{\eta}\ |_{(1,2)} + |\ \zeta\bar{\zeta}\ |_{(1,2)}]\ u(3)\ \alpha(3), \tag{3.16}$$

Eq. (3.4) gives

$$\Psi\left(t_2^2(^1A_1)e\ ^2E\ M{=}\frac{1}{2}\ u\right) = \frac{1}{\sqrt{3}}\,[|\ \xi\bar{\xi}u\ | + |\ \eta\bar{\eta}u\ | + |\ \zeta\bar{\zeta}u\ |]. \tag{3.17}$$

Similarly, since one has

$$\Psi'\left(t_2^2(^1E)e\ ^2E\ M{=}\frac{1}{2}\ u\right)$$
$$= -\frac{1}{2\sqrt{3}}\,[2|\ \zeta\bar{\zeta}\ |_{(1,2)} - |\ \xi\bar{\xi}\ |_{(1,2)} - |\ \eta\bar{\eta}\ |_{(1,2)}]\ u(3)\ \alpha(3)$$
$$+ \frac{1}{2}[|\ \xi\bar{\xi}\ |_{(1,2)} - |\ \eta\bar{\eta}\ |_{(1,2)}]\ v(3)\ \alpha(3), \tag{3.18}$$

Eq. (3.4) gives

$$\Psi\left(t_2^2(^1E)e\ ^2E\ M{=}\frac{1}{2}\ u\right) = \frac{1}{2}\left\{-\frac{1}{\sqrt{3}}\,[2\,|\ \zeta\bar{\zeta}u\ | - |\ \xi\bar{\xi}u\ | - |\ \eta\bar{\eta}u\ |]\right.$$
$$\left. + |\ \xi\bar{\xi}v\ | - |\ \eta\bar{\eta}v\ |\right\}. \tag{3.19}$$

Contrary to the case of t_2^3, the wavefunctions (3.17) and (3.19) are entirely different. All the wavefunctions of t_2^2e thus calculated are listed in Table 3.4.

Problem 3.2. Construct the wavefunctions of the t_2e^2 electron configuration. ◇

Now we will construct the wavefunctions of $t_2^2e^2$. Since the allowed terms are $^1A_1$, ^1E, $^1T_2$, and $^3T_1$ for t_2^2 and $^1A_1$, ^1E, $^3A_2$ for e^2, eighteen terms are expected for $t_2^2e^2$ as shown in Table 3.5. By denoting the wavefunctions of t_2^2 and e^2 by $\Psi(t_2^2S_1\Gamma_1M_1\gamma_1)$ and $\Psi(e^2S_2\Gamma_2M_2\gamma_2)$, respectively, the wavefunction of the $t_2^2(S_1\Gamma_1)\ e^2(S_2\Gamma_2)\ S\Gamma M\gamma$ state is obtained by antisymmetrizing

$$\Psi'(t_2^2(S_1\Gamma_1)\ e^2(S_2\Gamma_2)\ S\Gamma M\gamma) = \sum_{\substack{M_1M_2\\ \gamma_1\gamma_2}} \Psi(t_2^2S_1\Gamma_1M_1\gamma_1)\ \Psi(e^2S_2\Gamma_2M_2\gamma_2)$$
$$\times \langle S_1M_1S_2M_2\ |\ SM\rangle\langle\Gamma_1\gamma_1\Gamma_2\gamma_2\ |\ \Gamma\gamma\rangle. \tag{3.20}$$

TABLE 3.4

$\Psi(t_2^2(S_0\Gamma_0)eS\Gamma\, M\, \gamma)$

$S\Gamma$	$S_0\Gamma_0$	M	γ	Ψ
$^2A_1$	^1E	$\frac{1}{2}$	e_1	$\frac{1}{2}\left\{\frac{1}{\sqrt{3}}[2\,\|\zeta\bar{\zeta}u\| - \|\xi\bar{\xi}u\| - \|\eta\bar{\eta}u\|] + \|\xi\bar{\xi}v\| - \|\eta\bar{\eta}v\|\right\}$
$^2A_2$	^1E	$\frac{1}{2}$	e_2	$\frac{1}{2}\left\{\frac{1}{\sqrt{3}}[2\,\|\zeta\bar{\zeta}v\| - \|\xi\bar{\xi}v\| - \|\eta\bar{\eta}v\|] - \|\xi\bar{\xi}u\| + \|\eta\bar{\eta}u\|\right\}$
^2E	$^1A_1$	$\frac{1}{2}$	u	$\frac{1}{\sqrt{3}}[\|\xi\bar{\xi}u\| + \|\eta\bar{\eta}u\| + \|\zeta\bar{\zeta}u\|]$
			v	$\frac{1}{\sqrt{3}}[\|\xi\bar{\xi}v\| + \|\eta\bar{\eta}v\| + \|\zeta\bar{\zeta}v\|]$
^2E	^1E	$\frac{1}{2}$	u	$\frac{1}{2}\left\{-\frac{1}{\sqrt{3}}[2\,\|\zeta\bar{\zeta}u\| - \|\xi\bar{\xi}u\| - \|\eta\bar{\eta}u\|] + \|\xi\bar{\xi}v\| - \|\eta\bar{\eta}v\|\right\}$
			v	$\frac{1}{2}\left\{\frac{1}{\sqrt{3}}[2\,\|\zeta\bar{\zeta}v\| - \|\xi\bar{\xi}v\| - \|\eta\bar{\eta}v\|] + \|\xi\bar{\xi}u\| - \|\eta\bar{\eta}u\|\right\}$
$^2T_1$	$^1T_2$	$\frac{1}{2}$	α	$-\frac{1}{2\sqrt{2}}\{\sqrt{3}[\|\eta\bar{\zeta}u\| - \|\bar{\eta}\zeta u\|] + \|\eta\bar{\zeta}v\| - \|\bar{\eta}\zeta v\|\}$
			β	$\frac{1}{2\sqrt{2}}\{\sqrt{3}[\|\zeta\bar{\xi}u\| - \|\bar{\zeta}\xi u\|] - \|\zeta\bar{\xi}v\| + \|\bar{\zeta}\xi v\|\}$
			γ	$\frac{1}{\sqrt{2}}[\|\xi\bar{\eta}v\| - \|\bar{\xi}\eta v\|]$
$^2T_1$	$^3T_1$	$\frac{1}{2}$	α	$\frac{1}{2\sqrt{6}}\{[\|\bar{\eta}\zeta u\| + \|\eta\bar{\zeta}u\|] - \sqrt{3}[\|\bar{\eta}\zeta v\| + \|\eta\bar{\zeta}v\|] + 2[\|\eta\zeta\bar{u}\| - \sqrt{3}\,\|\eta\zeta\bar{v}\|]\}$
			β	$\frac{1}{2\sqrt{6}}\{[\|\bar{\zeta}\xi u\| + \|\zeta\bar{\xi}u\|] + \sqrt{3}[\|\bar{\zeta}\xi v\| + \|\zeta\bar{\xi}v\|] + 2[\|\zeta\xi\bar{u}\| + \sqrt{3}\,\|\zeta\xi\bar{v}\|]\}$
			γ	$\frac{1}{\sqrt{6}}\{-[\|\bar{\xi}\eta u\| + \|\xi\bar{\eta}u\|] + 2\,\|\xi\eta\bar{u}\|\}$
$^2T_2$	$^1T_2$	$\frac{1}{2}$	ξ	$\frac{1}{2\sqrt{2}}\{-[\|\eta\bar{\zeta}u\| - \|\bar{\eta}\zeta u\|] + \sqrt{3}[\|\eta\bar{\zeta}v\| - \|\bar{\eta}\zeta v\|]\}$
			η	$-\frac{1}{2\sqrt{2}}\{[\|\zeta\bar{\xi}u\| - \|\bar{\zeta}\xi u\|] + \sqrt{3}[\|\zeta\bar{\xi}v\| - \|\bar{\zeta}\xi v\|]\}$
			ζ	$\frac{1}{\sqrt{2}}[\|\xi\bar{\eta}u\| - \|\bar{\xi}\eta u\|]$

TABLE 3.4 (*continued*)

$S\Gamma$	$S_0\Gamma_0$	M	γ	Ψ
$^2T_2$	$^3T_1$	$\frac{1}{2}$	ξ	$\frac{1}{2\sqrt{6}}\{-\sqrt{3}[\mid \bar{\eta}\zeta u \mid + \mid \eta\bar{\zeta}u \mid] - [\mid \bar{\eta}\zeta v \mid + \mid \eta\bar{\zeta}v \mid]$ $+ 2[\sqrt{3} \mid \eta\zeta\bar{u} \mid + \mid \eta\zeta\bar{v} \mid]\}$
			η	$\frac{1}{2\sqrt{6}}\{\sqrt{3}[\mid \bar{\zeta}\xi u \mid + \mid \zeta\bar{\xi}u \mid] - [\mid \bar{\zeta}\xi v \mid + \mid \zeta\bar{\xi}v \mid]$ $+ 2[-\sqrt{3} \mid \zeta\xi\bar{u} \mid + \mid \zeta\xi\bar{v} \mid]\}$
			ζ	$\frac{1}{\sqrt{6}}[\mid \bar{\xi}\eta v \mid + \mid \xi\bar{\eta}v \mid - 2 \mid \xi\eta\bar{v} \mid]$
$^4T_1$	$^3T_1$	$\frac{3}{2}$	α	$\frac{1}{2}[-\mid \eta\zeta u \mid + \sqrt{3} \mid \eta\zeta v \mid]$
			β	$-\frac{1}{2}[\mid \zeta\xi u \mid + \sqrt{3} \mid \zeta\xi v \mid]$
			γ	$\mid \xi\eta u \mid$
$^4T_2$	$^3T_1$	$\frac{3}{2}$	ξ	$\frac{1}{2}[\sqrt{3} \mid \eta\zeta u \mid + \mid \eta\zeta v \mid]$
			η	$\frac{1}{2}[-\sqrt{3} \mid \zeta\xi u \mid + \mid \zeta\xi v \mid]$
			ζ	$-\mid \xi\eta v \mid$

In terms of Slater determinants, (3.20) can be expressed as

$$\Psi' = \sum_{k_1k_2k_3k_4} C_{k_1k_2k_3k_4} \mid \phi_{k_1}\phi_{k_2} \mid_{(1,2)} \times \mid \phi_{k_3}\phi_{k_4} \mid_{(3,4)}, \tag{3.21}$$

where

$$k_i = (t_2 m_i \gamma_i) \quad \text{for} \quad i = 1, 2$$
$$= (e m_i \gamma_i) \quad \text{for} \quad i = 3, 4.$$

Therefore, the antisymmetrization is achieved by making the following linear combination:

$$\frac{1}{\sqrt{6}} \sum_{k_1k_2k_3k_4} C_{k_1k_2k_3k_4}[\mid \phi_{k_1}\phi_{k_2} \mid_{(1,2)} \times \mid \phi_{k_3}\phi_{k_4} \mid_{(3,4)}$$
$$- \mid \phi_{k_1}\phi_{k_2} \mid_{(3,2)} \times \mid \phi_{k_3}\phi_{k_4} \mid_{(1,4)} - \mid \phi_{k_1}\phi_{k_2} \mid_{(4,2)} \times \mid \phi_{k_3}\phi_{k_4} \mid_{(3,1)}$$
$$- \mid \phi_{k_1}\phi_{k_2} \mid_{(1,3)} \times \mid \phi_{k_3}\phi_{k_4} \mid_{(2,4)} - \mid \phi_{k_1}\phi_{k_2} \mid_{(1,4)} \times \mid \phi_{k_3}\phi_{k_4} \mid_{(3,2)}$$
$$+ \mid \phi_{k_1}\phi_{k_2} \mid_{(3,4)} \times \mid \phi_{k_3}\phi_{k_4} \mid_{(1,2)}]. \tag{3.22}$$

TABLE 3.5

THE ALLOWED TERMS OF $t_2^2(S_1\Gamma_1)e^2(S_2\Gamma_2)S\Gamma$

$S_1\Gamma_1$	$S_2\Gamma_2$	$S\Gamma$
$^1A_1$		$^1A_1$
^1E	$^1A_1$	^1E
$^1T_2$		$^1T_2$
$^3T_1$		$^3T_1$
$^1A_1$		^1E
^1E	^1E	$^1A_1$, $^1A_2$, ^1E
$^1T_2$		$^1T_1$, $^1T_2$
$^3T_1$		$^3T_1$, $^3T_2$
$^1A_1$		$^3A_2$
^1E	$^3A_2$	^3E
$^1T_2$		$^3T_1$
$^3T_1$		$^1T_2$, $^3T_2$, $^5T_2$

Since $(1/\sqrt{6})[\cdots]$ in (3.22) is Laplace's expansion‡ of Slater determinant

‡ Laplace's expansion of the n-dimensional determinant D in terms of the r-dimensional ($r < n$) small determinants is given as

$$D = \sum_{(\alpha'\beta'\cdots\lambda')} D\begin{pmatrix} \overbrace{\alpha\ \beta\ \cdots\ \lambda}^{r} \\ \alpha'\beta'\ \cdots\ \lambda' \end{pmatrix} \times \operatorname{adj} D\begin{pmatrix} \alpha\ \beta\ \cdots\ \lambda \\ \alpha'\beta'\ \cdots\ \lambda' \end{pmatrix},$$

where

$$\operatorname{adj} D\begin{pmatrix} \alpha\ \beta\ \cdots\ \lambda \\ \alpha'\beta'\ \cdots\ \lambda' \end{pmatrix} = (-1)^{(\alpha+\beta+\cdots+\lambda)+(\alpha'+\beta'+\cdots+\lambda')} \operatorname{comp} D\begin{pmatrix} \alpha\ \beta\ \cdots\ \lambda \\ \alpha'\beta'\ \cdots\ \lambda \end{pmatrix}.$$

Here

$$D\begin{pmatrix} \alpha\ \beta\ \cdots\ \lambda \\ \alpha'\beta'\ \cdots\ \lambda' \end{pmatrix}$$

is an r-dimensional small determinant constructed by picking up elements at the α-, β-,..., λth rows (or columns) and the α'-, β'-,..., λ'th columns (or rows) of D. In this case the order of α, β,..., λ and also that of α', β',..., λ' are fixed as found in D. The summation runs over all possible sets of (α', β',..., λ'), so that the number of terms is given by ${}_nC_r$. Complementary minor,

$$\operatorname{comp} D\begin{pmatrix} \alpha\ \beta\ \cdots\ \lambda \\ \alpha'\beta'\ \cdots\ \lambda' \end{pmatrix},$$

is constructed by picking up elements at the remaining rows and columns with the order as found in D.

$|\phi_{k_1}\phi_{k_2}\phi_{k_3}\phi_{k_4}|$ in terms of the two-dimensional small determinants, (3.22) can be expressed as

$$\Psi(t_2^2(S_1\Gamma_1)\, e^2(S_2\Gamma_2)\, S\Gamma M\gamma) = \sum_{k_1k_2k_3k_4} C_{k_1k_2k_3k_4} |\phi_{k_1}\phi_{k_2}\phi_{k_3}\phi_{k_4}|, \tag{3.23}$$

which is the wavefunction of the $t_2^2(S_1\Gamma_1)\, e^2(S_2\Gamma_2)\, S\Gamma M\gamma$ state.

For example $\Psi(t_2^2(^1A_1)\, e^2(^1E)\, ^1Eu)$ is obtained as follows: Since one has

$$\Psi'(t_2^2(^1A_1)\, e^2(^1E)\, ^1Eu) = \frac{1}{\sqrt{6}} [|\xi\bar{\xi}|_{(1,2)} + |\eta\bar{\eta}|_{(1,2)} + |\zeta\bar{\zeta}|_{(1,2)}] \times [-|u\bar{u}|_{(3,4)} + |v\bar{v}|_{(3,4)}], \tag{3.24}$$

Eq. (3.23) gives

$$\Psi(t_2^2(^1A_1)\, e^2(^1E)\, ^1Eu) = \frac{1}{\sqrt{6}} [-|\xi\bar{\xi}u\bar{u}| + |\xi\bar{\xi}v\bar{v}| - |\eta\bar{\eta}u\bar{u}| + |\eta\bar{\eta}v\bar{v}| - |\zeta\bar{\zeta}u\bar{u}| + |\zeta\bar{\zeta}v\bar{v}|]. \tag{3.25}$$

The other ^1Eu state of $t_2^2(^1E)\, e^2(^1E)$ is obtained from

$$\begin{aligned}\Psi'(t_2^2(^1E)e^2(^1E)\, ^1Eu) = \frac{1}{2\sqrt{2}} \Big\{ -\frac{1}{\sqrt{3}} [-|\xi\bar{\xi}|_{(1,2)} - |\eta\bar{\eta}|_{(1,2)} \\ + 2|\zeta\bar{\zeta}|_{(1,2)}][-|u\bar{u}|_{(3,4)} + |v\bar{v}|_{(3,4)}] \\ + [|\xi\bar{\xi}|_{(1,2)} - |\eta\bar{\eta}|_{(1,2)}][|u\bar{v}|_{(3,4)} - |\bar{u}v|_{(3,4)}] \Big\}\end{aligned} \tag{3.26}$$

as

$$\begin{aligned}\Psi(t_2^2(^1E)e^2(^1E)\, ^1Eu) = \frac{1}{2\sqrt{2}} \Big\{ -\frac{1}{\sqrt{3}} [|\xi\bar{\xi}u\bar{u}| - |\xi\bar{\xi}v\bar{v}| \\ + |\eta\bar{\eta}u\bar{u}| - |\eta\bar{\eta}v\bar{v}| - 2|\zeta\bar{\zeta}u\bar{u}| + 2|\zeta\bar{\zeta}v\bar{v}|] \\ + |\xi\bar{\xi}u\bar{v}| - |\xi\bar{\xi}\bar{u}v| - |\eta\bar{\eta}u\bar{v}| + |\eta\bar{\eta}\bar{u}v| \Big\}.\end{aligned} \tag{3.27}$$

As expected (3.27) is entirely different from (3.25) with the same set of $S\Gamma M\gamma$.

The method of obtaining the wavefunctions of t_2^2e and $t_2^2e^2$ described here can easily be extended for obtaining those of $t_2^ne^m$ ($n \leqslant 6$, $m \leqslant 4$).

3.2 Formulas for Calculating Matrix Elements

3.2.1 Matrix Elements of One-Electron Operators

Let us denote operators acting on one electron such as the electric dipole moment $-e\mathbf{r}_i$, the crystalline field potential energy $V_c(\mathbf{r}_i)$ as f_i. In the N-electron system, a one-electron operator is given by

$$F = \sum_{i=1}^{N} f_i . \tag{3.28}$$

In this subsection we will give the formulas to reduce the integral,

$$\sum_{\sigma} \int d\boldsymbol{\tau} | \phi_{k_1}\phi_{k_2} \cdots \phi_{k_N} |^* F | \phi_{k_1'}\phi_{k_2'} \cdots \phi_{k_N'} |, \tag{3.29}$$

to one-electron integrals. In (3.29) $\sum_\sigma$ is the summation over the spin-coordinates of N electrons and the integration is carried out with respect to the space coordinates of N electrons. Inserting (3.28) into (3.29), we obtain

$$\begin{aligned}(3.29) &= \sum_{i=1}^{N} \sum_{\sigma} \int d\boldsymbol{\tau} | \phi_{k_1}\phi_{k_2} \cdots \phi_{k_N} |^* f_i \, | \phi_{k_1'}\phi_{k_2'} \cdots \phi_{k_N'} | \\ &= N \sum_{\sigma} \int d\boldsymbol{\tau} | \phi_{k_1}\phi_{k_2} \cdots \phi_{k_N} |^* f_1 \, | \phi_{k_1'}\phi_{k_2'} \cdots \phi_{k_N'} |. \end{aligned} \tag{3.30}$$

In deriving the last expression of (3.30), we have used the fact that the integral involving f_i does not change by renumbering i and 1 as 1 and i, respectively, as both the Slater determinants involved merely change their signs by this renumbering.

In (3.29) and (3.30) spin-orbitals ϕ_{k_i} and $\phi_{k_{i'}}$ in the Slater determinants are arranged in such a way that, if ϕ_{k_i} and $\phi_{k_i'}$ are the same, they are located at the same positions from the extreme right in the Slater determinants. In other words, $k_i \neq k_{j'}$ if $i \neq j$, $k_j = k_j'$ for all $j > i$ if $k_i = k_i'$, and $k_j \neq k_j'$ for all $j < i$ if $k_i \neq k_i'$. This arrangement can always be achieved by changing the order of columns in the Slater determinant. For example, when the matrix element between $| \eta\bar{\xi}\zeta |$ and $| \eta\zeta\bar{u} |$ is calculated, we first change the order of columns to give $-| \eta\bar{\xi}\zeta | = | \phi_{k_1}\phi_{k_2}\phi_{k_3} | = | \bar{\xi}\eta\zeta |$ and $(-1)^2 | \eta\zeta\bar{u} | = | \phi_{k_{1'}}\phi_{k_{2'}}\phi_{k_{3'}} | = | \bar{u}\eta\zeta |$, then calculate (3.29) and finally multiply the result by $(-1)^3$.

Now to simplify (3.30) further, we expand both the Slater determinants as

$$|\phi_{\alpha_1}\phi_{\alpha_2}\cdots\phi_{\alpha_N}| = N^{-1/2}\sum_{i=1}^{N}(-1)^{1+i}\phi_{\alpha_i}(1)$$
$$\times\,|\phi_{\alpha_1}\phi_{\alpha_2}\cdots\phi_{\alpha_{i-1}}\phi_{\alpha_{i+1}}\cdots\phi_{\alpha_N}|_{(2,3,\ldots,N)}$$
$$(\alpha_i = k_i \quad\text{and}\quad k_i'), \tag{3.31}$$

which is nothing but a particular case of Laplace's expansion. In (3.31) factor $N^{-1/2}$ comes from the normalization factors included in the Slater determinants. Inserting (3.31) into (3.30), one obtains

$$(3.29) = \sum_{i=1}^{N}\sum_{j=1}^{N}(-1)^{i+j}\langle\phi_{k_i}|f|\phi_{k_j}\rangle\, S_{ij}\,,$$
$$S_{ij} = \sum_{\sigma}{}'\int d\tau'\,|\phi_{k_1}\phi_{k_2}\cdots\phi_{k_{i-1}}\phi_{k_{i+1}}\cdots\phi_{k_N}|^*$$
$$\times\,|\phi_{k_1'}\phi_{k_2'}\cdots\phi_{k_{j-1}'}\phi_{k_{j+1}'}\cdots\phi_{k_N'}|, \tag{3.32}$$

where $\sum_\sigma' \int d\tau'$ is carried out for electrons, 2, 3,..., N. Because of the orthogonality relation between Slater determinants, S_{ij} is nonvanishing and is unity only when the two Slater determinants in the integrand are identical.

Case I $k_i = k_i'$ *for all* i. In this case, S_{ij} is nonvanishing and is unity only when $i = j$, and from (3.32) one obtains

$$\sum_{\sigma}\int d\tau\,|\phi_{k_1}\phi_{k_2}\cdots\phi_{k_N}|^* F|\phi_{k_1}\phi_{k_2}\cdots\phi_{k_N}| = \sum_{i=1}^{N}\langle\phi_{k_i}|f|\phi_{k_i}\rangle. \tag{3.33}$$

Case II $k_1 \neq k_1'$, $k_i = k_i'$ $(i \neq 1)$. In this case, S_{ij} is nonvanishing and unity only when $i = j = 1$. Thus, one obtains

$$\sum_{\sigma}\int d\tau\,|\phi_{k_1}\phi_{k_2}\cdots\phi_{k_N}|^* F|\phi_{k_1'}\phi_{k_2}\cdots\phi_{k_N}| = \langle\phi_{k_1}|f|\phi_{k_1'}\rangle \quad (k_1 \neq k_1'). \tag{3.34}$$

Case III $k_1 \neq k_1'$, $k_2 \neq k_2'$. In this case, S_{ij} is always zero, and one obtains

$$\sum_{\sigma}\int d\tau\,|\phi_{k_1}\phi_{k_2}\cdots\phi_{k_N}|^* F|\phi_{k_1'}\phi_{k_2'}\cdots\phi_{k_N'}| = 0$$
$$(k_1 \neq k_1', \quad k_2 \neq k_2'). \tag{3.35}$$

3.2.2 Matrix Elements of Two-Electron Operators

Let us denote operators acting on two electrons, i and j, such as the Coulomb interaction $1/r_{ij}$ as g_{ij}. In the N-electron system, a two-electron operator is given by

$$G = \sum_{j>i=1}^{N} g_{ij}\,. \tag{3.36}$$

In this subsection, we will derive the formulas to reduce the N-electron integral,

$$\sum_{\sigma} \int d\boldsymbol{\tau}\, |\phi_{k_1}\phi_{k_2} \cdots \phi_{k_N}|^* G |\phi_{k_1'}\phi_{k_2'} \cdots \phi_{k_N'}|, \tag{3.37}$$

to two-electron integrals. In (3.37) spin-orbitals in the Slater determinants are arranged in just the same way as that mentioned in the previous subsection. Inserting (3.36) into (3.37), one obtains

$$\begin{aligned}(3.37) &= \sum_{j>i=1}^{N} \sum_{\sigma} \int d\boldsymbol{\tau}\, |\phi_{k_1}\phi_{k_2} \cdots \phi_{k_N}|^* g_{ij} |\phi_{k_1'}\phi_{k_2'} \cdots \phi_{k_N'}| \\ &= \frac{N(N-1)}{2} \sum_{\sigma} \int d\boldsymbol{\tau}\, |\phi_{k_1}\phi_{k_2} \cdots \phi_{k_N}|^* g_{12} |\phi_{k_1'}\phi_{k_2'} \cdots \phi_{k_N'}|. \end{aligned} \tag{3.38}$$

For the derivation of the last expression of (3.38), we have used the fact that the integral involving g_{ij} is invariant to the renumbering of electrons, $i \to 1$, $1 \to i$, $j \to 2$, and $2 \to j$. Factor $N(N-1)/2$ is the number of terms, ${}_NC_2$, in (3.36).

By using the formula of Laplace's expansion, both the Slater determinants in (3.38) are expanded in terms of two-dimensional Slater determinants as follows:

$$\begin{aligned}|\phi_{\alpha_1}\phi_{\alpha_2} \cdots \phi_{\alpha_N}| &= [N(N-2)/2]^{-1/2} \sum_{j=1}^{N} (-1)^{i+j+3} \times |\phi_{\alpha_i}\phi_{\alpha_j}|_{(1,2)} \\ &\times |\phi_{\alpha_1}\phi_{\alpha_2} \cdots \phi_{\alpha_{i-1}}\phi_{\alpha_{i+1}} \cdots \phi_{\alpha_{j-1}}\phi_{\alpha_{j+1}} \cdots \phi_{\alpha_N}|_{(3,4,\ldots,N)}\,, \end{aligned} \tag{3.39}$$

where factor $[N(N-1)/2]^{-1/2}$ comes from the normalization factors in the Slater determinants. Inserting (3.39) into (3.38), one obtains

$$\begin{aligned}(3.37) = \sum_{q>p=1}^{N} \sum_{s>r=1}^{N} (-1)^{p+q+r+s} [&\langle \phi_{k_p}\phi_{k_q} | g | \phi_{k_r'}\phi_{k_s'} \rangle \\ &- \langle \phi_{k_p}\phi_{k_q} | g | \phi_{k_s'}\phi_{k_r'} \rangle]\, S_{pq,rs}\,, \end{aligned} \tag{3.40}$$

where

$$S_{pq,rs} = \sum_{\sigma}'' \int d\tau'' \, |\phi_{k_1}\phi_{k_2} \cdots \phi_{k_{p-1}}\phi_{k_{p+1}} \cdots \phi_{k_{q-1}}\phi_{k_{q+1}} \cdots \phi_{k_N}|^*$$

$$\times |\phi_{k_1'}\phi_{k_2'} \cdots \phi_{k'_{r-1}}\phi_{k'_{r+1}} \cdots \phi_{k'_{s-1}}\phi_{k'_{s+1}} \cdots \phi_{k_N'}|.$$

In factor $S_{pq,rs}$, $\sum_{\sigma}'' \int d\tau''$ is carried out for electrons, 3, 4,..., N, and $S_{pq,rs}$ is nonvanishing and is unity only when the two Slater determinants in the integrand are identical.

Case I $k_i = k_i'$ *for all* i. In this case $S_{pq,rs}$ is nonzero and unity only when $p = r$ and $q = s$. Therefore, (3.40) gives

$$\sum_{\sigma} \int d\tau \, |\phi_{k_1}\phi_{k_2} \cdots \phi_{k_N}|^* G |\phi_{k_1}\phi_{k_2} \cdots \phi_{k_N}|$$

$$= \sum_{j>i=1}^{N} [\langle \phi_{k_i}\phi_{k_j} | g | \phi_{k_i}\phi_{k_j} \rangle - \langle \phi_{k_i}\phi_{k_j} | g | \phi_{k_j}\phi_{k_i} \rangle]. \qquad (3.41)$$

Case II $k_1 \neq k_1'$, $k_i = k_i'$ $(i \neq 1)$. In this case, $S_{pq,rs}$ is nonzero and unity only when $p = r = 1$ and $q = s$. Thus, one obtains

$$\sum_{\sigma} \int d\tau \, |\phi_{k_1}\phi_{k_2} \cdots \phi_{k_N}|^* G |\phi_{k_1'}\phi_{k_2} \cdots \phi_{k_N}|$$

$$= \sum_{j=2,3,\ldots,N} [\langle \phi_{k_1}\phi_{k_j} | g | \phi_{k_1'}\phi_{k_j} \rangle - \langle \phi_{k_1}\phi_{k_j} | g | \phi_{k_j}\phi_{k_1'} \rangle]$$

$$(k_1 \neq k_1'). \qquad (3.42)$$

Case III $k_1 \neq k_1'$, $k_2 \neq k_2'$, $k_i = k_i'$ $(i \neq 1, 2)$. In this case, $S_{pq,rs}$ is nonzero and unity only when $p = r = 1$ and $q = s = 2$, and one obtains

$$\sum_{\sigma} \int d\tau \, |\phi_{k_1}\phi_{k_2}\phi_{k_3} \cdots \phi_{k_N}|^* G |\phi_{k_1'}\phi_{k_2'}\phi_{k_3} \cdots \phi_{k_N}|$$

$$= \langle \phi_{k_1}\phi_{k_2} | g | \phi_{k_1'}\phi_{k_2'} \rangle - \langle \phi_{k_1}\phi_{k_2} | g | \phi_{k_2'}\phi_{k_1'} \rangle \quad (k_1 \neq k_1', \quad k_2 \neq k_2'). \qquad (3.43)$$

Case IV $k_1 \neq k_1'$, $k_2 \neq k_2'$, $k_3 \neq k_3'$. In this case, $S_{pq,rs}$ is always zero, leading to

$$\sum_{\sigma} \int d\tau \, |\phi_{k_1}\phi_{k_2}\phi_{k_3} \cdots \phi_{k_N}|^* G |\phi_{k_1'}\phi_{k_2'}\phi_{k_3'} \cdots \phi_{k_N'}| = 0$$

$$(k_1 \neq k_1', \quad k_2 \neq k_2', \quad k_3 \neq k_3'). \qquad (3.44)$$

Let us calculate one example using (3.42):

$$\sum_\sigma \int d\tau \mid \bar{v}\eta\bar{\zeta} \mid G \mid \bar{u}\eta\bar{\zeta} \mid = [\langle v\beta\ \eta\alpha \mid g \mid u\beta\ \eta\alpha\rangle - \langle v\beta\ \eta\alpha \mid g \mid \eta\alpha\ u\beta\rangle] + [\langle v\beta\ \zeta\beta \mid g \mid u\beta\ \zeta\beta\rangle - \langle v\beta\ \zeta\beta \mid g \mid \zeta\beta\ u\beta\rangle]. \tag{3.45}$$

Because of the orthogonality between spin functions, the second term, $\langle v\beta\ \eta\alpha \mid g \mid \eta\alpha\ u\beta\rangle$, in (3.45) is zero.

3.3 Energy Matrices in the Three-Electron System

3.3.1 Term Energies in $t_2{}^3$

As mentioned in Section 3.1, the allowed terms of the $t_2{}^3$ electron configuration are $^4A_2$, ^2E, $^2T_1$ and $^2T_2$ and we see that no term appears more than once in this configuration. Therefore, the matrix of $\mathscr{H}_1$ is already diagonal if it is calculated by using the wavefunctions associated with these terms. Thus, the term energies within the $t_2{}^3$ configuration are obtained directly by using the wavefunctions in Table 3.2 and the formulas (3.41–3.44). Note that the matrix elements of $\mathscr{H}_0$ appearing in the diagonal are all the same within a fixed electron configuration. Since the term energies are independent of M and γ, it is convenient to choose the wavefunction of the simplest form associated with a particular set of M and γ.

For $^4A_2$, by using (3.41), one obtains

$$\langle t_2{}^3\,{}^4A_2\,M{=}\tfrac{3}{2} \mid \mathscr{H}_1 \mid t_2{}^3\,{}^4A_2\,M{=}\tfrac{3}{2}\rangle = \sum_\sigma \int d\boldsymbol{\tau} \mid \xi\eta\zeta \mid^* G \mid \xi\eta\zeta \mid$$
$$= \langle \xi\alpha\ \eta\alpha \mid g \mid \xi\alpha\ \eta\alpha\rangle - \langle \xi\alpha\ \eta\alpha \mid g \mid \eta\alpha\ \xi\alpha\rangle + \langle \xi\alpha\ \zeta\alpha \mid g \mid \xi\alpha\ \zeta\alpha\rangle - \langle \xi\alpha\ \zeta\alpha \mid g \mid \zeta\alpha\ \xi\alpha\rangle + \langle \eta\alpha\ \zeta\alpha \mid g \mid \eta\alpha\ \zeta\alpha\rangle - \langle \eta\alpha\ \zeta\alpha \mid g \mid \zeta\alpha\ \eta\alpha\rangle, \tag{3.46}$$

which, in terms of the Coulomb and the exchange integrals defined in Section 2.3, can simply be expressed as

$$\langle t_2{}^3\,{}^4A_2 M = \tfrac{3}{2} \mid \mathscr{H}_1 \mid t_2{}^3\,{}^4A_2\,M{=}\tfrac{3}{2}\rangle = 3[J(\xi\eta) - K(\xi\eta)]. \tag{3.47}$$

For ^2E, by using the wavefunction of the $t_2{}^3\,{}^2E\ M{=}\frac{1}{2}\,u$ state, one obtains

$$\langle t_2{}^3\,{}^2E\ M{=}\tfrac{1}{2}\,u|\ \mathscr{H}_1\ |\ t_2{}^3\,{}^2E\ M{=}\tfrac{1}{2}\,u\rangle$$
$$= \tfrac{1}{2}\sum_{\sigma}\int d\boldsymbol{\tau}[|\ \xi\bar{\eta}\zeta\ |^*\ G|\ \xi\bar{\eta}\zeta\ | + |\ \bar{\xi}\eta\zeta\ |^*\ G|\ \bar{\xi}\eta\zeta\ | - 2|\ \xi\bar{\eta}\zeta\ |^*\ G|\ \bar{\xi}\eta\zeta\ |]$$
$$= 3J(\xi\eta). \tag{3.48}$$

Similarly, one obtains

$$\langle t_2{}^3\,{}^2T_1\ |\ \mathscr{H}_1\ |\ t_2{}^3\,{}^2T_1\rangle = 2J(\xi\eta) + J(\zeta\zeta) - 2K(\xi\eta), \tag{3.49}$$

and

$$\langle t_2{}^3\,{}^2T_2\ |\ \mathscr{H}_1\ |\ t_2{}^3\,{}^2T_2\rangle = 2J(\xi\eta) + J(\zeta\zeta). \tag{3.50}$$

Now, considering the relations $K(\xi\eta) > 0$ and $J(\zeta\zeta) > J(\xi\eta)$ mentioned in Section 2.3, the order of the terms is found as

$$\begin{aligned} E(^2E) &> E(^4A_2), & E(^2T_2) &> E(^2T_1), \\ E(^2T_1) &> E(^4A_2), & E(^2T_2) &> E(^2E), \end{aligned} \tag{3.51}$$

which shows that the $^4A_2$ term is lowest in energy in agreement with the Hund rule. From (3.51) it is impossible to determine which is higher, $^2T_1$ or ^2E. However, if the t_2 orbital is assumed to be the *d*-function, one sees that the $^2T_1$ and ^2E levels are accidentally degenerate, as the term energies are given to this approximation as follows:

$$E(t_2{}^3\,{}^4A_2) = 3A - 15B, \tag{3.52a}$$
$$E(t_2{}^3\,{}^2E) = 3A - 6B + 3C, \tag{3.52b}$$
$$E(t_2{}^3\,{}^2T_1) = 3A - 6B + 3C, \tag{3.52c}$$
$$E(t_2{}^3\,{}^2T_2) = 3A + 5C. \tag{3.52d}$$

The relative positions of the terms with $t_2{}^3$ are visualized in Fig. 3.1 by using the result in (3.52).

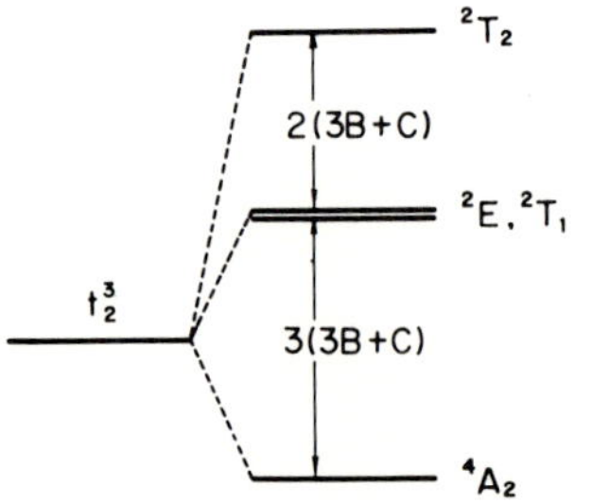

FIG. 3.1. Energy levels arising from the $t_2{}^3$ electron configuration.

3.3.2 Energy Matrix for ^2E

Two ^2E with $t_2{}^2(^1A_1)e$ and $t_2{}^2(^1E)e$ are allowed, besides the $t_2{}^3\ {}^2E$ term, as was shown in Section 3.1. In addition to these three ^2E terms, it will be shown in the next chapter that another ^2E term arises from the e^3 configuration. As in the case of the two-electron system, there are nondiagonal matrix elements of $\mathscr{H}_1$ among these four ^2E terms. Consequently, the exact term energies of ^2E are calculated by diagonalizing the energy matrix of $\mathscr{H}_0 + \mathscr{H}_1$. Such calculation taking into account the configuration mixing is particularly important when the cubic field splitting parameter $10Dq$ is not much greater than the Coulomb interaction.

Before calculating the nondiagonal elements, let us first calculate the diagonal elements for $t_2{}^2(^1A_1)e\ {}^2E$ and $t_2{}^2(^1E)e\ {}^2E$. Since the wavefunctions of these states were obtained in (3.17) and (3.19), one may calculate the diagonal elements of $\mathscr{H}_1$ by using the formulas given in (3.41)–(3.44) as follows:

$$\begin{aligned}\langle t_2{}^2(^1A_1)e\ {}^2E \mid \mathscr{H}_1 \mid t_2{}^2(^1A_1)e\ {}^2E\rangle\\ = J(\zeta\zeta) + J(\zeta u) + J(\zeta v) + 2K(\xi\eta) - \tfrac{1}{2}[K(\zeta u) + K(\zeta v)]\\ (= 3A + 8B + 6C).\end{aligned} \tag{3.53}$$

In deriving (3.53), the relation

$$J(\xi u) = J(\eta u) = \tfrac{1}{4}J(\zeta u) + \tfrac{3}{4}J(\zeta v) \tag{3.54}$$

was used. The last expression of (3.53) in brackets is the result obtained with the approximation where t_2 and e are the d-functions. Furthermore, one obtains

$$\begin{aligned}\langle t_2{}^2(^1E)e\ {}^2E \mid \mathscr{H}_1 \mid t_2{}^2(^1E)e\ {}^2E\rangle\\ = J(\zeta\zeta) + J(\zeta u) + J(\zeta v) - K(\xi\eta) - \tfrac{1}{2}[K(\zeta u) + K(\zeta v)]\\ (= 3A - B + 3C),\end{aligned} \tag{3.55}$$

in which new relations

$$\begin{aligned}\langle \xi u \mid\mid \xi v\rangle &= -\langle \eta u \mid\mid \eta v\rangle = \frac{\sqrt{3}}{4}[J(\zeta v) - J(\zeta u)],\\ \langle \xi u \mid\mid v\xi\rangle &= -\langle \eta u \mid\mid v\eta\rangle = \frac{\sqrt{3}}{4}[K(\zeta v) - K(\zeta u)]\end{aligned} \tag{3.56}$$

are used in addition to those already given. For example, the first relation in (3.56) can be proved by showing

$$\begin{aligned}\langle \xi u \mid\mid \xi v\rangle &= \langle C_4(y)\xi C_4(y)u \mid\mid C_4(y)\xi C_4(y)v\rangle\\ &= \frac{\sqrt{3}}{4}[\langle \zeta v \mid\mid \zeta v\rangle - \langle \zeta u \mid\mid \zeta u\rangle] + \frac{1}{2}\langle \zeta u \mid\mid \zeta v\rangle,\end{aligned} \tag{3.57}$$

in which the last term vanishes as follows:

$$\begin{aligned}\langle \zeta u \mid\mid \zeta v\rangle &= \langle C_4(z)\zeta C_4(z)u \mid\mid C_4(z)\zeta C_4(z)v\rangle \\ &= -\langle \zeta u \mid\mid \zeta v\rangle = 0.\end{aligned} \tag{3.58}$$

The remaining diagonal element is for $e^3\,{}^2E$. In the next chapter we will show that the wavefunctions of this term are given immediately as

$$\begin{aligned}\Psi(e^3\,{}^2E\ M{=}\tfrac{1}{2}\ u) &= |\, uv\bar{v}\,|,\\ \Psi(e^3\,{}^2E\ M{=}\tfrac{1}{2}\ v) &= |\, vu\bar{u}\,|,\end{aligned} \tag{3.59}$$

which may also be obtained from the wavefunction for $e^2\,{}^1A_1$ by using the method described in Section 3.1. By using (3.59) the diagonal element of $\mathscr{H}_1$ is calculated as

$$\langle e^3\,{}^2E \mid \mathscr{H}_1 \mid e^3\,{}^2E\rangle = 3J(uv) + K(uv) \qquad (= 3A - 8B + 4C), \tag{3.60}$$

in which the relation, $J(uu) = J(vv)$, is used.

Now we calculate the nondiagonal elements of $\mathscr{H}_1$. By using the following relations in addition to those already given,

$$\begin{aligned}\langle \eta\zeta \mid\mid \xi u\rangle &= \langle C_4(z)\eta C_4(z)\zeta \mid\mid C_4(z)\xi C_4(z)u\rangle = \langle \xi\zeta \mid\mid \eta u\rangle,\\ \langle \eta\zeta \mid\mid u\xi\rangle &= \langle \xi\zeta \mid\mid u\eta\rangle,\end{aligned} \tag{3.61}$$

$$\begin{aligned}\langle \xi\zeta \mid\mid \eta u\rangle &= \langle C_4(x)\xi C_4(x)\zeta \mid\mid C_4(x)\eta C_4(x)u\rangle \\ &= -\frac{\sqrt{3}}{2}\langle \xi\eta \mid\mid \zeta v\rangle - \frac{1}{2}\langle \xi\eta \mid\mid \zeta u\rangle \\ &= -2\langle \xi\eta \mid\mid \zeta u\rangle,\end{aligned} \tag{3.62}$$

$$\begin{aligned}\langle \xi\zeta \mid\mid u\eta\rangle &= -\frac{\sqrt{3}}{2}\langle \xi\eta \mid\mid v\zeta\rangle - \frac{1}{2}\langle \xi\eta \mid\mid u\zeta\rangle \\ &= \langle \xi\eta \mid\mid \zeta u\rangle,\end{aligned} \tag{3.63}$$

one obtains

$$\begin{aligned}&\langle t_2^3\,{}^2E\ M{=}\frac{1}{2}\,u \mid \mathscr{H}_1 \mid t_2^2({}^1A_1)e\,{}^2E\ M{=}\frac{1}{2}\,u\rangle \\ &= \frac{1}{\sqrt{6}}[2\langle \eta\zeta \mid\mid \xi u\rangle + 2\langle \xi\zeta \mid\mid \eta u\rangle - \langle \xi\zeta \mid\mid u\eta\rangle \\ &\quad - \langle \eta\zeta \mid\mid u\xi\rangle - \langle \xi\eta \mid\mid u\zeta\rangle - \langle \xi\eta \mid\mid \zeta u\rangle] \\ &= -2\sqrt{6}\langle \xi\eta \mid\mid \zeta u\rangle \qquad (= -6\sqrt{2}B).\end{aligned} \tag{3.64}$$

Also by using the relations

$$\langle \eta\zeta \,||\, \xi v\rangle = -\langle \xi\zeta \,||\, \eta v\rangle, \tag{3.65a}$$

$$\langle \eta\zeta \,||\, v\xi\rangle = -\langle \xi\zeta \,||\, v\eta\rangle, \tag{3.65b}$$

$$\langle \xi\zeta \,||\, \eta v\rangle = \frac{1}{2}\langle \xi\eta \,||\, \zeta v\rangle - \frac{\sqrt{3}}{2}\langle \xi\eta \,||\, \zeta u\rangle = 0, \tag{3.66}$$

$$\langle \xi\zeta \,||\, v\eta\rangle = -\sqrt{3}\langle \xi\eta \,||\, \zeta u\rangle, \tag{3.67}$$

one obtains

$$\langle t_2^3\,{}^2E\,M{=}\tfrac{1}{2}\,u \mid \mathscr{H}_1 \mid t_2^2({}^1E)e\,{}^2E\,M{=}\tfrac{1}{2}\,u\rangle = -\sqrt{6}\langle \xi\eta \,||\, \zeta u\rangle$$
$$(= -3\sqrt{2}B). \tag{3.68}$$

The nondiagonal element of $\mathscr{H}_1$ between $t_2^3\,{}^2E$ and $e^3\,{}^2E$ states is zero from (3.44). The remaining nondiagonal elements are

$$\langle t_2^2({}^1A_1)e\,{}^2E\,M{=}\frac{1}{2}\,u \mid \mathscr{H}_1 \mid t_2^2({}^1E)e\,{}^2E\,M{=}\frac{1}{2}\,u\rangle$$
$$= \frac{1}{2\sqrt{3}}\left\{-\frac{1}{\sqrt{3}}[-4J(\xi u) + 2K(\xi u) + 4J(\zeta u) - 2K(\zeta u)]\right.$$
$$\left. + 4\langle \xi u \,||\, \xi v\rangle - 2\langle \xi u \,||\, v\xi\rangle\right\}$$
$$= -J(\zeta u) + J(\zeta v) + \tfrac{1}{2}[K(\zeta u) - K(\zeta v)], \qquad (= 10B), \tag{3.69}$$

$$\langle t_2^2({}^1A_1)e\,{}^2E\,M{=}\tfrac{1}{2}\,u \mid \mathscr{H}_1 \mid e^3\,{}^2E\,M{=}\tfrac{1}{2}\,u\rangle$$
$$= \frac{\sqrt{3}}{2}[K(\zeta u) + K(\zeta v)] \qquad (= \sqrt{3}(2B + C)), \tag{3.70}$$

and

$$\langle t_2^2({}^1E)e\,{}^2E\,M{=}\tfrac{1}{2}\,u \mid \mathscr{H}_1 \mid e^3\,{}^2E\,M{=}\tfrac{1}{2}\,u\rangle$$
$$= \frac{\sqrt{3}}{2}[K(\zeta u) - K(\zeta v)] \qquad (= 2\sqrt{3}B). \tag{3.71}$$

The diagonal elements of $\mathscr{H}_0 = f_1 + f_2 + f_3$ are calculated by using (3.33) as follows:

$$\begin{aligned}
\langle t_2^3\,{}^2E \mid \mathscr{H}_0 \mid t_2^3\,{}^2E\rangle &= 3\langle \zeta \mid f \mid \zeta\rangle,\\
\langle e^3\,{}^2E \mid \mathscr{H}_0 \mid e^3\,{}^2E\rangle &= 3\langle v \mid f \mid v\rangle,\\
\langle t_2^2({}^1A_1)e\,{}^2E \mid \mathscr{H}_0 \mid t_2^2({}^1A_1)e\,{}^2E\rangle &= \langle t_2^2({}^1E)e\,{}^2E \mid \mathscr{H}_0 \mid t_2^2({}^1E)e\,{}^2E\rangle\\
&= 2\langle \zeta \mid f \mid \zeta\rangle + \langle v \mid f \mid v\rangle.
\end{aligned} \tag{3.72}$$

Now we have obtained all the necessary matrix elements. The term energies of ${}^2E$ are obtained by solving the four-dimensional secular equation expressed in terms of these matrix elements. The corresponding secular matrix can be found in Appendix IV.

Chapter IV ELECTRONS AND HOLES

4.1 Complementary States

In principle it is possible to calculate term energies of any N-electron system by using the method described in the previous chapter. However, the calculation becomes laborious as the number of electrons increases. In this chapter we will show that there is a simple relation between the matrix elements of operators in the states of $t_2{}^n e^m$ $(n + m = N)$ configurations and those in the states of $t_2^{6-n} e^{4-m}$. Consequently it is unnecessary to calculate energy matrices for $N = 6, 7, 8$, and 9 once those for $N = 4, 3, 2$, and 1 are calculated. The $t_2^{6-n}(S_1\Gamma_1)e^{4-m}(S_2\Gamma_2)S\Gamma - M\gamma$ state is called *complementary* to the $t_2{}^n(S_1\Gamma_1)\, e^m(S_2\Gamma_2)\; S\Gamma\, M\,\gamma$ state in the sense that these two states are complementary in constructing the closed shells $t_2{}^6e^4$, as shown later. For calculating energy matrices in the complementary states, it is not always necessary to obtain their wavefunctions, but we first construct the wavefunctions according to the prescription adopted so far.

4.1.1 Complementary States in the t_2 Shell

To consider the complementary states in the t_2 shell, we start from the $t_2{}^6\;{}^1A_1$ state of a closed-shell configuration, which is expressed by a single Slater determinant. By using the formula of Laplace's expansion, this determinant may be expanded in terms of n-dimensional $(n < 6)$

small determinants D_i^n including the normalization factor $(n!)^{-1/2}$ (Slater determinant) as follows:

$$\Psi(t_2^6\,{}^1A_1) = |\,\xi\bar{\xi}\eta\bar{\eta}\zeta\bar{\zeta}\,|$$

$$= q^{-1/2}\sum_{i=1}^{q} D_i^n \hat{D}_i^{6-n}, \qquad (4.1)$$

where D_i^n involves electrons 1, 2,..., n and $\hat{D}_i^{6-n}$ those $n+1$, $n+2$,..., 6. Here, $\hat{D}_i^{6-n}$ is the adjunct (or cofactor) of D_i^n also including the normalization factor $[(6-n)!]^{-1/2}$. Factor q is equal to ${}_6C_n = 6!/n!(6-n)!$ which comes from normalization factors in the Slater determinants. For example, in the case of $n = 2$, D_i^2 and $\hat{D}_i^{6-2}$ are given as

i	1	2	3	4	5	$\cdots$	
D_i^2	$\|\xi\bar{\xi}\|$	$\|\eta\bar{\eta}\|$	$\|\zeta\bar{\zeta}\|$	$\|\xi\eta\|$	$\|\eta\zeta\|$	$\cdots$	(4.2)
$\hat{D}_i^{6-2}$	$\|\eta\bar{\eta}\zeta\bar{\zeta}\|$	$\|\xi\bar{\xi}\zeta\bar{\zeta}\|$	$\|\xi\bar{\xi}\eta\bar{\eta}\|$	$-\|\bar{\xi}\bar{\eta}\zeta\bar{\zeta}\|$	$-\|\xi\bar{\xi}\bar{\eta}\bar{\zeta}\|$	$\cdots$.	

By introducing a row vector $\mathbf{D}^n$ of q components and a column vector $\hat{\mathbf{D}}^{6-n}$ of q components defined as follows,

$$\mathbf{D}^n = [D_1^n, D_2^n, \ldots, D_q^n],$$

$$\hat{\mathbf{D}}^{6-n} = \begin{bmatrix} \hat{D}_1^{6-n} \\ \hat{D}_2^{6-n} \\ \vdots \\ \hat{D}_q^{6-n} \end{bmatrix}, \qquad (4.3)$$

Eq. (4.1) can be expressed simply as

$$\Psi(t_2^6\,{}^1A_1) = q^{-1/2}\mathbf{D}^n\hat{\mathbf{D}}^{6-n}. \qquad (4.4)$$

As mentioned in Chapters II and III, there are q states of t_2^n. Denoting wavefunctions $\Psi(t_2^n S\Gamma\, M\gamma)$ as Ψ_i^n and defining a row vector $\mathbf{\Psi}^n$ of q components as

$$\mathbf{\Psi}^n = [\Psi_1^n, \Psi_2^n, \ldots, \Psi_q^n], \qquad (4.5)$$

one can express $\mathbf{\Psi}^n$ in terms of a unitary matrix $\mathbf{U}$ and $\mathbf{D}^n$ as follows:

$$\mathbf{\Psi}^n = \mathbf{D}^n\mathbf{U}. \qquad (4.6)$$

For example, by arranging Ψ_i^n for $n = 2$ as

$$\Psi_1^2 = \Psi(t_2^2\,{}^1A_1), \qquad \Psi_2^2 = \Psi(t_2^2\,{}^1Eu), \qquad \Psi_3^2 = \Psi(t_2^2\,{}^1Ev), \ldots, \qquad (4.7)$$

the unitary matrix $\mathbf{U}$ is found from Table 2.2 as follows:

$$\mathbf{U} = \left[\begin{array}{ccc|c} \frac{1}{\sqrt{3}} & -\frac{1}{\sqrt{6}} & \frac{1}{\sqrt{2}} & \\ \frac{1}{\sqrt{3}} & -\frac{1}{\sqrt{6}} & -\frac{1}{\sqrt{2}} & 0 \\ \frac{1}{\sqrt{3}} & \frac{2}{\sqrt{6}} & 0 & \\ \hline & 0 & & \begin{array}{c}\text{12-dimensional}\\ \text{matrix}\end{array} \end{array}\right]. \tag{4.8}$$

As seen in (4.8), $\mathbf{U}$ in our problem is real, so it is an orthogonal transformation matrix. However, to make possible a more general application of our argument, we will hereafter deal with $\mathbf{U}$ as if it were complex.

Now let the hermitian conjugate of $\mathbf{U}$ be $\mathbf{U}^{\dagger}$ which is the complex conjugate of transposed matrix $\tilde{\mathbf{U}}$, i.e., $(U^{\dagger})_{ij} = (\tilde{U})^{*}_{ij} = U^{*}_{ji}$. Then, one obtains $\mathbf{U}^{\dagger}\mathbf{U} = \mathbf{U}\mathbf{U}^{\dagger} = \mathbf{E}$ as $\mathbf{U}$ is unitary. Therefore, (4.4) can be reexpressed as

$$\begin{aligned} \Psi(t_2{}^{6}\,{}^{1}A_1) &= q^{-1/2}\mathbf{D}^{n}\mathbf{U}\mathbf{U}^{\dagger}\hat{\mathbf{D}}^{6-n} \\ &= q^{-1/2}\boldsymbol{\Psi}^{n}\mathbf{U}^{\dagger}\hat{\mathbf{D}}^{6-n}. \end{aligned} \tag{4.9}$$

In (4.9) $\mathbf{U}^{\dagger}\hat{\mathbf{D}}^{6-n}$ is a column vector of q components whose elements are the antisymmetric functions of electrons, $n+1, n+2, \ldots, 6$. We express $\mathbf{U}^{\dagger}\hat{\mathbf{D}}^{6-n}$ in the form

$$\mathbf{U}^{\dagger}\hat{\mathbf{D}}^{6-n} = \begin{bmatrix} c_1\hat{\Psi}_1^{6-n} \\ c_2\hat{\Psi}_2^{6-n} \\ \vdots \\ c_q\hat{\Psi}_q^{6-n} \end{bmatrix}, \tag{4.10}$$

where the c_i's are numerical constants. For a moment we assume that for $i = (S\Gamma M\gamma)$, c_i is given as

$$c_i = \alpha\,\langle SMS\;-M \mid 00\rangle\langle \Gamma\gamma\Gamma\gamma \mid A_1e_1\rangle, \tag{4.11}$$

where in general α is a constant depending upon S, Γ, M, and γ. By denoting $\hat{\Psi}_i^{6-n}$ as $\hat{\Psi}(t_2^{6-n}S\Gamma\;-M\gamma)$, (4.9) is now written as

$$\begin{aligned} \Psi(t_2{}^{6}\,{}^{1}A_1) = q^{-1/2} &\sum_{\substack{S\Gamma \\ M\gamma}} \alpha\,\langle SMS\;-M \mid 00\rangle\langle \Gamma\gamma\Gamma\gamma \mid A_1e_1\rangle \\ &\times \Psi(t_2{}^{n}S\Gamma M\gamma)\,\hat{\Psi}(t_2^{6-n}S\Gamma\;-M\gamma), \end{aligned} \tag{4.12}$$

which tells us that, if α is a constant $\alpha_{S\Gamma}$, independent of $M\gamma$, $\hat{\Psi}(t_2^{6-n}S\Gamma\, -M\gamma)$ may be the γ base of irreducible representation Γ of the O_h-group as well as the eigenfunction of $\mathbf{S}^2$ and S_z with eigenvalues $S(S+1)$ and $-M$, respectively. In fact, if $\hat{\Psi}(t_2^{6-n}S\Gamma\, -M\gamma)$ has the above-mentioned properties, then

$$\sum_{M\gamma} \langle SMS\, -M \mid 00\rangle\langle \Gamma\gamma\Gamma\gamma \mid A_1e_1\rangle\, \Psi(t_2{}^nS\Gamma M\gamma)\, \hat{\Psi}(t_2^{6-n}S\Gamma\, -M\gamma)$$

is the base of A_1 of the O_h-group with $S=0$ irrespective of S and Γ, so that the right-hand side of (4.12) with $\alpha = \alpha_{S\Gamma}$ also has the property of ${}^1A_1$.

Considering that $\hat{\Psi}(t_2^{6-n}S\Gamma\, -M\gamma)$ is the antisymmetric function of electrons, $n+1$, $n+2$,..., 6, we now see that $\hat{\Psi}(t_2^{6-n}S\Gamma\, -M\gamma)$ can be the wavefunction associated with the $S\Gamma\, -M\gamma$ state of the t_2^{6-n} electron configuration if $\alpha = \alpha_{S\Gamma}$. The constant $\alpha_{S\Gamma}$ is determined so that $\hat{\Psi}(t_2^{6-n}S\Gamma\, -M\gamma)$ is normalized;

$$|\, \alpha_{S\Gamma} \langle SMS\, -M \mid 00\rangle\langle \Gamma\gamma\Gamma\gamma \mid A_1e_1\rangle|^2 = 1. \tag{4.13}$$

Since the Wigner coefficient and the C–G coefficient appearing in (4.12) are given as

$$\langle SMS\, -M \mid 00\rangle = (-1)^{S-M}(2S+1)^{-1/2}, \tag{4.14a}$$

and

$$\langle \Gamma\gamma\Gamma\gamma \mid A_1e_1\rangle = (\Gamma)^{-1/2}\epsilon_\Gamma, \tag{4.14b}$$

with

$$\epsilon_\Gamma{}^2 = 1,$$

Eq. (4.13) is satisfied if $\alpha_{S\Gamma}$ is given as

$$\alpha_{S\Gamma} = \epsilon_\Gamma[(2S+1)(\Gamma)]^{1/2}. \tag{4.15}$$

Of course (4.13) does not determine $\alpha_{S\Gamma}$ uniquely as (4.15), but the choice of the phase of $\alpha_{S\Gamma}$ as given in (4.15) is necessary for later discussions concerning the complementary states of a half-filled shell. Replacing α in (4.12) by $\alpha_{S\Gamma}$ in (4.15), we finally obtain

$$\Psi(t_2{}^6\, {}^1A_1) = q^{-1/2} \sum_{\substack{S\Gamma\\ M\gamma}} (-1)^{S-M}\Psi(t_2{}^nS\Gamma M\gamma)\, \hat{\Psi}(t_2^{6-n}S\Gamma\, -M\gamma), \tag{4.16}$$

which is the expression connecting $\hat{\Psi}(t_2^{6-n}S\Gamma\, -M\gamma)$ with $\Psi(t_2{}^nS\Gamma M\gamma)$. As seen in (4.16) they are complementary in constructing the ${}^1A_1$ state of the $t_2{}^6$ closed shell. It is interesting to note that $\Psi(t_2{}^6\, {}^1A_1)$ is totally antisymmetric while each term in (4.16) is antisymmetric only with respect to the electron exchange within each set of electrons (1, 2,..., n) and ($n + 1, n + 2,..., 6$). The wavefunction $\Psi(t_2{}^n S\Gamma M\gamma)$ with $n \leqslant 3$ directly calculated by using the methods described in the previous chapters is called the wavefunction of the L-state and denoted by Ψ_L, and $\hat{\Psi}(t_2^{6-n} S\Gamma -M\gamma)$ with $n \leqslant 3$ connected with $\Psi(t_2{}^n S\Gamma M\gamma)$ by (4.16) is called the wavefunction of the R-state Ψ_R. This labeling is important when the state has a half-filled configuration ($n = 3$), as shown later. The labeling may be extended to the cases of $\Psi(t_2{}^n S\Gamma M\gamma)$ and $\hat{\Psi}(t_2^{6-n} S\Gamma M\gamma)$ with $n > 3$ by introducing $\Psi_L(t_2{}^n S\Gamma M\gamma)$ with $n > 3$ and $\Psi_R(t_2{}^n S\Gamma M\gamma)$ with $n < 3$ which satisfy the relation,

$$\begin{aligned} \Psi_L(t_2{}^n S\Gamma M\gamma) &= \Psi_R(t_2{}^n S\Gamma M\gamma) \qquad (n > 3), \\ \Psi_R(t_2{}^n S\Gamma M\gamma) &= \Psi_L(t_2{}^n S\Gamma M\gamma) \qquad (n < 3). \end{aligned} \tag{4.17}$$

As seen from (4.10), (4.11), and (4.15), $\hat{\Psi}(t_2^{6-n} S\Gamma -M\gamma)$ appearing in (4.16) can simply be obtained from $\Psi(t_2{}^n S\Gamma M\gamma)$ by using the following procedures: (1) Change the coefficients of D_i^n in $\Psi(t_2{}^n S\Gamma M\gamma)$ into their complex conjugates‡; (2) Replace the Slater determinants D_i^n in $\Psi(t_2{}^n S\Gamma M\gamma)$ by their adjunct $\hat{D}_i^{6-n}$; (3) Multiply the factor $(-1)^{S-M}$. For example, from (4.2) and (4.8) one obtains

$$\hat{\Psi}(t_2{}^4\,{}^1A_1e_1) = \frac{1}{\sqrt{3}}[|\,\eta\bar{\eta}\zeta\bar{\zeta}\,| + |\,\zeta\bar{\zeta}\xi\bar{\xi}\,| + |\,\xi\bar{\xi}\eta\bar{\eta}\,|], \tag{4.18a}$$

$$\hat{\Psi}(t_2{}^4\,{}^1Eu) = \frac{1}{\sqrt{6}}[-|\,\eta\bar{\eta}\zeta\bar{\zeta}\,| - |\,\zeta\bar{\zeta}\xi\bar{\xi}\,| + 2|\,\xi\bar{\xi}\eta\bar{\eta}\,|], \tag{4.18b}$$

$$\hat{\Psi}(t_2{}^4\,{}^1Ev) = \frac{1}{\sqrt{2}}[|\,\eta\bar{\eta}\zeta\bar{\zeta}\,| - |\,\zeta\bar{\zeta}\xi\bar{\xi}\,|]. \tag{4.18c}$$

In deriving (4.18), procedures (1) and (3) give no effect on the results.

Problem 4.1. Derive $\hat{\Psi}(t_2{}^4\,{}^3T_1M\gamma)$. ◇

When $N = 3$, the phases of $\Psi_L(t_2{}^3 S\Gamma M\gamma)$ and $\Psi_R(t_2{}^3 S\Gamma M\gamma)$ are not necessarily the same. For example, from Table 3.2 one can derive

$$\begin{aligned} \Psi_L\left(t_2{}^3\,{}^4A_2\, M{=}-\frac{3}{2}\, e_2\right) &= -|\,\bar{\xi}\bar{\eta}\bar{\zeta}\,|, \\ \Psi_L\left(t_2{}^3\,{}^2T_2\, M{=}-\frac{1}{2}\,\zeta\right) &= -\frac{1}{\sqrt{2}}[|\,\zeta\bar{\xi}\xi\,| + |\,\zeta\bar{\eta}\eta\,|], \end{aligned} \tag{4.19}$$

‡ In the present problem this procedure is unnecessary, as the coefficients are real.

from which one obtains

$$\Psi_R\left(t_2{}^3\,{}^4A_2\,M{=}\frac{3}{2}\,e_2\right) = |\,\xi\eta\zeta\,|,$$

$$\Psi_R\left(t_2{}^3\,{}^2T_2\,M{=}\frac{1}{2}\,\zeta\right) = \frac{1}{\sqrt{2}}\,[|\,\zeta\xi\bar{\xi}\,| + |\,\zeta\eta\bar{\eta}\,|]. \tag{4.20}$$

Thus, comparing (4.20) with Table 3.2, one sees that

$$\Psi_R(t_2{}^3\,{}^4A_2) = -\Psi_L(t_2{}^3\,{}^4A_2),$$
$$\Psi_R(t_2{}^3\,{}^2T_2) = \Psi_L(t_2{}^3\,{}^2T_2). \tag{4.21}$$

Similarly, for all the remaining terms, ${}^2E$ and ${}^2T_1$, we can show that Ψ_L has the phase opposite to that of Ψ_R. All the results may be summarized as follows:

$$\Psi_R(t_2{}^nS\Gamma) = \mu_1\Psi_L(t_2{}^nS\Gamma), \tag{4.22}$$

where $\mu_1 = -1$ for $n = 3$ and $S\Gamma = {}^4A_2$, ${}^2E$, ${}^2T_1$, and $\mu_1 = 1$ for all the other cases including $n = 3$ and $S\Gamma = {}^2T_2$.

The above-mentioned arguments can be applied as well to the wavefunctions of e^m, and the results may be summarized as

$$\Psi_R(e^mS\Gamma) = \mu_2\Psi_L(e^mS\Gamma), \tag{4.23}$$

where $\mu_2 = -1$ for $m = 2$ and $S\Gamma = {}^1E$, ${}^3A_2$, and $\mu_2 = 1$ for all the other cases including $m = 2$ and $S\Gamma = {}^1A_1$.

Relations (4.22) and (4.23) will conveniently be used in calculating matrix elements of various operators in the states of the half-filled shell electron configurations.

4.1.2 Complementary States in the $(t_2\,, e)$ Shell

Starting from the $t_2{}^6e^4\,{}^1A_1$ state, one can derive complementary states in the $(t_2\,, e)$ shell in a similar way to that mentioned in the previous subsection. Expanding a ten-dimensional Slater determinant of the $t_2{}^6e^4\,{}^1A_1$ state in terms of $n + m = N$ dimensional small determinants, and using a similar procedure to that mentioned in the previous subsection, one obtains the expression corresponding to (4.16) as follows:

$$\begin{aligned}\Psi(t_2{}^6e^4\,{}^1A_1) &= |\,\xi\bar{\xi}\eta\bar{\eta}\zeta\bar{\zeta}u\bar{u}v\bar{v}\,|\\ &= {}_{10}C_N^{-1/2} \sum_{\substack{S_1\Gamma_1S_2\Gamma_2\\ S\Gamma M\gamma\\ n,m(n+m=N)}} (-1)^{S-M}\Psi(t_2{}^n(S_1\Gamma_1)\,e^m(S_2\Gamma_2)\,S\Gamma M\gamma)\\ &\quad\times \Psi_{S_1\Gamma_1,S_2\Gamma_2}(t_2^{6-n}e^{4-m}S\Gamma\,-M\gamma),\end{aligned} \tag{4.24}$$

in which ${}_{10}C_N = 10!/N!(10-N)!$ is the number of all the allowed states of $t_2^n e^m$ $(n+m=N)$ and $\Psi(t_2^n(S_1\Gamma_1)\, e^m(S_2\Gamma_2)\, S\Gamma M\gamma)$ is the wavefunction of the $t_2^n(S_1\Gamma_1)\, e^m(S_2\Gamma_2)\, S\Gamma M\gamma$ state involving electrons, 1, 2,..., N. And $\Psi_{S_1\Gamma_1 S_2\Gamma_2}(t_2^{6-n} e^{4-m} S\Gamma -M\gamma)$ is the normalized wavefunction of the $S\Gamma -M\gamma$ state of the $t_2^{6-n} e^{4-m}$ electron configuration involving electrons, $N+1$, $N+2$,..., 10, but we cannot conclude, from the previously mentioned argument, that it is the wavefunction of the $t_2^{6-n}(S_1\Gamma_2)\, e^{4-m}(S_2\Gamma_2)\, S\Gamma -M\gamma$ state. Since there are several $S\Gamma -M\gamma$ states of $t_2^{6-n} e^{4-m}$, it is, in general, given by linear combination of the wavefunctions of the $S\Gamma -M\gamma$ states of the $t_2^{6-n} e^{4-m}$ electron configuration, for example, $\Psi_R(t_2^{6-n}(S_1'\Gamma_1')\, e^{4-m}(S_2'\Gamma_2')\, S\Gamma -M\gamma)$, as

$$\begin{aligned}\Psi_{S_1\Gamma_2, S_2\Gamma_2}&(t_2^{6-n} e^{4-m} S\Gamma -M\gamma)\\ &= \sum_{\substack{S_1'\Gamma_1'\\ S_2'\Gamma_2'}} \alpha_{nm}(S_1\Gamma_1\,,\, S_2\Gamma_2 : S_1'\Gamma_1', S_2'\Gamma_2')\\ &\quad \times \Psi_R(t_2^{6-n}(S_1'\Gamma_1')\, e^{4-m}(S_2'\Gamma_2')\, S\Gamma -M\gamma),\end{aligned} \tag{4.25}$$

where $\Psi_R(t_2^{6-n}(S_1\Gamma_1)\, e^{4-m}(S_2\Gamma_2)\, S\Gamma M\gamma)$ is defined by

$$\begin{aligned}\Psi_R(t_2^{6-n}&(S_1\Gamma_1)\, e^{4-m}(S_2\Gamma_2)\, S\Gamma M\gamma)\\ &= (-1)^{nm} \sum_{\substack{M_1M_2\\ \gamma_1\gamma_2}} \mathscr{A}\Psi_R(t_2^{6-n} S_1\Gamma_1 M_1\gamma_1)\, \Psi_R(e^{4-m} S_2\Gamma_2 M_2\gamma_2)\\ &\quad \times \langle S_1M_1S_2M_2 \mid SM\rangle\langle \Gamma_1\gamma_1\Gamma_2\gamma_2 \mid \Gamma\gamma\rangle,\end{aligned} \tag{4.26}$$

with the inclusion of the phase factor $(-1)^{nm}$ in the wavefunction of the R-state. In (4.26), $\mathscr{A}$ is an operator which makes the right-hand side of (4.26) antisymmetric with respect to the exchange of electrons in each of the t_2 and e shells. However, after a somewhat complicated manipulation, one can prove that

$$\begin{aligned}\alpha_{nm}(S_1\Gamma_1\,,\, S_2\Gamma_2 : S_1'\Gamma_1', S_2'\Gamma_2') &= \delta(S_1S_1')\, \delta(S_2S_2')\\ &\quad \times \delta(\Gamma_1\Gamma_1')\, \delta(\Gamma_2\Gamma_2').\end{aligned} \tag{4.27}$$

The proof of (4.27) will be given in Appendix V.

Then, one obtains

$$\begin{aligned}\Psi(t_2^6 e^4\, {}^1A_1) = {}_{10}C_N^{-1/2} &\sum_{\substack{S_1\Gamma_1S_2\Gamma_2\\ S\Gamma M\gamma\\ n,m(n+m=N)}} (-1)^{S-M}\\ &\times \Psi_L(t_2^n(S_1\Gamma_1)\, e^m(S_2\Gamma_2)\, S\Gamma M\gamma)\\ &\times \Psi_R(t_2^{6-n}(S_1\Gamma_1)\, e^{4-m}(S_2\Gamma_2)\, S\Gamma -M\gamma),\end{aligned} \tag{4.28}$$

which should be compared with (4.16). The phase relation between the wavefunctions of the R and L states is given by

$$\Psi_R(t_2{}^n(S_1\Gamma_1)\, e^m(S_2\Gamma_2)\, S\Gamma M\gamma) = (-1)^{nm}\mu_1\mu_2 \times \Psi_L(t_2{}^n(S_1\Gamma_1)\, e^m(S_2\Gamma_2)\, S\Gamma M\gamma), \quad (4.29)$$

in which μ_1 and μ_2 are already defined in (4.22) and (4.23). This phase relation can be obtained by inserting first (4.27) into (4.25), then using (4.26), and finally applying the phase relations (4.22) and (4.23).

Equation (4.28) tells us that, quite similarly to the case of complementary states in the t_2 shell, $\Psi_R(t_2^{6-n}(S_1\Gamma_1)\, e^{4-m}(S_2\Gamma_2)\, S\Gamma -M\gamma)$ is obtained from $\Psi_L(t_2{}^n(S_1\Gamma_1)\, e^m(S_2\Gamma_2)\, S\Gamma M\gamma)$ by using the following procedures: (1) Change the coefficients of Slater determinants in

$$\Psi_L(t_2{}^n(S_1\Gamma_1)\, e^m(S_2\Gamma_2)\, S\Gamma M\gamma)$$

into their complex conjugates; (2) Replace the Slater determinants by their adjuncts; (3) multiply the factor $(-1)^{S-M}$.

Problem 4.2. Derive $\Psi_{L,R}(t_2{}^3({}^4A_2)\, e^2({}^3A_2)\, {}^6A_1\; M=\frac{5}{2})$ and

$$\Psi_{L,R}(t_2{}^3({}^4A_2)\, e^2({}^1A_1)\, {}^4A_2\; M=\tfrac{3}{2})$$

by using the methods described here. Then confirm the phase relation in (4.29). ◇

4.2 Matrix Elements in Complementary States

4.2.1 Connection between $\Psi(\alpha S\Gamma -M\gamma)$ and $\Psi(\alpha S\Gamma M\gamma)$

In the previous section, we discussed the relation between the wavefunctions of the $t_2{}^n(S_1\Gamma_1)\, e^m(S_2\Gamma_2)\, S\Gamma M\gamma$ and

$$t_2^{6-n}(S_1\Gamma_1)\, e^{4-m}(S_2\Gamma_2)\, S\Gamma -M\gamma$$

states. However, in order to discuss the relation between the matrix elements in the $t_2{}^n(S_1\Gamma_1)\, e^m(S_2\Gamma_2)\, S\Gamma M\gamma$ and $t_2^{6-n}(S_1\Gamma_1)e^{4-m}(S_2\Gamma_2)S\Gamma M\gamma$ states, it is convenient to derive a simple method of obtaining the wavefunction of the $t_2{}^n(S_1\Gamma_1)\, e^m(S_2\Gamma_2)\, S\Gamma M\gamma$ state from that of the $t_2{}^n(S_1\Gamma_1)\, e^m(S_2\Gamma_2)\, S\Gamma -M\gamma$ state.

For this purpose let us introduce an operator,

$$K = K_S K_O\,, \quad (4.30)$$

where K_O is the operator which acting on an orbital part changes a function into its complex conjugate and K_S the operator acting on a spin part defined by

$$K_S\theta(\tfrac{1}{2}, m) = (-1)^{(1/2)-m}\theta(\tfrac{1}{2}, -m). \tag{4.31}$$

Operator K is called the *time reversal operator*. At present it is sufficient merely to know its definition. The physical implication of this operator will be fully discussed in a later chapter.

From (4.30) one can derive

$$K| a_1a_2 \cdots a_{n_1}\bar{a}_{n_1+1}\bar{a}_{n_1+2} \cdots \bar{a}_{n_1+n_2} |$$
$$= (-1)^{n_2} | \bar{a}_1\bar{a}_2 \cdots \bar{a}_{n_1}a_{n_1+1}a_{n_1+2} \cdots a_{n_1+n_2} |, \tag{4.32}$$

where the orbital functions a_i's are assumed to be real as in the present case. By using (4.32) one can prove the relation

$$\Psi(\alpha S\Gamma M\gamma) = (-1)^{S+M}K\Psi(\alpha S\Gamma\ -M\gamma), \tag{4.33}$$

in which the coefficients of Slater determinants are assumed to be real. The proof of (4.33) is given in Problem 4.3. By using (4.33) and (4.32), it is now easy to derive $\Psi(\alpha S\Gamma M\gamma)$ from $\Psi(\alpha S\Gamma\ -M\gamma)$.

Problem 4.3. First prove a special case of (4.33),

$$\Psi(\alpha S\Gamma\ M{=}-S\ \gamma) = K\Psi(\alpha S\Gamma\ M{=}S\ \gamma),$$

by using (4.31), and the relations

$$S_-\Psi(\alpha S\Gamma M\gamma) = [S(S+1) - M(M-1)]^{1/2}\Psi(\alpha S\Gamma\ M{-}1\ \gamma)$$

and

$$S_- | a_1a_2 \cdots a_n | = \sum_i | a_1a_2 \cdots \bar{a}_i \cdots a_n |.$$

Then, assuming (4.33) to be valid, prove

$$\Psi(\alpha S\Gamma\ M{+}1\ \gamma) = (-1)^{S+M+1}K\Psi(\alpha S\Gamma\ -M{-}1\ \gamma). \quad \diamond$$

4.2.2 Matrix Elements of One-Electron Operators

In order to derive the relation between the matrix elements of one-electron operator in the $t_2{}^n(S_1\Gamma_1)\, e^m(S_2\Gamma_2)\, S\Gamma M\gamma$ and

$$t_2^{6-n}(S_1\Gamma_1)\, e^{4-m}(S_2\Gamma_2)\, S\Gamma M\gamma$$

states, we first examine the matrix elements between $D_k{}^N$ and those between $(-1)^{S-M}(-1)^{S+M}K\hat{D}_k^{10-N} = (-1)^N K\hat{D}_k^{10-N}$: Here it should be borne in mind that $\Psi(t_2^n(S_1\Gamma_1)\, e^m(S_2\Gamma_2)\, S\Gamma M\gamma)$ is given by a linear combination of D_k^N and $\Psi(t_2^{6-n}(S_1\Gamma_1)\, e^{4-m}(S_2\Gamma_2)\, S\Gamma M\gamma)$ by that of $(-1)^{S-M}(-1)^{S+M}K\hat{D}_k^{10-N}$, where $\hat{D}_k^{10-N}$ is the adjunct of D_k^N.

By assuming that

$$D_k^N = | \alpha_1\alpha_2 \cdots \alpha_N |, \tag{4.34}$$

where the α_i's are the abbreviation of spin-orbitals, the diagonal element of operator F between D_k^N is given from (3.33) as

$$\begin{aligned} F_{kk}^N &= \sum_\sigma \int d\tau\, D_k^{N*} F D_k^N \\ &= \sum_{i=1}^{N} \langle \alpha_i | f | \alpha_i \rangle. \end{aligned} \tag{4.35}$$

On the other hand, since one has

$$D_k^{10-N} = | \alpha_{N+1}\alpha_{N+2} \cdots \alpha_{10} |, \tag{4.36}$$

the diagonal element of F between $(-1)^N K\hat{D}_k^{10-N}$ is given as

$$\begin{aligned} F_{kk}^{10-N} &= \sum_\sigma \int d\tau\, [(-1)^N K\hat{D}_k^{10-N}]^* F[(-1)^N K\hat{D}_k^{10-N}] \\ &= \sum_\sigma \int d\tau\, [K | \alpha_{N+1}\alpha_{N+2} \cdots \alpha_{10} |]^* F[K | \alpha_{N+1}\alpha_{N+2} \cdots \alpha_{10} |] \\ &= \sum_{i=N+1}^{10} \langle K\alpha_i | f | K\alpha_i \rangle. \end{aligned} \tag{4.37}$$

If operator f is independent of spin as in the case of the ligand-field potential energy, (4.37) can be simplified as

$$F_{kk}^{10-N} = \sum_{i=N+1}^{10} \langle \alpha_i | f | \alpha_i \rangle. \tag{4.38}$$

Introducing a quantity F_0,

$$\begin{aligned} F_0 &= \sum_\sigma \int d\tau\, | \alpha_1\alpha_2 \cdots \alpha_{10} |^* F | \alpha_1\alpha_2 \cdots \alpha_{10} | \\ &= \sum_{i=1}^{10} \langle \alpha_i | f | \alpha_i \rangle, \end{aligned} \tag{4.39}$$

which is independent of state k we are considering, one obtains from (4.35) and (4.38)

$$F_{kk}^{10-N} = F_0 - F_{kk}^N. \tag{4.40}$$

The nondiagonal element between $D_k{}^N$ and $D^N_{k'}$, the latter of which is assumed to be

$$D^N_{k'} = | \alpha_1\alpha_2 \cdots \alpha_p{}' \cdots \alpha_N | \qquad (\alpha_p{}' = \alpha_{N+q}), \tag{4.41}$$

is calculated from (3.34) as

$$\begin{aligned} F^N_{kk'} &= \sum_\sigma \int d\boldsymbol{\tau}\, D^{N*}_k F D^N_{k'} \\ &= \langle \alpha_p | f | \alpha_p{}' \rangle = \langle \alpha_p | f | \alpha_{N+q} \rangle. \end{aligned} \tag{4.42}$$

On the other hand, the nondiagonal element between $(-1)^N K\hat{D}^{10-N}_k$ and $(-1)^N K\hat{D}^{10-N}_{k'}$, in which $\hat{D}^{10-N}_{k'}$ is given as

$$\hat{D}^{10-N}_{k'} = -| \alpha_{N+1}\alpha_{N+2} \cdots \alpha'_{N+q} \cdots \alpha_{10} | \qquad (\alpha'_{N+q} = \alpha_p), \tag{4.43}$$

is calculated as

$$\begin{aligned} F^{10-N}_{kk'} &= \sum_\sigma \int d\boldsymbol{\tau}\, [(-1)^N K\hat{D}^{10-N}_k]^* F[(-1)^N K\hat{D}^{10-N}_{k'}] \\ &= -\sum \int d\boldsymbol{\tau}\, [K | \alpha_{N+1}\alpha_{N+2} \cdots \alpha_{N+q} \cdots \alpha_{10} |]^* F \\ &\qquad \times [K | \alpha_{N+1}\alpha_{N+2} \cdots \alpha'_{N+q} \cdots \alpha_{10} |] \\ &= -\langle K\alpha_{N+q} | f | K\alpha'_{N+q} \rangle \\ &= -\langle K\alpha_{N+q} | f | K\alpha_p \rangle. \end{aligned} \tag{4.44}$$

The minus sign of (4.43) comes from the relation,

$$\begin{aligned} &\text{adj} | \alpha_1\alpha_2 \cdots \alpha_{p-1}\alpha_{p+1} \cdots \alpha_N\alpha_{N+q} | \\ &\quad = (-1)^{N+q-p} | \alpha_p\alpha_{N+1}\alpha_{N+2} \cdots \alpha_{N+q-1}\alpha_{N+q+1} \cdots \alpha_{10} |; \end{aligned} \tag{4.45}$$

consequently

$$\begin{aligned} &\text{adj} | \alpha_1\alpha_2 \cdots \alpha_p{}' \cdots \alpha_N | \\ &\quad = (-1)^{N+q-p}(-1)^{N-p} | \alpha_p\alpha_{N+1}\alpha_{N+2} \cdots \alpha_{N+q-1}\alpha_{N+q+1} \cdots \alpha_{10} | \\ &\quad = (-1)^{N+q-p}(-1)^{N-p}(-1)^{q-1} | \alpha_{N+1}\alpha_{N+2} \cdots \alpha'_{N+q} \cdots \alpha_{10} |. \end{aligned} \tag{4.46}$$

If operator f is independent of spin, (4.44) can be simplified as

$$F^{10-N}_{kk'} = -\langle \alpha_{N+q} | f | \alpha_p \rangle. \tag{4.47}$$

Furthermore, if the matrix of f is Hermitian, namely $\langle i | f | j \rangle = \langle j | f | i \rangle^*$, and is real as in the case of the ligand-field potential energy, one obtains from (4.42) and (4.47)

$$F_{kk'}^{10-N} = -F_{kk'}^{N} . \tag{4.48}$$

Now we are ready to discuss the relation between the matrix element in the $t_2{}^n(S_1\Gamma_1)\, e^m(S_2\Gamma_2)\, S\Gamma M\gamma$ states and that in the

$$t_2^{6-n}(S_1\Gamma_1)\, e^{4-m}(S_2\Gamma_2)\, S\Gamma M\gamma$$

states. From the results of the previous section, the wavefunctions of these states are given as

$$\Psi_i{}^N = \sum_k D_k{}^N U_{ki} , \tag{4.49a}$$

$$\begin{aligned} \Psi_i^{10-N} &= (-1)^N \sum_k (K\hat{D}_k^{10-N})\, K_0 U_{ki}^* \\ &= (-1)^N \sum_k (K\hat{D}_k^{10-N})\, U_{ki} , \end{aligned} \tag{4.49b}$$

where $\Psi_i{}^N$ and Ψ_i^{10-N} with $i = t_2{}^n(S_1\Gamma_1)\, e^m(S_2\Gamma_2)\, S\Gamma M\gamma$ are the abbreviations of $\Psi_L(t_2{}^n(S_1\Gamma_1)e^m(S_2\Gamma_2)S\Gamma M\gamma)$ and $\Psi_R(t_2^{6-n}(S_1\Gamma_1)e^{4-m}(S_2\Gamma_2)S\Gamma M\gamma)$, respectively. Therefore, one can show that the matrix elements,

$$\sum_\sigma \int d\tau\, \Psi_i^{N*} F \Psi_j{}^N = \sum_{kl} U_{ki}^* U_{lj} F_{kl}^N \tag{4.50}$$

and

$$\sum_\sigma \int d\tau\, \Psi_i^{10-N*} F \Psi_j^{10-N} = \sum_{kl} U_{ki}^* U_{lj} F_{kl}^{10-N}, \tag{4.51}$$

are related to each other as

$$\sum_\sigma \int d\tau\, \Psi_i^{10-N*} F \Psi_j^{10-N} = F_0\, \delta_{ij} - \sum_\sigma \int d\tau\, \Psi_i^{N*} F \Psi_j{}^N, \tag{4.52}$$

if the matrix of f is spin-independent, hermitian and real. Eq. (4.52) shows that the system with $10 - N$ electrons in the (t_2 , e) shell behaves just like the system with N particles having positive charges with respect to the ligand field, except an energy shift as a whole due to the diagonal term F_0 . These particles are called *holes.*

4.2.3 Matrix Elements of Two-Electron Operators

For the purpose of deriving the relation between the matrix elements of two-electron operators in the $t_2{}^n(S_1\Gamma_1)\, e^m(S_2\Gamma_2)\, S\Gamma M\gamma$ states and those in the $t_2^{6-n}(S_1\Gamma_1)\, e^{4-m}(S_2\Gamma_2)\, S\Gamma M\gamma$ states, let us first consider the matrix elements between the $D_k{}^N$'s and those between the $(-1)^N K\hat{D}_k^{10-N}$'s.

By using the expression of $D_k{}^N$ in (4.34), the diagonal element of G in $D_k{}^N$ is given from (3.42) as

$$G_{kk}^{N} = \sum_{\sigma} \int d\boldsymbol{\tau}\, D_k^{N*} G D_k{}^N$$

$$= \sum_{j>i=1}^{N} [\langle \alpha_i\alpha_j \mid g \mid \alpha_i\alpha_j\rangle - \langle \alpha_i\alpha_j \mid g \mid \alpha_j\alpha_i\rangle], \tag{4.53}$$

while, by using (4.36), the diagonal element of G in $(-1)^N K\hat{D}_k^{10-N}$ is calculated as

$$G_{kk}^{10-N} = \sum_{\sigma} \int d\boldsymbol{\tau}\, (K\hat{D}_k^{10-N})^* G (K\hat{D}_k^{10-N})$$

$$= \sum_{j>i=N+1}^{10} [\langle K\alpha_i K\alpha_j \mid g \mid K\alpha_i K\alpha_j\rangle - \langle K\alpha_i K\alpha_j \mid g \mid K\alpha_j K\alpha_i\rangle]. \tag{4.54}$$

If operator g is independent of spin as in the case of the Coulomb interaction operator $1/r_{12}$, operator K may be eliminated in (4.54), and one obtains

$$G_{kk}^{10-N} = \sum_{j>i=N+1}^{10} [\langle \alpha_i\alpha_j \mid g \mid \alpha_i\alpha_j \rangle - \langle \alpha_i\alpha_j \mid g \mid \alpha_j\alpha_i\rangle]$$

$$= \left(\sum_{j>i=1}^{10} - \sum_{j=1}^{10}\sum_{i=1}^{N} + \sum_{j>i=1}^{N} \right) [\langle \alpha_i\alpha_j \mid g \mid \alpha_i\alpha_j\rangle - \langle \alpha_i\alpha_j \mid g \mid \alpha_j\alpha_i\rangle], \tag{4.55}$$

in which the decomposition of the sum is explained in Fig. 4.1. Note that

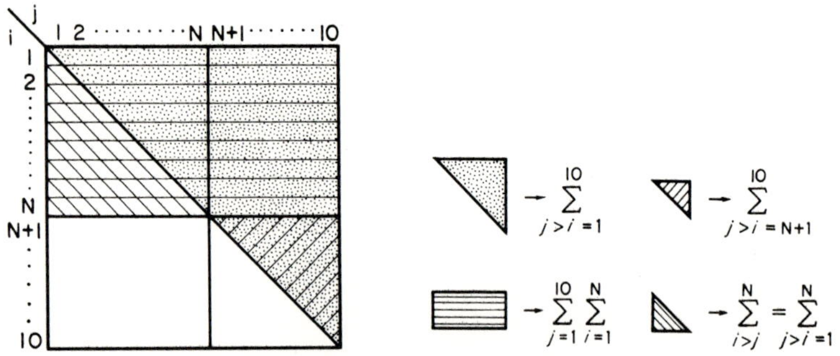

FIG. 4.1. Decomposition of the summation,

$$\sum_{j>i=N+1}^{10} = \sum_{j>i=1}^{10} - \sum_{j=1}^{10}\sum_{i=1}^{N} + \sum_{j>i=1}^{N}.$$

the expression in the square bracket in (4.55) is zero when $i = j$. The first sum in the last expression of (4.55) is equal to

$$G_0 = \sum_{\sigma} \int d\tau \mid \alpha_1\alpha_2 \cdots \alpha_{10} \mid^* G \mid \alpha_1\alpha_2 \cdots \alpha_{10} \mid, \tag{4.56}$$

which is independent of state k. In the second sum one can show that the term

$$G_i = \sum_{j=1}^{10} [\langle \alpha_i\alpha_j \mid g \mid \alpha_i\alpha_j \rangle - \langle \alpha_i\alpha_j \mid g \mid \alpha_j\alpha_i \rangle] \tag{4.57}$$

is independent of i as long as α_i is one of the spin-orbitals involving ξ, η, and ζ, and may be denoted as $G(t_2)$. The same statement is applied when α_i is one of the spin-orbitals involving u and v and term G_i may be denoted as $G(e)$. The proof is as follows: Let us denote G_i as $G(a_i)$ when α_i involves orbital function a_i. Since operator g and $\sum_{j=1}^{10} \alpha_j{}^*\alpha_j$ are invariant to any symmetry operation of the O_h-group, one obtains

$$\begin{aligned} G(\xi) = G(C_4(z)\xi) &= G(\eta) \\ &= G(C_4(x)\eta) = G(\zeta) \\ &= G(t_2), \end{aligned} \tag{4.58}$$

and similarly

$$\begin{aligned} G(u) &= G(C_3(xyz)u) = G(C_3{}^2(xyz)u), \\ G(v) &= G(C_3(xyz)v) = G(C_3{}^2(xyz)v). \end{aligned} \tag{4.59}$$

Furthermore, by using $\mathbf{D}^{(E)}((C_3(xyz))$ in (1.52), it is easy to show the relation

$$\begin{aligned} &G(u) + G(C_3(xyz)u) + G(C_3{}^2(xyz)u) \\ &\quad = G(v) + G(C_3(xyz)v) + G(C_3{}^2(xyz)v), \end{aligned} \tag{4.60}$$

which together with (4.59) leads to

$$G(u) = G(v) = G(e). \tag{4.61}$$

As a result of (4.58) and (4.61), the second sum in the last expression of (4.55) is given as

$$nG(t_2) + mG(e). \tag{4.62}$$

The physical implication of (4.58) and (4.61) is that the interaction energy of an electron in a shell with closed shells is independent of the state of the electron. In general, $G(t_2) \neq G(e)$ but, if the t_2 and e functions are the d-functions, one can show the relation

$$G(t_2) = G(e) = G_1, \tag{4.63}$$

and the second sum is equal to NG_1. The relation (4.63) is understandable from the above-mentioned physical argument, as the (t_2, e) shells can be regarded as a single shell, d-shell, to this approximation.

Problem 4.4. If the t_2 and e functions are the d-functions, $\sum_{j=1}^{10} \alpha_j{}^* \alpha_j$ as well as operator g is invariant to any symmetry operation of the continuous rotation group. Using this fact, prove (4.63). ◇

The third term in the last expression of (4.55) is nothing but G_{kk}^N. Therefore, we finally obtain

$$G_{kk}^{10-N} = G_0 - [nG(t_2) + mG(e)] + G_{kk}^N. \tag{4.64}$$

The nondiagonal element of G between $D_k{}^N$ and $D_{k'}^N$, which are given in (4.34) and (4.41), is calculated from (3.42) as

$$\begin{aligned} G_{kk'}^N &= \sum_\sigma \int d\boldsymbol{\tau} \mid \alpha_1\alpha_2 \cdots \alpha_p \cdots \alpha_N \mid^* G \mid \alpha_1\alpha_2 \cdots \alpha_p{}' \cdots \alpha_N \mid \\ &= \sum_{j=1}^{N} [\langle \alpha_p\alpha_j \mid g \mid \alpha_{N+q}\alpha_j \rangle - \langle \alpha_p\alpha_j \mid g \mid \alpha_j\alpha_{N+q} \rangle], \end{aligned} \tag{4.65}$$

in which the term with $j = p$ is formally included as it is zero. On the other hand, the nondiagonal element of G between $(-1)^N K\hat{D}_k^{10-N}$ and $(-1)^N K\hat{D}_{k'}^{10-N}$, in which $\hat{D}_k^{10-N}$ and $\hat{D}_{k'}^{10-N}$ are given in (4.36) and (4.43), is calculated as

$$\begin{aligned} G_{kk'}^{10-N} &= -\sum_\sigma \int d\boldsymbol{\tau}\, (K \mid \alpha_{N+1}\alpha_{N+2} \cdots \alpha_{N+q} \cdots \alpha_{10} \mid)^* G \\ &\quad \times (K \mid \alpha_{N+1}\alpha_{N+2} \cdots \alpha_{N+q}' \cdots \alpha_{10} \mid) \\ &= -\sum_{j=N+1}^{10} [\langle \alpha_{N+q}\alpha_j \mid g \mid \alpha_p\alpha_j \rangle - \langle \alpha_{N+q}\alpha_j \mid g \mid \alpha_j\alpha_p \rangle] \\ &= -\sum_{j=N+1}^{10} [\langle \alpha_p\alpha_j \mid g \mid \alpha_{N+q}\alpha_j \rangle - \langle \alpha_p\alpha_j \mid g \mid \alpha_j\alpha_{N+q} \rangle]. \end{aligned} \tag{4.66}$$

In (4.66) operator g is assumed to be the Coulomb interaction operator and spin-orbitals to be real. Now let us examine the difference between (4.65) and (4.66),

$$G_{kk'}^N - G_{kk'}^{10-N} = \sum_{j=1}^{10} [\langle \alpha_p\alpha_j \mid g \mid \alpha_{N+q}\alpha_j \rangle - \langle \alpha_p\alpha_j \mid g \mid \alpha_j\alpha_{N+q} \rangle]. \tag{4.67}$$

This difference can be shown to be zero, as operator g and $\sum_{j=1}^{10} \alpha_j{}^* \alpha_j$ are invariant to any symmetry operation of the O_h-group, and in addition to this one can always find a symmetry operation which transforms $\alpha_p{}^*\alpha_{N+q}$ into $-\alpha_p{}^*\alpha_{N+q}$ ($a_p \neq a_{N+q}$: a_i's are orbital functions in spin-orbitals of the α_i's). For example, when $a_p = \xi$ and $a_{N+q} = v$, one can show that the relation, $C_4{}^2(y)\xi C_4{}^2(y)v = -\xi v$. Then one obtains

$$G_{kk'}^{10-N} = G_{kk'}^{N} . \tag{4.68}$$

The nondiagonal element of G between $D_k{}^N$ and $D_{k''}^N$, in which $D_{k''}^N$ is given by

$$D_{k''}^N = | \alpha_1\alpha_2 \cdots \alpha_p' \cdots \alpha_q' \cdots \alpha_N | \tag{4.69}$$

with

$$\alpha_p' = \alpha_{N+r} \qquad \text{and} \qquad \alpha_q' = \alpha_{N+s} ,$$

is calculated from (3.43) as

$$\begin{aligned} G_{kk''}^N &= \sum_\sigma \int d\tau \, | \alpha_1\alpha_2 \cdots \alpha_p \cdots \alpha_q \cdots \alpha_N |^* G \\ &\quad \times | \alpha_1\alpha_2 \cdots \alpha_p' \cdots \alpha_q' \cdots \alpha_N | \\ &= \langle \alpha_p\alpha_q | g | \alpha_{N+r}\alpha_{N+s} \rangle - \langle \alpha_p\alpha_q | g | \alpha_{N+s}\alpha_{N+r} \rangle . \end{aligned} \tag{4.70}$$

On the other hand, the nondiagonal element of G between $(-1)^N K\hat{D}_k^{10-N}$ and (-1) $K\hat{D}_{k''}^{10-N}$, in which $\hat{D}_{k''}^{10-N}$ is given from (4.69) as

$$\hat{D}_{k''}^{10-N} = | \alpha_{N+1}\alpha_{N+2} \cdots \alpha_{N+r}' \cdots \alpha_{N+s}' \cdots \alpha_{10} | \tag{4.71}$$

with

$$\alpha_{N+r}' = \alpha_p \qquad \text{and} \qquad \alpha_{N+s}' = \alpha_p ,$$

is calculated as

$$\begin{aligned} G_{kk''}^{10-N} &= \sum_\sigma \int d\tau \, (K| \alpha_{N+1}\alpha_{N+2} \cdots \alpha_{N+r} \cdots \alpha_{N+s} \cdots \alpha_{10} |)^* G \\ &\quad \times (K | \alpha_{N+1}\alpha_{N+2} \cdots \alpha_{N+r}' \cdots \alpha_{N+s}' \cdots \alpha_{10} |) \\ &= \langle \alpha_{N+r}\alpha_{N+s} | g | \alpha_p\alpha_q \rangle - \langle \alpha_{N+r}\alpha_{N+s} | g | \alpha_q\alpha_p \rangle \\ &= \langle \alpha_p\alpha_q | g | \alpha_{N+r}\alpha_{N+s} \rangle - \langle \alpha_p\alpha_q | g | \alpha_{N+s}\alpha_{N+r} \rangle . \end{aligned} \tag{4.72}$$

In (4.72) operator g is assumed to be $1/r_{12}$ and the spin-orbitals to be real. Comparing (4.70) with (4.72), one obtains

$$G_{kk''}^{10-N} = G_{kk''}^{N} . \tag{4.73}$$

All the other kinds of nondiagonal elements in $D_k{}^N$ and $\hat{D}_k^{10-N}$ are zero because of the formula in (3.44).

By using (4.64), (4.68), and (4.73) and remembering that Ψ_i^N and Ψ_i^{10-N} are given as in (4.49), we finally obtain

$$\sum_\sigma \int d\tau\, \Psi_i^{10-N*} G \Psi_j^{10-N} = \{G_0 - [nG(t_2) + mG(e)]\}\, \delta_{ij} + \sum_\sigma \int d\tau\, \Psi_i^{N*} G \Psi_j^N. \qquad (4.74)$$

In (4.74) operator G is assumed to be independent of spin and orbitals involved are to be real. If the t_2 and e functions are the d-functions, $nG(t_2) + mG(e)$ may be replaced by NG_1 which is independent of the individual values of n and m.

Except for an energy shift as a whole due to the diagonal term $G_0 - NG_1$ (the t_2 and e functions are assumed to be the d-functions), the result in (4.74) also suggests that the $(10 - N)$-electron system can be regarded as the N-hole system.

Problem 4.5. Express $G(t_2)$ and $G(e)$ in terms of the two-electron integrals given in Chap. II. Assuming the t_2 and e functions to be the d-functions, express $G(t_2)$ and $G(e)$ in terms of the Racah parameters and confirm (4.64). ◇

4.3 Energy Matrices

By using (4.74) the matrix elements of $\mathscr{H}_1 = \sum_{i>j} 1/r_{ij}$ in the $(10 - N)$-electron system of $t_2^{6-n}e^{4-m}$ $(n + m = N;\ 10 - N = 6, 7, 8)$ are immediately obtained from those in the N-electron system of $t_2^n e^m$ $(N = 2, 3, 4,)$. In particular, if the t_2 and e functions are assumed to be the d-functions, the same matrix elements of $\mathscr{H}_1$ can be used for both the $(10 - N)$- and N-electron systems, as the additional common term, $G_0 - NG_1$, in the diagonal elements of the $(10 - N)$-electron system only shifts the terms as a whole.

The matrix elements of $\mathscr{H}_1$ in the systems with $N \leqslant 5$ are obtained as follows. Those for $N = 2$ and 3 have already been obtained in Chaps. II and III, respectively. In the system with $N = 4$, we have states of t_2^4, t_2^3e, $t_2^2e^2$, t_2e^3, and e^4 electron configurations. The wavefunctions for t_2^4 are obtained from those for t_2^2 by the use of the result in Section 4.1.1. The wavefunctions for t_2e^3 are obtained by calculating first those for e^3 from those for e, and then by combining the wavefunctions of t_2 and e^3 as mentioned in Section 3.1.2. The wavefunctions of the other configurations are easily obtained. Then, it is straightforward to calculate the matrix elements of $\mathscr{H}_1$ by using formulas given in (3.41)–(3.44).

As in the case of $N = 4$, the wavefunctions of the $N = 5$ system are calculated from those with $t_2{}^n$ ($n \leqslant 3$) and e^m ($m \leqslant 2$) according to the methods described in Sections 4.1.1 and 3.1. Therefore, in principle, there is no difficulty in calculating the matrix elements of $\mathscr{H}_1$. However, the number of the matrix elements to be calculated can be greatly reduced if one uses the following facts:

(i) As easily seen from (3.44), one has

$$\langle t_2{}^n(S_1\Gamma_1)\, e^m(S_2\Gamma_2)\, S\Gamma |\, G \,|\, t_2^{n-k}(S_1'\Gamma_1')\, e^{m+k}(S_2'\Gamma_2')\, S\Gamma\rangle = 0 \qquad (4.75)$$

for $|\, k \,| > 2$. This fact can also be used in the case of $N = 4$. Examples are:

$$\begin{aligned} &\langle t_2{}^4\, {}^3T_1 \,|\, G \,|\, t_2 e^3\, {}^3T_1\rangle = 0, \\ &\langle t_2{}^5\, {}^2T_2 \,|\, G \,|\, t_2{}^2({}^1T_2)\, e^3\, {}^2T_2\rangle = 0, \qquad \text{etc.} \end{aligned} \qquad (4.76)$$

(ii) In the case of $N = 5$ the $t_2^{6-n}(S_1\Gamma_1)\, e^{4-m}(S_2\Gamma_2)\, S\Gamma$ and

$$t_2{}^n(S_1\Gamma_1)\, e^m(S_2\Gamma_2)\, S\Gamma$$

terms ($n \neq 3$, $m \neq 2$) are allowed as a pair. By using the results that $\Psi_R(t_2^{6-n}(S_1\Gamma_1)\, e^{4-m}(S_2\Gamma_2)\, S\Gamma M\gamma)$ is related to

$$\Psi_L(t_2^{6-n}(S_1\Gamma_1)\, e^{4-m}(S_2\Gamma_2)\, S\Gamma M\gamma)$$

by (4.29) and the matrix elements of $\mathscr{H}_1$ are the same in the complementary states except the constant‡ $G_0 - NG_1$, appearing in diagonals, one can show

$$M' = (-1)^{nm}(-1)^{n'm'}\mu_1\mu_2\mu_1'\mu_2' M, \qquad (4.77)$$

where

$$M = \langle t_2{}^n(S_1\Gamma_1)\, e^m(S_2\Gamma_2)\, S\Gamma |\, G \,|\, t_2^{n'}(S_1'\Gamma_1')\, e^{m'}(S_2'\Gamma_2')\, S\Gamma\rangle, \qquad (4.78)$$

$$M' = \langle t_2^{6-n}(S_1\Gamma_1)\, e^{4-m}(S_2\Gamma_2)\, S\Gamma |\, G \,|\, t_2^{6-n'}(S_1'\Gamma_1')\, e^{4-m'}(S_2'\Gamma_2')\, S\Gamma\rangle. \qquad (4.79)$$

In (4.77) μ_1' and μ_2' are defined for n', m', $S_1'\Gamma_1'$, and $S_2'\Gamma_2'$ just in the same way as μ_1 and μ_2 for n, m, $S_1\Gamma_1$ and $S_2\Gamma_2$, i.e.,

$$\begin{aligned} \mu_1 \ (\text{or } \mu_1') &= -1 \qquad \text{for} \quad n \ (\text{or } n') = 3 \quad \text{and} \\ &\qquad\qquad S_1\Gamma_1 \ (\text{or } S_1'\Gamma_1') = {}^4A_2,\, {}^2E,\, {}^2T_1, \\ &= 1 \qquad \text{otherwise,} \end{aligned}$$

‡ The t_2 and e are assumed to be the d-functions.

and

$$\mu_2 \quad (\text{or } \mu_2') = -1 \qquad \text{for} \quad m \quad (\text{or } m') = 2 \qquad \text{and}$$
$$S_2\Gamma_2 \quad (\text{or } S_2'\Gamma_2') = {}^1E, {}^3A_2,$$
$$= 1 \qquad \text{otherwise.}$$

In the $n + m = 5$ system, $(-1)^{nm}(-1)^{n'm'}$ is always plus unity. Therefore, (4.77) can be expressed as

$$M' = \mu_1\mu_2\mu_1'\mu_2'M. \tag{4.80}$$

In the present case of $n \neq 3$ and $n' \neq 3$ (accordingly $m \neq 2$ and $m' \neq 2$) (4.80) is

$$M' = M \qquad (n \neq 3, \quad n' \neq 3). \tag{4.81}$$

Because of (4.81), it is unnecessary to calculate M' if M is known. Examples are:

$$\langle t_2^5\ {}^2T_2 |\ G \mid t_2^4({}^3T_1)e\ {}^2T_2\rangle = \langle t_2e^4\ {}^2T_2 |\ G \mid t_2^2({}^3T_1)\ e^3\ {}^2T_2\rangle$$
$$= -3\sqrt{6}B, \tag{4.82}$$
$$\langle t_2^5\ {}^2T_2 \mid G \mid t_2^5\ {}^2T_2\rangle = \langle t_2e^4\ {}^2T_2 \mid G \mid t_2e^4\ {}^2T_2\rangle$$
$$= -20B + 10C, \qquad \text{etc.} \tag{4.83}$$

(iii) If $n = 3$ and $n' \neq 3$ (accordingly $m = 2$ and $m' \neq 2$), (4.80) is given as

$$M' = \mu_1\mu_2M, \tag{4.84}$$

which gives M' immediately from M. Examples are:

$$\langle t_2^3({}^2T_1)\ e^2({}^3A_2)\ {}^2T_2 \mid G \mid t_2^4({}^1T_2)e\ {}^2T_2\rangle$$
$$= \langle t_2^3({}^2T_1)\ e^2({}^3A_2)\ {}^2T_2 \mid G \mid t_2^2({}^1T_2)\ e^3\ {}^2T_2\rangle$$
$$= -\ 3\sqrt{6}B/2, \tag{4.85}$$

$$\langle t_2^3({}^2T_2)\ e^2({}^1E)\ {}^2T_2 \mid G \mid t_2^5\ {}^2T_2\rangle$$
$$= -\ \langle t_2^3({}^2T_2)\ e^2({}^1E)\ {}^2T_2 \mid G \mid t_2e^4\ {}^2T_2\rangle$$
$$= 2B, \qquad \text{etc.} \tag{4.86}$$

(iv) If $n = 3$ and $n' = 3$, M' should be equal to M as they are the matrix elements in the complementary states. Therefore, one can conclude from (4.80) that

$$M = M' = 0 \qquad \text{if} \quad \mu_1\mu_2\mu_1'\mu_2' = -1. \tag{4.87}$$

An example is

$$\langle t_2^3(^2T_1)\, e^2(^3A_2)\, ^2T_2 \mid G \mid t_2^3(^2T_2)\, e^2(^1E)\, ^2T_2 \rangle = 0. \tag{4.88}$$

All the matrices of $\mathscr{H}_1$ calculated in this way for $N = 2, 3, 4$, and 5 are given in Appendix IV, in which the t_2 and e functions ar eassumed to be the d-functions. To this approximation the matrices for $N = 8, 7, 6$ are, respectively, just those for $N = 2, 3, 4$ with the bases obtained by replacing $t_2{}^n(S_1\Gamma_1)\, e^m(S_2\Gamma_2)\, S\Gamma$ for $N = 2, 3, 4$ by $t_2^{6-n}(S_1\Gamma_1)\, e^{4-m}(S_2\Gamma_2)\, S\Gamma$. Common terms appearing in diagonals for a fixed value of N are neglected. When these matrices are diagonalized, they have to give the energies of the terms of free atoms or ions indicated in brackets above the corresponding matrices.

To obtain energy matrices of $\mathscr{H}_0 + \mathscr{H}_1$ of (2.5), we must add

$$(-4n + 6m)\, Dq \tag{4.89}$$

to the diagonal element in the state of $t_2{}^n e^m$. For the state of $t_2^{6-n} e^{4-m}$, (4.89) should be

$$[-4(6 - n) + 6(4 - m)]\, Dq = -(-4n + 6m)\, Dq, \tag{4.90}$$

in agreement with the result in (4.52). This shows that the term energies for $N = 6, 7, 8$ are obtained, respectively, from those for $N = 4, 3, 2$ by changing the sign of Dq.

Chapter V MULTIPLETS IN OPTICAL SPECTRA

5.1 Energy Level Diagrams

5.1.1 Racah Parameters B and C

In the previous chapter all the energy matrices for the systems of the $t_{2g}^n e_g{}^m$ $(n + m = N = 1, 2,..., 9)$ electron configurations were calculated in terms of Racah parameters B and C and the cubic field splitting parameter 10 Dq. To apply the theory to the analysis of experiments, it is convenient to plot the energies of the states as a function of the parameters involved in the energy matrices. For this purpose let us first point out that the ratio $C/B = \gamma$ is almost independent of both the atomic number and the number of electrons in the iron-group ions, thus making it possible to reduce the number of parameters.

This fact may be understood from the following arguments: If the radial part of the wavefunction $R_d(r)$ is assumed to be that of a hydrogen-like, or Slater-type wavefunction with an appropriate effective nuclear charge, that is,

$$R_d(r) = Nr^2e^{-\kappa r}, \tag{5.1}$$

where

$$N = [(2\kappa)^7/6!]^{1/2}, \tag{5.2}$$

then Slater integral $F^k(dd)$ given in (2.101) is calculated as

$$\begin{aligned} F^k(dd) &= \int_0^\infty dr_1 r_1{}^2 \int_0^\infty dr_2 r_2{}^2 R_d{}^2(r_1)\, R_d{}^2(r_2)\, \frac{r_<{}^k}{r_>^{k+1}} \\ &= 2N^4 \int_0^\infty dr_1 r_1^{-(k-1)} R_d{}^2(r_1) \int_0^{r_1} dr_2 r_2^{k+2} R_d{}^2(r_2) \\ &= \frac{4\kappa(6+k)!}{(6!)^2}\left[(5-k)! - \sum_{n=1}^{7+k} \frac{(12-n)!}{2^{13-n}(7+k-n)!}\right], \end{aligned} \tag{5.3}$$

which is always proportional to κ. Therefore, one easily sees that the ratio of B to C, which is given in terms of $F^k(dd)$ as in (2.104) and (2.106), is independent of κ.

The values of Racah parameters in the free iron-group ions may be determined directly from the spectroscopic data,‡ or, if the data are not available, by interpolating or extrapolating from those experimentally determined. The values of Racah parameters and γ thus determined are listed in Table 5.1 for the divalent and trivalent positive ions. In Table 5.1 one sees that the values of γ fall in a relatively narrow range of values 4 to 5.

TABLE 5.1

RACAH PARAMETERS FOR FREE IONS[a]

M^{2+}	B	C	γ	M^{3+}	B	C	γ
Ti^{2+}	695 cm^{-1}	2910 cm^{-1}	4.19				
V^{2+}	755	3257	4.31	V^{3+}	862	3815	4.43
Cr^{2+}	810	3565	4.40	Cr^{3+}	918	4133	4.50
Mn^{2+}	860	3850	4.78	Mn^{3+}	965	4450	4.61
Fe^{2+}	917	4040	4.41	Fe^{3+}	1015	4800	4.73
Co^{2+}	971	4497	4.63	Co^{3+}	1065	5120	4.81
Ni^{2+}	1030	4850	4.71	Ni^{3+}	1115	5450	4.89

[a] These values are taken from Y. Tanabe and S. Sugano, *J. Phys. Soc. Japan* **9**, 766 (1954).

As emphasized in Section 2.3.6, the radial functions $R_d(r)$ in crystals and complex ions are not necessarily equal to the radial functions of the $3d$-atomic orbitals. However, it seems reasonable to assume that the deviation from the atomic orbitals would not be large. In particular the deviation of γ in crystals from that in free ions would be small. Therefore, in what follows, we express the energy matrices in terms of two parameters, B and $10Dq$ assuming the free-ion value for γ.

5.1.2 Illustration of the Diagrams

Now, dividing all the matrix elements in the energy matrices by B, one can calculate the energy eigenvalues in unit of B as a function of Dq/B. The important low-lying energy levels thus calculated are shown in Figs. 5.1–5.7. The values of γ used in the calculation are indicated in

‡ "Atomic Energy Levels," National Bureau of Standards (1952).

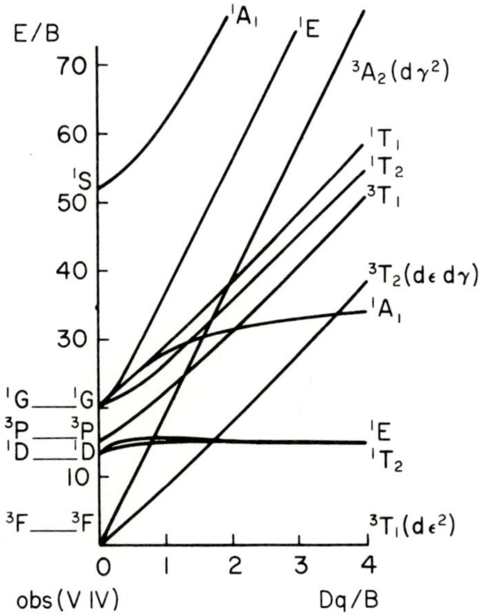

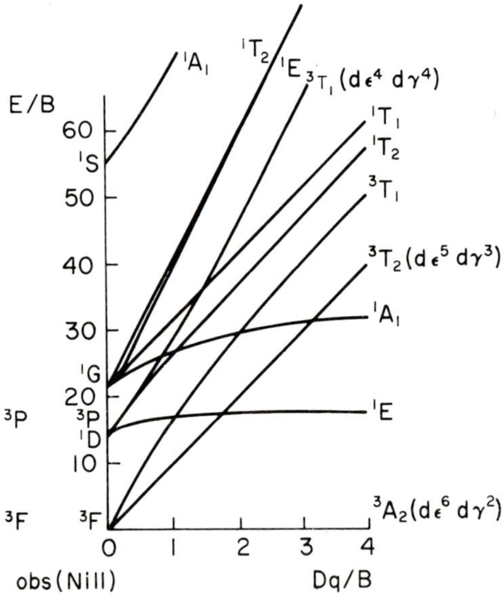

Fig. 5.1. The energy level diagram for the $N = 2$ system (V IV : $3d4s\ ^3D \sim 111$; $\gamma = 4.42$; $B = 860$).

Fig. 5.2. The energy level diagram for the $N = 8$ system (Ni III : $3d^7(^4F)4s\ ^5F \sim 51$; $\gamma = 4.71$; $B = 1030$).

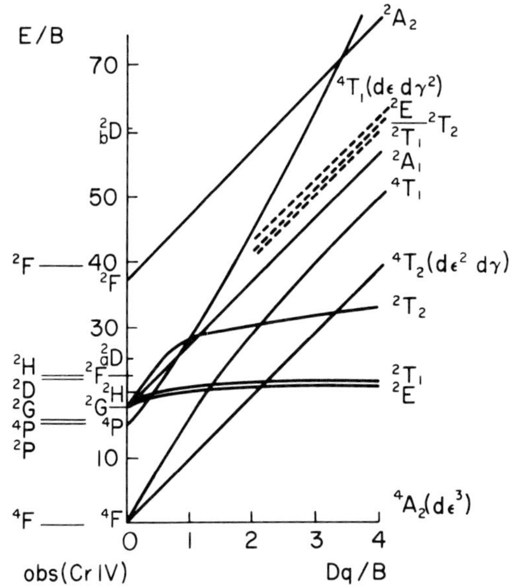

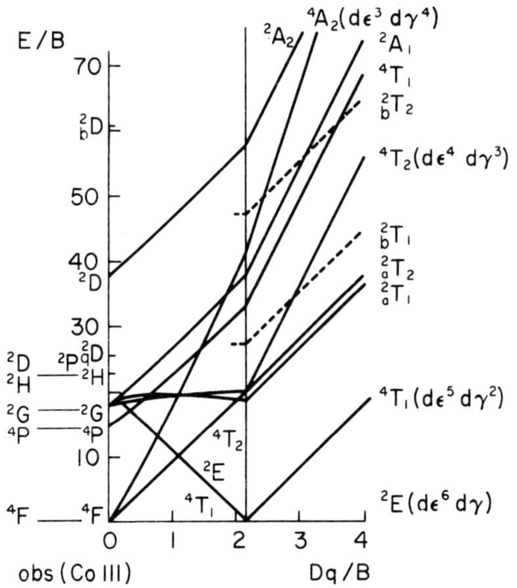

FIG. 5.3. The energy level diagram for the $N = 3$ system (Cr IV : $3d^2(^3F)\,4s\,^4F \sim 113$; $\gamma = 4.50$; $B = 918$).

FIG. 5.4. The energy level diagram for the $N = 7$ system (Co III : $3d^6(^5D)4s\,^6D \sim 48$; $\gamma = 4.63$; $B = 971$).

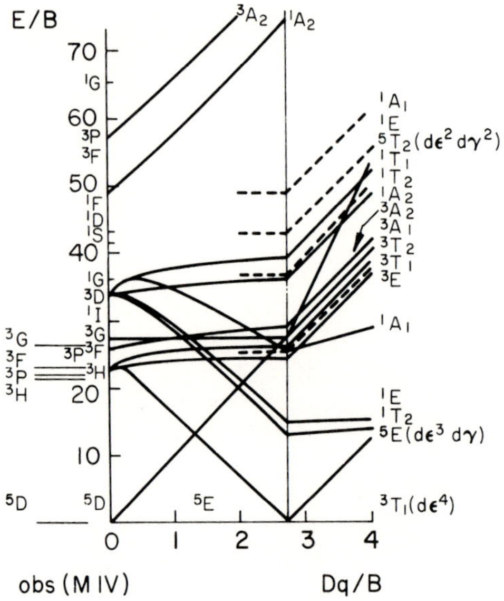

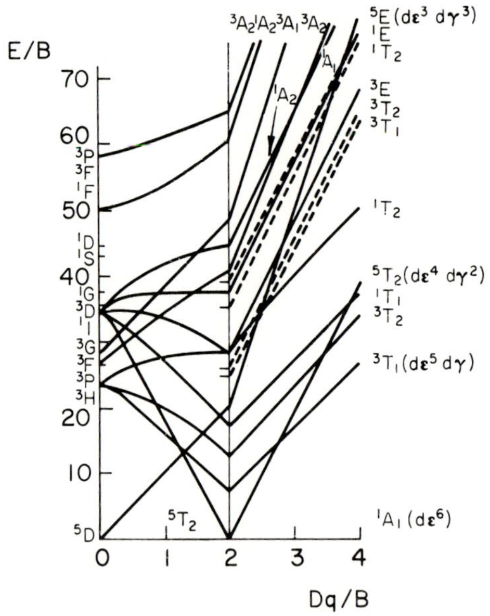

FIG. 5.5. The energy level diagram for the $N = 4$ system (Mn IV : $3d^3(^4F)4s\,^5F \sim 116$; $\gamma = 4.61$; $B = 965$).

FIG. 5.6. The energy level diagram for the $N = 6$ system (Co IV : $3d^5(^6S)\,4s\,^7S \sim$?; $\gamma = 4.81$; $B = 1065$).

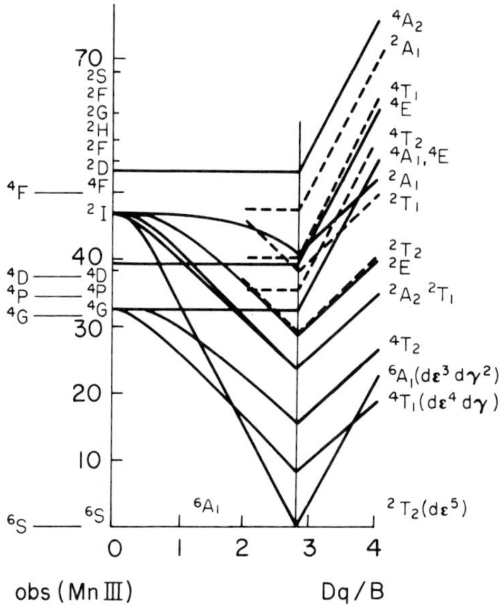

FIG. 5.7. The energy level diagram for the $N = 5$ system (Mn III : $3d^4(^5D)\ 4s\ ^6D \sim$?; $\gamma = 4.48$; $B = 860$).

the figures. Notations, such as Cr IV and Ni III mean Cr^{3+} and Ni^{2+}, respectively. In each figure the energies are always measured from the lowest energy levels.

As seen in the figures the curves representing energies of the terms of the same electron configuration‡ are almost parallel to each other in the range of large values of Dq/B. This is due to the fact that the effects of the configuration mixing are small when the cubic field splitting is large as compared with the Coulomb interaction.

At the left of the vertical axes, the energy levels observed in the free ions are indicated by the horizontal lines, for which the values of B shown in the figures are assumed. The positions of the energy levels at $Dq/B = 0$ should coincide with the horizontal lines of corresponding terms of the free ions, but because of the approximate nature of the theory for free ions, the coincidence is imperfect as seen in the figures. To show the fact that the energy of the lowest term of the $d^{N-1}s$ configuration is much higher than the energies of the terms of d^N, it is indicated in the figures in unit of B.

‡ In the figures the electron configurations are denoted by $d\epsilon^n\ d\gamma^m$ instead of $t_{2g}^n e_g^m$. Notations $d\epsilon$ and $d\gamma$ are sometimes used in place of t_{2g} and e_g, respectively, when the t_{2g} and e_g orbitals are assumed to be the d-functions.

5.1.3 Breakdown of the Hund Rule

As seen in Figs. 5.4–5.7, the interchange of the ground state occurs at a certain value of $Dq/B = (Dq/B)_0$ in the 4, 5, 6, and 7-electron systems. In the range of $Dq/B > (Dq/B)_0$, therefore, the Hund rule with respect to spin which is valid for free atoms is no longer applicable to the ions in a cubic field. This is explained by the fact that maximum spin multiplicity is attained at the expense of excitation energies due to the cubic field splitting. This situation is clearly seen in Fig. 5.8. Figure 5.8a shows the spin arrangements in which the Hund rule with respect to spin is valid. Figure 5.8b shows the spin arrangements in

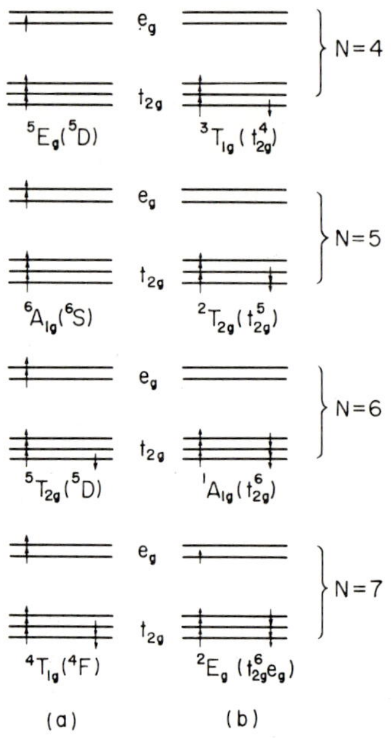

Fig. 5.8. Spin arrangements in the ground states: (a) high-spin systems, (b) low-spin systems.

which breakdown of the Hund rule occurs due to the large cubic field splitting between the t_{2g} and e_g levels. It is clear that in 2, 3, and 8-electron systems the Hund rule is always valid.

The systems in which the Hund rule with respect to spin is valid are sometimes called *high-spin systems*, and those in which it is no longer valid are called *low-spin systems*.

5.2 Optical Transitions

In order to prepare for the comparison between the present theory and optical experiments, qualitative discussions will be given in this section on the intensities and the line widths of optical transitions between the terms, i.e., of the multiplets in a cubic field. More detailed arguments on these subjects will be found in later chapters.

5.2.1 Intensities

Let us first consider transitions between the terms of the same spin-multiplicity which are called *intrasystem combinations*. Since all the terms we are considering belong to even-parity, the electric-dipole transitions which are proportional to the absolute square of the matrix elements of the electric-dipole moment, $\mathbf{P} = -e\sum_i \mathbf{r}_i$, are forbidden as the parity of $\mathbf{P}$ is odd. This selection rule is called the *parity selection rule*. The parity selection rule is slightly released if cubic symmetry of the system is slightly distorted either by the presence of a weak low-symmetry field of odd-parity or if it is instantaneously distorted by the presence of lattice vibrations of certain modes: in the latter case, the instantaneous distortion also brings in a weak low-symmetry field of odd-parity. The odd-parity field admixes even-parity states with odd-parity states, resulting in nonvanishing matrix elements of the electric dipole moment. Let us denote $\langle V_{\text{odd}}\rangle$ as the matrix element of the static or instantaneous odd-parity field, V_{odd}, between even- and odd-parity states. The degree of the admixture of an even-parity state with an odd-parity state is given as $\langle V_{\text{odd}}\rangle/\Delta E_{\text{eo}}$, where ΔE_{eo} is the energy separation between the even-parity and odd-parity states. Therefore, the oscillator strength $f_{\text{el}}^{\text{forb}}$ of our parity forbidden transition is given approximately as

$$f_{\text{el}}^{\text{forb}} \sim f_{\text{el}}^{\text{allow}} \times \left(\frac{\langle V_{\text{odd}}\rangle}{\Delta E_{\text{eo}}}\right)^2, \tag{5.4}$$

where $f_{\text{el}}^{\text{allow}}$ is the oscillator strength of the parity allowed transition given as

$$f_{\text{el}}^{\text{allow}} = \frac{8\pi^2 m}{3he^2}\,\nu\,(|\,p_x\,|^2 + |\,p_y\,|^2 + |\,p_z\,|^2). \tag{5.5}$$

In (5.5) p_i ($i = x, y, z$) is the matrix element of the i component of the electric dipole moment P_i between the initial and final states of the transition. In many cases $f_{\text{el}}^{\text{allow}}$ is of the order of unity. To obtain the magnitude of $f_{\text{el}}^{\text{forb}}$ it is necessary to estimate the magnitude of V_{odd}.

Although a more detailed treatment will be given later, a rough estimate of V_{odd} due to lattice vibrations may be made as follows: Since V_{odd} is caused by a small nuclear displacement Q from an equilibrium position where only a cubic field V_c is present, one sees that

$$V_{odd} \sim V_c \times Q/R, \tag{5.6}$$

where R is the distance between the metal and ligand ions. The instantaneous nuclear displacement Q may be estimated from the zero-point amplitude of the vibration by equating the classical energy of a harmonic oscillator with amplitude Q, mass m_0, and frequency ν_0 to the quantum-mechanical zero-point energy as follows:

$$2\pi^2 m_0 \nu_0{}^2 Q^2 = \tfrac{1}{2} h\nu_0. \tag{5.7}$$

Assuming that $\nu_0 \sim 10^{13}$ sec^{-1} and $m_0 \sim 10^{-23}$ g which is the mass of the ligand ion, we obtain from (5.7) the zero-point amplitude of the order of 10^{-9} cm. Then, assuming the values of $R \sim 10^{-8}$ cm and $V_c \sim 10^4$ cm^{-1}, one obtains from (5.6) $V_{odd} \sim 10^3$ cm^{-1}. In view of experimental data on the absorption spectra in the ultraviolet region, it seems reasonable to assume $\Delta E_{eo} \sim 10^5$ cm^{-1}. Then (5.4) gives

$$f_{e1}^{forb} \sim f_{e1}^{allow} \times 10^{-4} \sim 10^{-4}. \tag{5.8}$$

The weak low-symmetry ligand fields of odd-parity also come from nuclear displacements, but in this case the displacements are static and associated with the goemetrical structure of a crystal or molecule. In many cases the magnitudes of the static displacements are of the same order of magnitude as that of the amplitudes of the zero-point vibrations. Therefore, the static low-symmetry fields of odd-parity are considered to be also of the order of 1000 cm^{-1}, which leads us to the same result for f_{e1}^{forb} as given in (5.8).

For our intrasystem combinations, magnetic-dipole transitions are generally allowed, as the magnetic-dipole moment

$$\mathbf{M} = -\frac{e\hbar}{2mc} \sum_i (\mathbf{l}_i + 2\mathbf{s}_i)$$

is of even parity. The oscillator strength of the magnetic-dipole transition, f_{m1}, is given by the formula,

$$f_{m1} = \frac{8\pi^2 m}{3he^2} \nu \left(|\, m_x \,|^2 + |\, m_y \,|^2 + |\, m_z \,|^2 \right), \tag{5.9}$$

in which m_i ($i = x, y, z$) is the matrix element of the magnetic-dipole moment M_i between the initial and final states of the transition. Since m_i is considered to be of the order of one Bohr magneton, (5.9) gives

$$f_{m1} \sim 10^{-6}. \tag{5.10}$$

In deriving (5.10) ν is assumed to be 6×10^{14} sec^{-1} as we are concerned with the transitions in the visible region.

Besides magnetic-dipole transitions, electric-quadrupole transitions are allowed between the states with the same parity, although they are considered to be very weak. The oscillator strength of the electric-quadrupole transition f_{e2} is given by the formula,

$$f_{e2} = \frac{4\pi^4 m}{5hc^2e^2}\nu^3 \sum_{i,j,=x,y,z} |\, q_{ij} \,|^2, \tag{5.11}$$

in which q_{ij} is the matrix element of the electric-quadrupole moment

$$Q_{ij} = -e \sum_{n:\text{electrons}} (r_{i,n}r_{j,n} - \tfrac{1}{3}r_n{}^2\, \delta_{ij})$$

between the initial and final states of the transition. The magnitude of $|q_{ij}|$ ($\approx q$) may be estimated from the approximate relation, $q \sim e\langle r\rangle^2$, where $\langle r\rangle$ is the average of r. Then, assuming $\langle r\rangle \sim 10^{-8}$ cm, one obtains from (5.11)

$$f_{e2} \sim 10^{-7}. \tag{5.12}$$

Although these estimates of the order of the oscillator strengths are very crude and may contain an error of factor $10^{\pm 1}$, it seems reasonable to conclude from (5.8), (5.10), and (5.12) that for intrasystem combinations the electric-dipole transitions slightly allowed by the presence of the odd-parity field or the odd-parity nuclear vibration predominate over other kinds of transitions.

One thing to be remarked here is that so far we have not considered selection rules other than the parity-selection rule. Transitions are sometimes forbidden by the rotational symmetries associated with both the initial and final states and also with the transition moment including the odd-parity perturbation. Such a selection rule will be discussed in later chapters. The above-mentioned arguments, of course, will be applicable only to the case in which the transitions are not forbidden by other selection rules.

Now let us consider the transitions between the terms of different spin-multiplicities, which are called *intersystem combinations*. Since none of the electric-dipole, magnetic-dipole, and electric-quadrupole moments has matrix elements between the states of different spin-multiplicities, the intersystem combinations are forbidden. This selection rule is called *spin-selection rule*. The spin-selection rule is slightly released if the spin-orbit interaction is taken into account. As explained in detail later, the spin-orbit interaction connects the terms with resultant spins S and S', where $| S - S' | = 0, 1$. Therefore, the term with S may have small components of the terms with $S \pm 1$ if the spin-orbit interaction is taken into account, and the presence of these small components slightly allows the spin-forbidden transitions $S \rightleftarrows S \pm 1$. The degree of the admixture of the S term with the $S \pm 1$ terms is approximately given by $\langle V_{\text{so}} \rangle / \Delta E_{\text{ee}}$, where $\langle V_{\text{so}} \rangle$ is the matrix element of the spin-orbit interaction between the S and $S \pm 1$ terms and ΔE_{ee} is the energy separation between them. In our problem of iron-group ions in a cubic field, $\langle V_{\text{so}} \rangle$ is of the order of $100\ \text{cm}^{-1}$ and ΔE_{ee} is of $3000\ \text{cm}^{-1}$. Therefore, the oscillator strengths of intersystem combinations for the parity-forbidden electric-dipole, magnetic dipole, and electric-quadrupole transitions are, respectively, given as

$$f'^{\text{forb}}_{\text{e1}} \sim 10^{-7}, \tag{5.13}$$

$$f'_{\text{m1}} \sim 10^{-9}, \tag{5.14}$$

$$f'_{\text{e2}} \sim 10^{-10}. \tag{5.15}$$

More detailed arguments on the intersystem combinations will be given after the spin-orbit interaction is fully discussed.

For the comparison of the present theory with experiments it is also of some help to point out the fact that, if the terms are well specified by electron configuration $t_{2g}^n e_g{}^m$, optical transitions are forbidden between the terms of $t_{2g}^n e_g{}^m$ and $t_{2g}^{n-k} e_g^{m+k}$ where $| k | \geqslant 2$. This selection rule is called *configuration-selection rule*, and is expressed by saying that *k-electron jumps* ($| k | \geqslant 2$) are forbidden. The explanation of this selection rule is as follows: Since the transition moments are one-electron operators, their matrix elements between the terms of $t_{2g}^n e_g{}^m$ and $t_{2g}^{n-k} e_g^{m+k}$ are zero from (3.35). Even in the case of the electric-dipole transitions slightly allowed by the odd-parity perturbation where two one-electron operators, the electric dipole moment and the odd-parity perturbation V_{odd}, are incorporated in the transition matrix elements, one can prove the validity of this selection rule. Clearly, the selection rule cannot be applied when the configuration mixing is appreciable.

Problem 5.1. Assuming V_{odd} as a small perturbation, show that the configuration-selection rule may be applied to the electric-dipole transitions slightly allowed by perturbation V_{odd}. ◇

5.2.2 Line Widths

As pointed out in Section 5.1.2, the separations of the energy levels of the same configuration are almost independent of the cubic field splitting parameter in the range of large Dq/B, while those of the levels belonging to different electron configurations are almost proportional to $10Dq$. This fact tells us that the transition energies between the terms of the same electron configurations is independent of the fluctuation of the cubic field which is caused by the nuclear vibration, while those between the terms belonging to different configurations have a certain spread corresponding to the same fluctuation. The magnitude of this fluctuation may be considered to be of the same order of $V_{\text{odd}} \sim 1000 \text{ cm}^{-1}$ at $T = 0°\text{K}$, as one can apply the same method of the estimation to the present case as that used in estimating V_{odd}. Therefore, we expect that the spectral lines connecting the terms of different configurations should have the spectral widths of $\sim 1000 \text{ cm}^{-1}$; consequently they should be observed as broad bands even at $T = 0°\text{K}$. On the other hand, the spectral lines connecting the terms of the same configuration are expected to be observed as sharp lines.

In the present arguments, we have considered only a single mode of the nuclear vibration which keeps the cubic symmetry of the system. This mode is sometimes called a *breathing mode* of the vibration. However, if one takes into account the effects of other vibrational modes, the spectral lines connecting the terms of the same electron configuration may be broadened due to the vibrational fluctuation of low-symmetry fields. This point will be discussed in the next chapter.

In the next section, the qualitative arguments on the line widths as given here will be found very useful in assigning both the sharp and broad structures in the absorption spectra observed in the crystals and complex ions containing the iron-group ions.

5.3 Comparison between Theory and Experiments

In order to show how our theory may be compared with experimental data, we will mention two examples, the comparison with the absorption spectra of Al_2O_3 containing Cr^{3+} impurity ions called ruby and of MnF_2 crystals which have been studied in detail both experimentally and theoretically.

5.3.1 Multiplets in Ruby

To the first approximation Cr^{3+} ions are considered to be in a cubic field in ruby, although the actual symmetry at the Cr^{3+} site is trigonal C_3.

The effects of the trigonal field and the spin-orbit interaction will be discussed later. Here we completely neglect the fine structure of the multiplets.

The observed absorption spectrum around the visible region is shown in Fig. 5.9. In the visible region there are two broad bands having

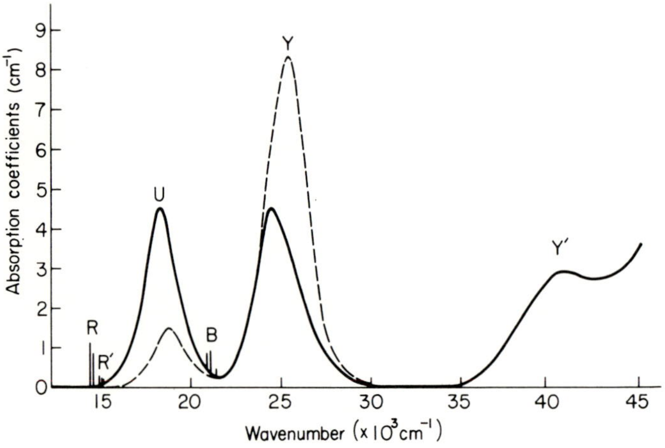

FIG. 5.9. Absorption spectrum of ruby (by A. Misu, unpublished). ——, $E \perp C_3$ (σ); - - - -, $E // C_3$ (π) (0.28 wt % Cr_2O_3; room temperature).

the widths of $\sim 3000\ \text{cm}^{-1}$ which are called U and Y bands, and three groups of sharp lines called R, R', and B. The spectral widths of these lines range from 0.1 to $10\ \text{cm}^{-1}$. In addition to them, one relatively weak broad band is observed in the untraviolet region, which will be called Y'. The oscillator strengths of these absorption bands and lines are estimated from the observed absorption coefficients $k(\nu)$ by using the relation‡

$$f = \frac{mc}{\pi N e^2} \int k(\nu)\, d\nu, \tag{5.16}$$

where N is the number of the absorption centers per cubic centimeter,

‡ In solution the absorption intensity is given in terms of the extinction coefficient $\epsilon(\nu)$ defined by the relation $I = I_0 10^{-\epsilon c_0 l}$, where l is the thickness of the solution in centimeters, c_0 is the concentration of the absorption center in moles per liter, and I_0 and I are the intensities of the incident and transmitted light, respectively. Then, the oscillator strength is given as

$$f = \frac{2.3 \times 10^3}{N_{\text{Avog}}} \frac{mc}{\pi e^2} \int \epsilon(\nu)\, d\nu,$$

where N_{Avog} is the Avogadro number.

and ν the frequency measured per second. For a rough estimate of f, one can replace the integral in (5.16) by $k_{\max}\,\Delta\nu$ assuming the gaussian shape for the absorption curve: here $k_{\max}$ is the absorption coefficient at the absorption peak and $\Delta\nu$ the half-width. Then, one finds that the oscillator strengths of the broad absorption bands are of the order of 10^{-4}, and those of the sharp lines of the order of 10^{-6} to 10^{-7}. From polarization measurements, the transitions responsible for the absorption are known to be of the electric-dipole type. The observed integrated intensities are almost independent of temperature.

Considering these experimental facts, one may conclude that the absorption bands are due to the intrasystem combinations, and the sharp lines due to the intersystem combinations. Actually, looking at Fig. 5.3 one can find three intrasystem combinations,

$$t_{2g}^3\ {}^4A_{2g} \rightarrow t_{2g}^2e_g\ {}^4T_{2g}\,, \qquad t_{2g}^2e_g\ {}^4T_{1g}\,, \qquad \text{and} \qquad t_{2g}e_g^2\ {}^4T_{1g}\,,$$

and three intersystem combinations,

$$t_{2g}^3\ {}^4A_{2g} \rightarrow t_{2g}^3\ {}^2E_g\,, \qquad t_{2g}^3\ {}^2T_{1g}\,, \qquad \text{and} \qquad t_{2g}^3\ {}^2T_{2g}$$

with E/B less than 30. Since these intrasystem combinations connect the terms of different electron configurations and the intersystem ones connect the terms of the same configuration, it is also reasonable, in view of spectral widths, to assign these intrasystem combinations to the broad bands and the intersystem combinations to the groups of the sharp lines.

Actually, one can fit the observed spectrum with the calculated energy level diagram, if the assignments shown in the tabulation are made.

Absorption	Transitions
R lines	$t_{2g}^3\ {}^4A_{2g} \rightarrow t_{2g}^3\ {}^2E_g$
R' lines	$\rightarrow t_{2g}^3\ {}^2T_{1g}$
U band	$\rightarrow t_{2g}^2e_g\ {}^4T_{2g}$
B lines	$\rightarrow t_{2g}^3\ {}^2T_{2g}$
Y band	$\rightarrow t_{2g}^2e_g\ {}^4T_{1g}$
Y' band	$\rightarrow t_{2g}e_g^2\ {}^4T_{1g}$

To obtain a quantitative agreement, it is found that one has to choose the values of parameter, $10Dq = 17{,}000\ \text{cm}^{-1}$, $B = 700\ \text{cm}^{-1}$, $\gamma = C/B$

= 4.0. The assignment of the Y' band indicates that it is due to a two-electron jump. The weak, but observable, intensity of the Y' band may be explained by taking into account the configuration mixing of the $t_{2g}^2e_g\ {}^4T_{1g}$ and $t_{2g}e_g^2\ {}^4T_{1g}$ terms. Temperature-independent intensities of these absorption bands and lines indicate that all the transitions are the electric-dipole transitions slightly allowed by a static odd-parity field. Actually, the Cr^{3+} site has no inversion symmetry so that the presence of a static odd-parity field may be expected.

In ruby the $t_{2g}^3\ {}^2E_g$ and $t_{2g}^3\ {}^2T_{1g}$ excited states can be populated as much as in the ground state by using a strong optical excitation called optical pumping, and optical absorption from these excited states can be observed as shown in Figs. 5.10 and 5.11, in which the absorption curves for the light polarized perpendicular (σ) and parallel (π) to the crystal axis are given separately.

The observed locations of the α and β groups of relatively sharp lines shown in Fig. 5.10 suggest the assignments shown in the tabulation.

Absorption	Transitions
α lines	$t_{2g}^3\ {}^2E_g \rightarrow t_{2g}^3\ {}^2T_{2g}$
β lines	$t_{2g}^3\ {}^2T_{1g} \rightarrow t_{2g}^3\ {}^2T_{2g}$

These assignments have been confirmed by detailed studies of the fine structure and the temperature dependence of the intensities of these groups of lines. The sharpness of the lines also support the above assignments, as these transitions connect the terms of the same electron configuration t_{2g}^3. Although these transitions are spin-allowed, the observed oscillator strengths are of the order of 10^{-6} which seems too small: The explanation has not yet been found.

The oscillator strengths of the I, II, III, and IV bands in Fig. 5.11 are comparable to those of the U and Y bands. This suggests that these bands are due to the spin-allowed transitions mainly from the $t_{2g}^3\ {}^2E_g$ term to the $t_{2g}^2e_g\ {}^2\Gamma_g$ terms: they should be ascribed to the one-electron jumps as the band widths are broad. There are eight $t_{2g}^2e_g\ {}^2\Gamma_g$ terms,

$$t_{2g}^2({}^3T_{1g})\ e_g\ {}^2T_{1g}, \qquad t_{2g}^2({}^1A_{1g})\ e_g\ {}^2E_g,$$

$$t_{2g}^2({}^1T_{2g})\ e_g\ {}^2T_{1g}, \qquad t_{2g}^2({}^1E_g)\ e_g\ {}^2E_g,$$

$$t_{2g}^2({}^3T_{1g})\ e_g\ {}^2T_{2g}, \qquad t_{2g}^2({}^1E_g)\ e_g\ {}^2A_{1g},$$

$$t_{2g}^2({}^1T_{2g})\ e_g\ {}^2T_{2g}, \qquad t_{2g}^2({}^1E_g)\ e_g\ {}^2A_{2g},$$

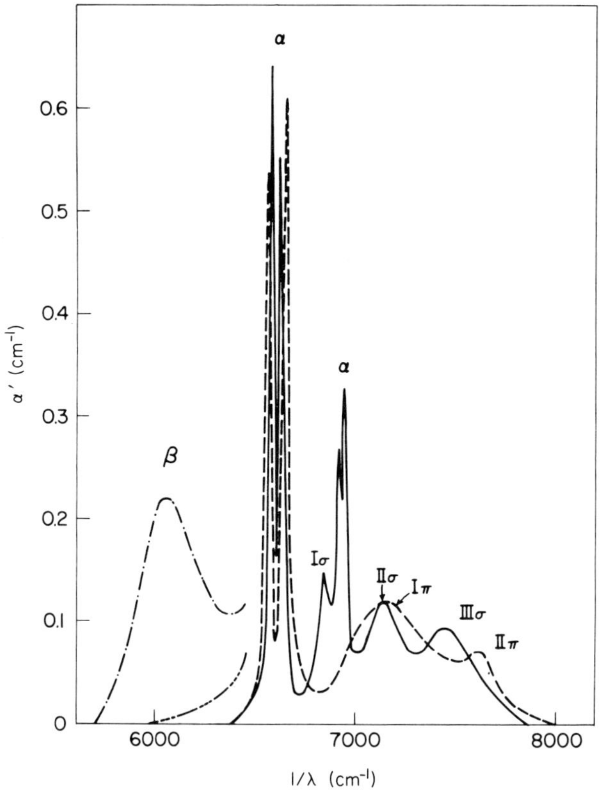

FIG. 5.10. Absorption spectrum of optically pumped ruby (——, σ, 86°K; - - - -, π, 86°K; — · — · —, σ, 300°K; — ··· — ··· —, π, 300°K). [T. Kushida, *J. Phys. Soc. Japan* **21**, 1331 (1966).]

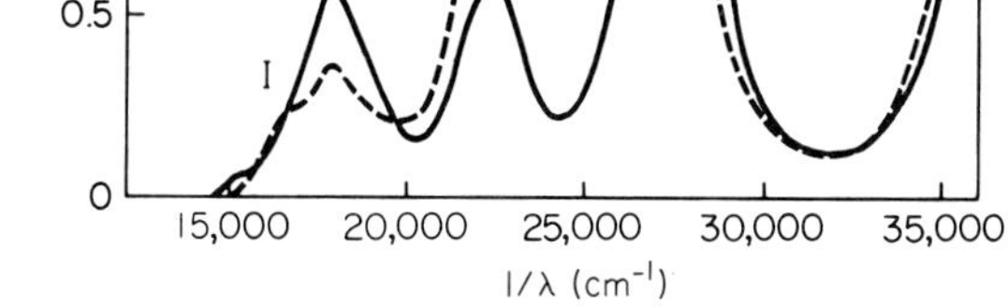

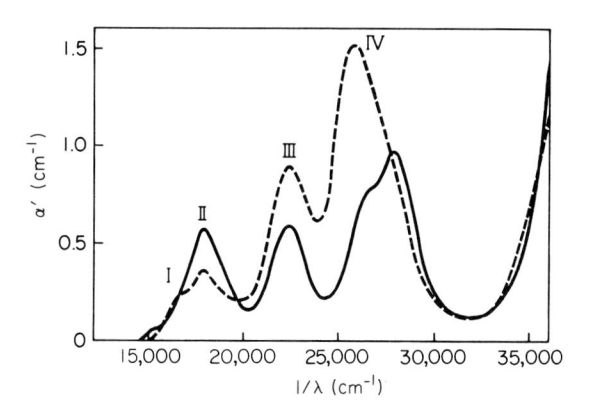

FIG. 5.11. Absorption spectrum of optically pumped ruby (——, σ; - - -, π). [T. Kushida, *J. Phys. Soc. Japan* **21**, 1331 (1966).]

but one can show that the transitions from the $t_{2g}^3\ {}^2E_g$ term to the four ${}^2E_g$, ${}^2E_g$, ${}^2A_{1g}$, ${}^2A_{2g}$ terms are forbidden (see Problem 5.2). Therefore, the observed bands are expected to be due to the transitions to the four $t_{2g}^2 e_g\ {}^2\Gamma_g$ ($\Gamma_g = T_{1g}, T_{2g}$) terms. By using the same values of the parameters as those determined from the absorption spectrum of unpumped ruby, it has been found that, if one uses the assignments in the tabulation, the observed peak energies of these bands may reasonably

Absorption	Transitions
I band	$t_{2g}^3\ {}^2E_g \rightarrow t_{2g}^2({}^3T_{1g})\ e_g\ {}^2T_{2g}$
II band	$\rightarrow t_{2g}^2({}^1T_{2g})\ e_g\ {}^2T_{1g}$
III band	$\rightarrow t_{2g}^2({}^3T_{1g})\ e_g\ {}^2T_{1g}$
IV band	$\rightarrow t_{2g}^2({}^1T_{2g})\ e_g\ {}^2T_{2g}$

be fitted to the calculated transition energies. In the absorption experiment of unpumped ruby, the transitions to these $t_{2g}^2 e_g\ {}^2\Gamma_g$ terms are spin-forbidden and correspond to the one-electron jumps. Therefore, these transitions are unobservable because of the weak intensities and the large spectral widths expected for these transitions.

Problem 5.2. By using the wavefunctions of the $t_{2g}^3\ {}^2E_g$ and $t_{2g}^2 e_g\ {}^2\Gamma_g$ ($\Gamma_g = E_g, A_{1g}, A_{2g}$) terms, show that the matrix elements of any one-electron operator between the $t_{2g}^3\ {}^2E_g$ and $t_{2g}^2 e_g\ {}^2\Gamma_g$ ($\Gamma_g = E_g, A_{1g}, A_{2g}$) terms are zero. ◇

5.3.2 Multiplets in MnF_2 Crystals

A MnF_2 crystal has rutile structure, but to the first approximation Mn^{2+} ions are surrounded octahedrally by six F^- ions, which makes possible the application of our theory.‡ The observed absorption spectrum of MnF_2 is illustrated in Fig. 5.12. In Mn^{2+} ions we have only one $S = 5/2$ term which is the ground state as seen in Fig. 5.7. Therefore, all the transitions from the ground state are expected to be spin-

‡ Since Mn^{2+} ions are placed periodically in MnF_2, the exact wavefunctions should be the bases of the irreducible representations of a space-group involving translational symmetry and the energy levels in general have dispersion with respect to the wave-vector forming energy bands. However, if the widths of the energy bands are much smaller than the term separations, it is possible to apply the ligand field theory without taking into consideration the periodicity in crystals.

forbidden. Actually, all the observed absorption intensities are very weak. As mentioned in the previous section, the spin-forbidden transitions from the ground state to the $S = 3/2$ terms are slightly allowed by the spin-orbit interaction. Therefore, one expects that the final states of the observed transitions are spin-quartets. As seen from a relatively small value of $10Dq$ determined later, the effects of the configuration mixing are appreciable in this case, so that to specify the terms it is convenient to indicate, in addition to the main electron configurations $t_{2g}^n e_g{}^m$, the free ion terms ${}^{2S+1}L$ from which the terms ${}^{2S+1}\Gamma$ arise.

It has been found that, if one uses the values of parameters, $B = 675\ \text{cm}^{-1}$, $C = 3750\ \text{cm}^{-1}$, and $10Dq = 7000\ \text{cm}^{-1}$, the observed absorption peaks are nicely explained by the theory as shown in the tabulation. In Fig. 5.12 the final states of the transitions are indicated above the corresponding absorption peaks.

Transitions	Observed peak energies	Calculated
$t_{2g}^3 e_g{}^2\ {}^6A_{1g}({}^6S)$		
$\rightarrow {}^4T_{1g}(t_{2g}^4 e_g : {}^4G)$	$19.4 \times 10^3\ \text{cm}^{-1}$	$21.1 \times 10^3\ \text{cm}^{-1}$
$\rightarrow {}^4T_{2g}(t_{2g}^4 e_g : {}^4G)$	23.5	23.9
$\rightarrow {}^4A_{1g}(t_{2g}^3 e_g{}^2 : {}^4G)$ $\rightarrow {}^4E_g(t_{2g}^3 e_g{}^2 : {}^4G)$	25.2 25.3 25.5	25.5
$\rightarrow {}^4T_{2g}(t_{2g}^3 e_g{}^2 : {}^4D)$	28.1 28.4	28.7
$\rightarrow {}^4E_g(t_{2g}^3 e_g{}^2 : {}^4D)$	30.2	30.2
$\rightarrow {}^4T_{1g}(t_{2g}^3 e_g{}^2 : {}^4P)$	33.1	34.6
$\rightarrow {}^4A_{2g}(t_{2g}^3 e_g{}^2 : {}^4F)$	39.0	41.1
$\rightarrow {}^4T_{1g}(t_{2g}^3 e_g{}^2 : {}^4F)$	41.4[a]	41.9

[a] Observed by H. J. Hrostowski.

5.3.3 Empirical Values of B and $10Dq$

As seen in the examples given in the previous subsection, a suitable choice of the parameter values B, C, and $10Dq$ in our theory explains the observed optical spectra in many insulating crystals and complex ions involving metal ions with d-electrons. The values of B and Dq chosen

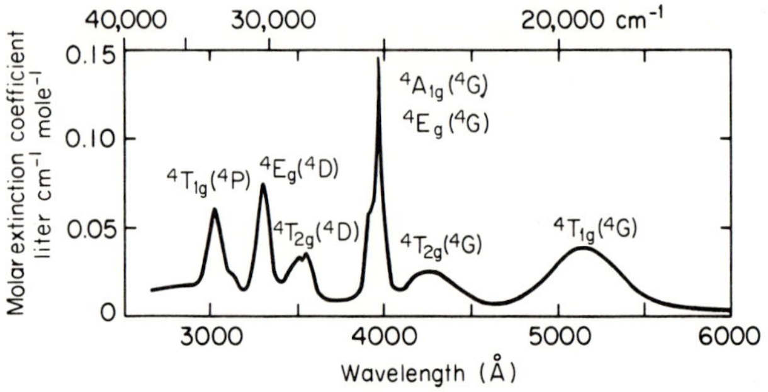

FIG. 5.12. Absorption spectrum of MnF_2 at room temperature (molar extinction coefficient, liter cm^{-1} $mole^{-1}$). [J. W. Stout, *J. Chem. Phys.* **31**, 709 (1959).]

for iron-group metal complexes with H_2O ligands are summarized in Table 5.2.

TABLE 5.2

EXPERIMENTAL VALUES[a] OF Dq AND B

	Ti^{3+}	V^{3+}	Cr^{3+}	Mn^{3+}	Fe^{3+}	Co^{3+}	
Dq (cm^{-1})	2030	1860	1720	2100	1350	1920	
B (cm^{-1})		642 (862)	765 (918)		820 (1015)		

	Cr^{2+}	Mn^{2+}	Fe^{2+}	Co^{2+}	Ni^{2+}	Cu^{2+}
Dq	1390	1230	1030	840	820	1220
B	810 (810)	860 (860)	917 (917)	971 (971)	1030 (1030)	

[a] Y. Tanabe and S. Sugano, *J. Phys. Soc. Japan* **9**, 766 (1954). Values in the parentheses are those of B of free ions.

As for the cubic field splitting parameter $10Dq$, the following empirical rules have been found: Irrespective of the ligand and the central metal ions, $10Dq$ in the systems with divalent metal ions is around 10,000 cm^{-1} and in those with trivalent metal ions around 20,000 cm^{-1}. The change of the ligand elements gives the variation of $10Dq$ which is of the order of 1000 cm^{-1}. It has been found that, when the metal element is fixed and the ligand is varied, the magnitudes of $10Dq$ may be arranged in the following order:

$$I < Br < Cl < S < F < O < N < C,$$

where the elements are those in ligands attached directly to the metal ion. This order is called Tsuchida's *spectrochemical series*. On the other hand, Jørgensen[‡] has pointed out that, when the ligand is fixed and the metal ion is varied, the magnitudes of $10Dq$ may be arranged in the following order:

$$Mn^{2+} < Ni^{2+} < Co^{2+} < Fe^{2+} < V^{2+} < Fe^{3+} < Cr^{3+} < V^{3+} < Co^{3+}$$
$$< Mn^{4+} < Mo^{3+} < Rh^{3+} \sim Ru^{3+} < Pd^{4+} < Ir^{3+} < Re^{4+} < Pt^{4+}.$$

Detailed discussion on the origin of $10Dq$ will be given in Chapter X on the molecular orbital and the Heitler–London theories.

As for the Racah parameter B, the empirically determined values, in particular for trivalent metal ions, are smaller than those of free ions as shown in Table 5.2. This means that radial functions $R_d(r)$ in crystals and complex ions are slightly different from those of the free ion. Probably the reduction from the free ion values would mean the expansion of the free ion radial functions in the presence of ligands. This problem will also be discussed in Chapter X.

[‡] C. K. Jørgensen, "Absorption Spectra and Chemical Bonding in Complexes. Pergamon Press, New York, 1962.

Chapter VI **LOW-SYMMETRY FIELDS**

So far we have been concerned only with the systems of cubic symmetry. There are many crystals and complex ions in which, to the first approximation, transition metal ions are surrounded octahedrally by ligands. However, a more detailed examination of the site symmetry of the metal ion very often reveals that the site symmetry is lower than cubic. For example, the site symmetry of Cr^{3+} ions in ruby is trigonal, C_3, as mentioned in the previous chapter, and the site symmetry of Mn^{2+} ions in MnF_2 is orthorhombic, D_{2h}. In these systems the electrons associated with the metal ions are exposed to the fields of low symmetry in addition to a cubic field. The presence of a low-symmetry field causes splittings and shifts of the terms in a cubic system, and gives, together with the spin-orbit interaction discussed in the next chapter, fine structure of the multiplets. This chapter presents the methods of deriving the ligand-field potentials of low symmetry and calculating their matrix elements between the components of the cubic terms.

6.1 Single Electron in Fields of Low Symmetry

6.1.1 Derivation of the Ligand-Field Potentials

As discussed in Chapter I, the ligand-field potential is invariant to any symmetry operation in the group to which symmetry of the system belongs. Therefore, it has the same symmetry property as that of the base function of the identity representation of the group. For example, the cubic-field potential has the same symmetry property as that of the base function of irreducible representation A_{1g} of the O_h-group. If the base function of A_{1g} of the O_h-group is expressed in the form

$$\sum_{km} A_{km}(r)\, C_m^{(k)}(\theta\varphi), \tag{6.1}$$

the terms of the lowest order k, except $k = 0$, should be those of $k = 4$. This is because, as seen in Table 1.2, A_{1g} first appears when one reduces representation $D^{(l)}$ with $l = 4$ (except $l = 0$) whose base functions are given in terms of $C_m^{(k)}(\theta\varphi)$ with $k = 4$. The next higher terms in (6.1) have to be those with $k = 6$, as A_{1g} appears when $D^{(l)}$ with $l = 6$ is reduced.

Instead of reducing representation $D^{(l)}$, we will construct the terms of $k = 4$ according to the consideration in Section 2.2.2. The explicit form of these terms may be obtained by using (2.32) as proportional to

$$\begin{aligned}\varphi_u(\mathbf{r})\, \varphi_u(\mathbf{r}) \langle EuEu \mid A_1 e_1\rangle + \varphi_v(\mathbf{r})\, \varphi_v(\mathbf{r}) \langle EvEv \mid A_1 e_1\rangle \\ = \frac{1}{\sqrt{2}} [\varphi_u(\mathbf{r})\, \varphi_u(\mathbf{r}) + \varphi_v(\mathbf{r})\, \varphi_v(\mathbf{r})],\end{aligned} \tag{6.2}$$

in which φ_u and φ_v are already given in (1.32a) and (1.32b). According to the arguments given in Section 2.2.2, it is clear that (6.2) is the base function of irreducible representation A_{1g} of the O_h-group.

Since (6.2) is given by a linear combination of $Y_{k'm'}(\theta\varphi) Y_{k''m''}(\theta\varphi)$ with $k' = k'' = 2$, it is necessary for obtaining an expression like (6.1) to express the product of two spherical harmonics by a linear combination of spherical harmonics as

$$Y_{k'm'}(\theta\varphi)\, Y_{k''m''}(\theta\varphi) = \sum_{k,m} a_{km}(k'm', k''m'')\, C_m^{(k)}(\theta\varphi). \tag{6.3}$$

By using the orthogonality relation,

$$\int d\varphi\, d\theta \sin\theta\, C_{m'}^{(k')}(\theta\varphi)^*\, C_m^{(k)}(\theta\varphi) = \frac{4\pi}{2k+1} \delta_{kk'}\, \delta_{mm'}\,, \tag{6.4}$$

coefficient $a_{km}(k'm', k''m'')$ is obtained as

$$\begin{aligned}a_{km}(k'm', k''m'') &= \frac{2k+1}{4\pi} \int d\varphi\, d\theta \sin\theta\, C_m^{(k)*}(\theta\varphi)\, Y_{k'm'}(\theta\varphi)\, Y_{k''m''}(\theta\varphi) \\ &= \frac{2k+1}{4\pi} (-1)^{m''} c^k(k'm', k'' -m''),\end{aligned} \tag{6.5}$$

which is nonvanishing only when

$$m = m' + m''. \tag{6.6}$$

The values of $c^k(k'm', k'' -m'')$ are given in Table 1.2.

According to (6.3), (6.5), and (6.6),

$$Y_{20}(\theta\varphi)\, Y_{20}(\theta\varphi) = \frac{1}{4\pi}\left[C_0^{(0)}(\theta\varphi) + \frac{10}{7}\, C_0^{(2)}(\theta\varphi) + \frac{18}{7}\, C_0^{(4)}(\theta\varphi)\right], \tag{6.7}$$

$$Y_{2\pm2}(\theta\varphi)\, Y_{2\pm2}(\theta\varphi) = \frac{1}{4\pi}\, 3\sqrt{\frac{10}{7}}\, C_{\pm4}^{(4)}, \tag{6.8}$$

and

$$Y_{2\pm2}(\theta\varphi)\, Y_{2\mp2}(\theta\varphi) = \frac{1}{4\pi}\left[C_0^{(0)}(\theta\varphi) - \frac{10}{7}\, C_0^{(2)}(\theta\varphi) + \frac{3}{7}\, C_0^{(4)}(\theta\varphi)\right]. \tag{6.9}$$

Then, one finds that (6.2) is proportional to

$$\frac{2}{3}\, C_0^{(0)}(\theta\varphi) + \left\{C_0^{(4)}(\theta\varphi) + \sqrt{\frac{5}{14}}\, [C_4^{(4)}(\theta\varphi) + C_{-4}^{(4)}(\theta\varphi)]\right\}. \tag{6.10}$$

The angular dependence of the second term is in agreement with that of the fourth-order term in (1.11) derived from the point-charge model.

Similarly, we may derive the explicit form of the potential energies of low-symmetry fields. Let us first consider the system of tetragonal symmetry illustrated in Fig. 1.7. As pointed out in (1.95) this system has symmetry D_{4h}. The character table for the D_{4h}-group is easily obtained from Table 1.3 for the D_4-group. As seen from Table 1.3, the base of the identity representation of the D_{4h}-group is the u base of the E_g irreducible representation of the O_h-group. Therefore, the potential of the D_{4h}-field with $k = 2$ in the form of (6.1) is immediately found to be proportional to

$$C_0^{(2)}(\theta\varphi). \tag{6.11}$$

Since E_g of the O_h-group also appears when $D^{(4)}$ is reduced as shown in Table 1.2, there should be the term with $k = 4$ in the potential which transforms like the u base of E_g. This term is obtained by using (2.37) as

$$\begin{aligned}\varphi_u(\mathbf{r})\, \varphi_u(\mathbf{r}) \langle EuEu \mid Eu\rangle + \varphi_v(\mathbf{r})\, \varphi_v(\mathbf{r}) \langle EvEv \mid Eu\rangle \\ = \frac{1}{\sqrt{2}}\, [-\varphi_u(\mathbf{r})\, \varphi_u(\mathbf{r}) + \varphi_v(\mathbf{r})\, \varphi_v(\mathbf{r})].\end{aligned} \tag{6.12}$$

Again, using (6.7), (6.8), and (6.9), one finds that (6.12) is proportional to

$$\frac{4}{\sqrt{3}}\, C_0^{(2)}(\theta\varphi) + \left\{C_0^{(4)}(\theta\varphi) - \sqrt{\frac{7}{10}}\, [C_4^{(4)}(\theta\varphi) + C_{-4}^{(4)}(\theta\varphi)]\right\}. \tag{6.13}$$

Putting the first term of (6.13) and (6.11) together, one obtains the D_{4h} ligand-field potential, V_{tet}, up to $k = 4$ as

$$V_{\text{tet}} = A(r)\, C_0^{(2)}(\theta\varphi) + B'(r) \left\{ C_0^{(4)}(\theta\varphi) - \sqrt{\frac{7}{10}}\, [C_4^{(4)}(\theta\varphi) + C_{-4}^{(4)}(\theta\varphi)] \right\}, \quad (6.14)$$

where $A(r)$ and $B'(r)$ are the functions of r only. We can express (6.14) as the sum of cubic and axial fields as follows:

$$V_{\text{tet}}(\mathbf{r}) = V_{\text{c}}{}^0(\mathbf{r}) + V_{\text{ax},z}(\mathbf{r}), \quad (6.15)$$

where

$$V_{\text{ax},z} = A(r)\, C_0^{(2)}(\theta\varphi) + B(r)\, C_0^{(4)}(\theta\varphi). \quad (6.16)$$

In (6.15) and (6.16)

$$V_c{}^0(\mathbf{r}) = -\frac{7}{5} B'(r) \left\{ C_0^{(4)}(\theta\varphi) + \sqrt{\frac{5}{14}}\, [C_4^{(4)}(\theta\varphi) + C_{-4}^{(4)}(\theta\varphi)] \right\} \quad (6.17)$$

and

$$B(r) = \frac{12}{5} B'(r). \quad (6.18)$$

In (6.16) $V_{\text{ax},z}$ is invariant to rotations around the z-axis by any angle φ as it is independent of φ. Therefore, it is the potential of an axially symmetric field.

Problem 6.1. Calculate the potential energy of an electron due to the ligand point-charges in the system of the D_{4h}-symmetry as shown in Fig. 1.7. ◇

In contrast to the D_{4h}-group, the D_4-group has no inversion symmetry so that the base function of the identity representation A_1 of the D_4-group is given by a linear combination of the u components of E_g and E_u of the O_h-group. Accordingly, the D_4-field potential has to have the odd-parity part in addition to the even-parity part given in (6.15). Since E_u first appears when $D^{(5)}$ is reduced as seen in Table 1.2, the odd-parity potential of the lowest order is given in terms of $C_m^{(5)}(\theta\varphi)$. The even-parity terms of ligand fields are responsible for the term splittings and shifts, but the odd-parity terms are responsible for allowing slightly the parity-forbidden electric dipole transitions.

Problem 6.2. Show that the even-parity potentials in the C_{4v}, C_{4h}, and C_4 systems are given by V_{tet} in (6.15), and the odd-parity potentials of the lowest order for C_{4v} and C_4 are given in terms of $C_m^{(1)}(\theta\varphi)$, while for C_{4h} it is given in terms of $C_m^{(5)}(\theta\varphi)$. ◇

At this point, it is important to make the following statement. In many textbooks, ligand-field potentials are given as

$$\sum_{km} A_{km} r^k C_m^{(k)}(\theta\varphi), \tag{6.19}$$

which is derived by assuming that the potential $V(r)$ satisfies the Laplace equation

$$\Delta V(\mathbf{r}) = 0. \tag{6.20}$$

The solution of the Laplace equation, which is analytic near the origin, is given in the form of (6.19) in which the A_{km}'s are numerical constants. However, if one takes into account the spatial distribution of ligand electrons extended toward the metal ion, one has the equation

$$\Delta V(\mathbf{r}) = 4\pi e \rho(\mathbf{r}), \tag{6.21}$$

where $\rho(\mathbf{r})$ is the electron density of ligands. The solution of (6.21) is given in the form of (6.1) rather than (6.19).

Now let us next consider the system in which ligands 1, 2, 3 are displaced by δ along the [111] direction and ligands 4, 5, 6 are displaced by the same amount along the $[\bar{1}\bar{1}\bar{1}]$ direction as shown in Fig. 6.1. This system has symmetry D_{3d}, which is the direct product of D_3 and C_i. The D_3-group contains six elements: E, $C_3(xyz)$, $C_3{}^2(xyz) = C_3(\bar{x}\bar{y}\bar{z})$, $C_2(x\bar{y})$, $C_2(y\bar{z})$, and $C_2(z\bar{x})$. These elements are classified into three

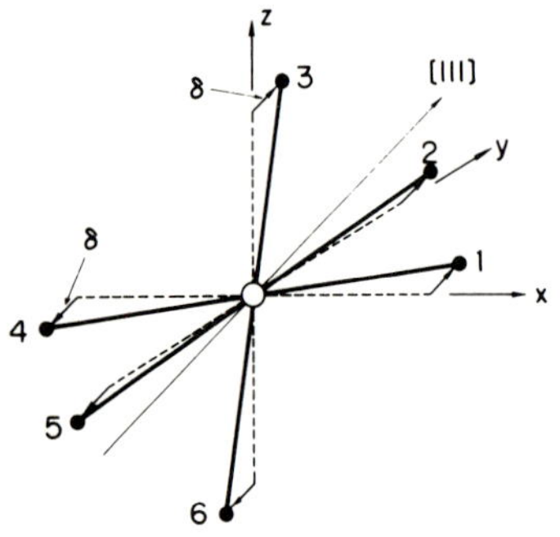

FIG. 6.1. An MX_6 system of D_{3d} symmetry.

classes, $\hat{E}$, $\hat{C}_3$, and $\hat{C}_2$. For this simple group, the character table can be constructed only by the use of the orthogonality relations of the first and second kinds given, respectively, in (1.67) and (1.69b). The character table for D_3 thus obtained is given in Table 6.1. Comparing this table with Table 1.1, one may derive the following reduction of the irreducible representations of the O-group when symmetry is reduced to D_3:

$$\begin{aligned} &A_1 \to A_1, \quad A_2 \to A_2, \quad E \to E, \\ &T_1 \to A_2 + E, \quad T_2 \to A_1 + E. \end{aligned} \tag{6.22}$$

TABLE 6.1

CHARACTER TABLE OF THE D_3-GROUP

Irred. repres.	$\hat{E}$	$2\hat{C}_3$	$3\hat{C}_2$
A_1	1	1	1
A_2	1	1	-1
E	2	-1	0

We see from (6.22) that the base of the A_{1g} of the D_{3d}-group reduced from T_{2g} of the O_h-group is $(\xi + \eta + \zeta)/\sqrt{3}$, consequently the angular dependence of the ligand-field potential in the D_{3d}-system with the lowest k ($k \neq 0$) should be the same as that of $(yz + zx + xy)/\sqrt{3}$. The base of the A_{1g} of D_{3d} reduced from A_{1g} of O_h clearly gives the cubic-field potential with $k = 4$.

Before deriving the explicit form of the D_{3d} potential, it is convenient to introduce a new coordinate (XYZ), in which the Z and Y axes are

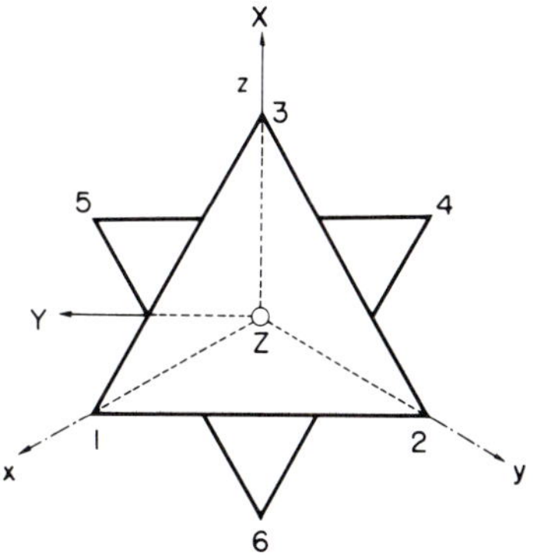

FIG. 6.2. Coordinates (X, Y, Z) and (x, y, z).

chosen along the [111] and [1$\bar{1}$0] directions, respectively, as shown in Fig. 6.2. The relation between the (xyz) and (XYZ) coordinates is given as follows:

$$x = -\frac{1}{\sqrt{6}}X + \frac{1}{\sqrt{2}}Y + \frac{1}{\sqrt{3}}Z,$$

$$y = -\frac{1}{\sqrt{6}}X - \frac{1}{\sqrt{2}}Y + \frac{1}{\sqrt{3}}Z, \qquad (6.23)$$

$$z = \frac{\sqrt{2}}{\sqrt{3}}X + \frac{1}{\sqrt{3}}Z.$$

It is straightforward to show, by calculating the characters from the transformation matrices and comparing them with Table 1.3, that the angular functions tabulated in Table 6.2 may be the bases of the

TABLE 6.2

TRIGONAL BASES

Irred. repres.	Components	Bases $\varphi_{\Gamma M}$
T_1	a_+	$Y_{11}(\theta\varphi)$
	a_-	$Y_{1-1}(\theta\varphi)$
	a_0	$Y_{10}(\theta\varphi)$
E	u_+	$-[Y_{2-2}(\theta\varphi) - \sqrt{2}Y_{21}(\theta\varphi)]/\sqrt{3}$
	u_-	$[Y_{22}(\theta\varphi) + \sqrt{2}Y_{2-1}(\theta\varphi)]/\sqrt{3}$
T_2	x_+	$-[\sqrt{2}Y_{2-2}(\theta\varphi) + Y_{21}(\theta\varphi)]/\sqrt{3}$
	x_-	$[\sqrt{2}Y_{22}(\theta\varphi) - Y_{2-1}(\theta\varphi)]/\sqrt{3}$
	x_0	$Y_{20}(\theta\varphi)$

irreducible representations of the O-group: In the table spherical harmonics are referred to the new coordinate system. One may also show after elementary, but lengthy, calculation that the bases, which will be denoted by $\varphi_{\Gamma M}$, in Table 6.2 are obtained from the linear combination of $\varphi_{\Gamma\gamma}$'s referred to the (xyz) coordinate system as follows:

$$\varphi_{\Gamma M} = \sum_{\gamma} \varphi_{\Gamma\gamma} \langle \Gamma\gamma \mid \Gamma M \rangle, \tag{6.24}$$

in which the $\langle \Gamma\gamma \mid \Gamma M \rangle$'s are the numerical coefficients: The unitary matrices whose elements are $\langle \Gamma\gamma \mid \Gamma M \rangle$ are given as

$$\begin{array}{c} {}_{\Gamma\gamma}\diagdown^{\Gamma M} \\ u \\ v \end{array} \begin{array}{c} \begin{matrix} u_+ & u_- \end{matrix} \\ \begin{bmatrix} -1 & 1 \\ -i & -i \end{bmatrix} \end{array} \times \frac{1}{\sqrt{2}} \tag{6.25}$$

and

$$\begin{array}{c} {}_{\Gamma\gamma}\diagdown^{\Gamma M} \\ \alpha\ (\xi) \\ \beta\ (\eta) \\ \gamma\ (\zeta) \end{array} \begin{array}{c} \begin{matrix} a_+ & a_- & a_0 \\ (x_+) & (x_-) & (x_0) \end{matrix} \\ \begin{bmatrix} -\omega & \bar{\omega} & 1 \\ -\bar{\omega} & \omega & 1 \\ -1 & 1 & 1 \end{bmatrix} \end{array} \times \frac{1}{\sqrt{3}}, \tag{6.26}$$

where $\omega = e^{2\pi i/3}$ and $\bar{\omega} = \omega^2 = e^{-2\pi i/3}$.

Problem 6.3. Confirm (6.25) and (6.26). ◇

By applying symmetry operations in the D_3-group, the functions in each set of (u_+, u_-), (a_+, a_-), and (x_+, x_-) are shown to be the bases of E, the a_0 function the base of A_2, and the x_0 the base of A_1 of the D_3-group. Therefore these functions are called *trigonal bases*. For further discussions on the trigonal systems, it is convenient to have Clebsch–Gordan coefficients for the trigonal bases, $\langle \Gamma_1 M_1 \Gamma_2 M_2 \mid \Gamma M \rangle$. As easily seen from the definition of C–G coefficients, the C–G coefficients for the trigonal bases are calculated from those for the cubic bases as

$$\langle \Gamma_1 M_1 \Gamma_2 M_2 \mid \Gamma M \rangle = \sum_{\gamma_1 \gamma_2 \gamma} \langle \Gamma_1 M_1 \mid \Gamma_1 \gamma_1 \rangle \langle \Gamma_2 M_2 \mid \Gamma_2 \gamma_2 \rangle \langle \Gamma_1 \gamma_1 \Gamma_2 \gamma_2 \mid \Gamma \gamma \rangle \langle \Gamma \gamma \mid \Gamma M \rangle, \tag{6.27}$$

in which necessary $\langle \Gamma M \mid \Gamma \gamma \rangle$'s are already given in (6.25) and (6.26). The C-G coefficients for the trigonal bases calculated by (6.27) are given in Appendix VI.

Now returning to the problem of deriving the explicit form of the D_{3d} potential, the $k = 2$ term, which is found to be proportional to $x_0 = (\xi + \eta + \zeta)/\sqrt{3}$, is given from Table 6.2 as proportional to $C_0^{(2)}(\theta\varphi)$, in which the spherical harmonic is referred to as the (XYZ) coordinate system. The $k = 4$ term, which transforms like the x_0 component of T_{2g}, may be obtained from

$$\varphi_{x_+}\varphi_{x_-}\langle T_2 x_+ T_2 x_- \mid T_2 x_0 \rangle + \varphi_{x_-}\varphi_{x_+}\langle T_2 x_- T_2 x_+ \mid T_2 x_0 \rangle + \varphi_{x_0}\varphi_{x_0}\langle T_2 x_0 T_2 x_0 \mid T_2 x_0 \rangle$$

$$= (\tfrac{2}{3})^{1/2} (\varphi_{x_+}\varphi_{x_-} + \varphi_{x_0}\varphi_{x_0}), \tag{6.28}$$

which, by using Table 6.2, is given in terms of the products of spherical harmonics referred to the (XYZ) coordinate system. The products of spherical harmonics other than (6.7)–(6.9) are expressed by the linear combinations of spherical harmonics as follows:

$$Y_{21}(\theta\varphi)\, Y_{2-1}(\theta\varphi) = -\frac{1}{4\pi}\left[C_0^{(0)}(\theta\varphi) + \frac{5}{7}\, C_0^{(2)}(\theta\varphi) - \frac{12}{7}\, C_0^{(4)}(\theta\varphi)\right] \tag{6.29}$$

$$Y_{2\pm2}(\theta\varphi)\, Y_{2\pm1}(\theta\varphi) = \frac{3}{4\pi}\left(\frac{5}{7}\right)^{1/2} C_{\pm3}^{(4)}(\theta\varphi). \tag{6.30}$$

Using (6.9), (6.29), and (6.30), one finds that (6.28) is proportional to

$$\frac{3}{4}\, C_0^{(2)}(\theta\varphi) + \left\{C_0^{(4)}(\theta\varphi) - \frac{1}{2}\left(\frac{7}{10}\right)^{1/2} [C_3^{(4)}(\theta\varphi) - C_{-3}^{(4)}(\theta\varphi)]\right\}. \tag{6.31}$$

On the other hand the term of cubic symmetry, which of course is invariant to the D_{3d}-symmetry operations, is obtained from

$$\varphi_{x_+}\varphi_{x_-}\langle T_2x_+T_2x_- \mid A_1e_1\rangle + \varphi_{x_-}\varphi_{x_+}\langle T_2x_-T_2x_+ \mid A_1e_1\rangle + \varphi_{x_0}\varphi_{x_0}\langle T_2x_0T_2x_0 \mid A_1e_1\rangle$$

$$= \frac{1}{\sqrt{3}}(-2\varphi_{x_+}\varphi_{x_-} + \varphi_{x_0}\varphi_{x_0}), \tag{6.32}$$

which, by using (6.9), (6.29), and (6.30) again, is found to be proportional to

$$\frac{3}{2}C_0^{(0)}(\theta\varphi) + \left\{C_0^{(4)}(\theta\varphi) + \left(\frac{10}{7}\right)^{1/2}[C_3^{(4)}(\theta\varphi) - C_{-3}^{(4)}(\theta\varphi)]\right\}. \tag{6.33}$$

Neglecting the first term in (6.33) and putting together (6.31) and (6.33), one finally obtains the potential in the D_{3d}-symmetry system as

$$V_{\text{trig}} = V_{C^0} + V_{\text{ax},Z}, \tag{6.34}$$

where

$$V_{C^0} = D(r)\left\{C_0^{(4)}(\theta\varphi) + \left(\frac{10}{7}\right)^{1/2}[C_3^{(4)}(\theta\varphi) - C_{-3}^{(4)}(\theta\varphi)]\right\}, \tag{6.35}$$

$$V_{\text{ax},Z} = A(r)\,C_0^{(2)}(\theta\varphi) + B(r)\,C_0^{(4)}(\theta\varphi). \tag{6.36}$$

The apparent forms of $V_{\text{ax},Z}$ in (6.36) and $V_{\text{ax},z}$ in (6.16) are the same: The only difference is that in the latter the quantization axis (z-axis) is along the fourfold symmetry axis of the octahedron.

In contrast to the D_{3d} case, the D_3-group has no inversion symmetry so that the ligand-field potential in the D_3 system has to involve the odd-parity part in addition to the even-parity part given in (6.34)–(6.36). As seen in Table 1.2, T_{2u} first appears when $D^{(3)}$ is reduced. Therefore, the odd-parity potential of the lowest order is given in terms of $C_m^{(3)}(\theta\varphi)$ and it is proportional to

$$\varphi_{a_+}\varphi_{x_-}\langle T_1a_+T_2x_- \mid T_2x_0\rangle + \varphi_{a_-}\varphi_{x_+}\langle T_1a_-T_2x_+ \mid T_2x_0\rangle$$

$$= \frac{i}{\sqrt{2}}(\varphi_{a_+}\varphi_{x_-} - \varphi_{a_-}\varphi_{x_+}), \tag{6.37}$$

where the explicit forms of $\varphi_{a_\pm}$ are given in Table 6.2. By using the relations derived from (6.3), (6.5), and

$$Y_{1\pm1}(\theta\varphi)\,Y_{2\pm2}(\theta\varphi) = \frac{3}{4\pi}\,C_{3\pm}^{(3)}(\theta\varphi), \tag{6.38}$$

$$Y_{1\pm1}(\theta\varphi)\,Y_{2\mp1}(\theta\varphi) = -\frac{3}{4\pi\sqrt{5}}[C_0^{(1)}(\theta\varphi) - C_0^{(3)}(\theta\varphi)], \tag{6.39}$$

Eq. (6.37) is found to be proportional to

$$V_{\text{trig}}^{\text{odd}} = E(r)[C_3^{(3)}(\theta\varphi) + C_{-3}^{(3)}(\theta\varphi)], \tag{6.40}$$

which is nothing but the odd-parity potential in the D_3-symmetry system.

By using the above mentioned analytical method based on the group theory, it is possible to obtain the explicit form of the ligand-field potential in any symmetry system. However, in the analysis of experiments, the derivation of the explicit form of the potential is not always necessary and a more general treatment of the potential can be applied as discussed in the next section.

6.1.2 Term Splittings and Shifts

Let us first calculate the splittings of the cubic $^2T_{2g}$ and ^2E_g terms in the system of a single electron in a tetragonal field. The wavefunctions associated with these terms are assumed to be those given in (1.31) and (1.32), i.e., the *d*-functions. Due to the cubic part of the ligand-field potential, the $^2T_{2g}$ and ^2E_g terms are separated by $10Dq$.

By using (6.16), (1.31), (1.32), and (1.15), the matrix elements of $V_{\text{ax},z}$ between the components of the $^2T_{2g}$ term are given as

$$\begin{aligned}\langle \xi \mid V_{\text{ax},z} \mid \xi \rangle &= \langle \eta \mid V_{\text{ax},z} \mid \eta \rangle \\ &= \langle A(r) \rangle\, c^2(21, 21) + \langle B(r) \rangle\, c^4(21, 21) \\ &= \frac{1}{7}\langle A(r) \rangle - \frac{4}{21}\langle B(r) \rangle,\end{aligned} \tag{6.41}$$

$$\begin{aligned}\langle \zeta \mid V_{\text{ax},z} \mid \zeta \rangle &= \langle A(r) \rangle\, c^2(22, 22) + \langle B(r) \rangle\, c^4(22, 22) \\ &= -\frac{2}{7}\langle A(r) \rangle + \frac{1}{21}\langle B(r) \rangle,\end{aligned} \tag{6.42}$$

$$\langle \xi \mid V_{\text{ax},z} \mid \eta \rangle = \langle \eta \mid V_{\text{ax},z} \mid \zeta \rangle = \langle \zeta \mid V_{\text{ax},z} \mid \xi \rangle = 0, \tag{6.43}$$

where for any function $f(r)$ of r

$$\langle f(r) \rangle = \int_0^\infty dr\, r^2 f(r)\, R_d^2(r). \tag{6.44}$$

Therefore, the energy eigenvalues of the three-dimensional secular equation are obtained as follows:

$$\begin{aligned}E(B_{2g}) &= \langle \zeta \mid V_{\text{ax},z} \mid \zeta \rangle, \\ E(E_g) &= \langle \xi \mid V_{\text{ax},z} \mid \xi \rangle,\end{aligned} \tag{6.45}$$

where state B_{2g} is nondegenerate and state E_g is doubly degenerate. Thus, the splitting Q_1 of the $^2T_{2g}$ is given as

$$\begin{aligned} Q_1 &= E(E_g) - E(B_{2g}) \\ &= \frac{3}{7}\langle A(r)\rangle - \frac{5}{21}\langle B(r)\rangle. \end{aligned} \tag{6.46}$$

Similarly the matrix elements of $V_{\mathrm{ax},z}$ between the components of the ^2E_g term are given as

$$\langle u \mid V_{\mathrm{ax},z} \mid u\rangle = \frac{2}{7}\langle A(r)\rangle + \frac{6}{21}\langle B(r)\rangle, \tag{6.47}$$

$$\langle v \mid V_{\mathrm{ax},z} \mid v\rangle = -\frac{2}{7}\langle A(r)\rangle + \frac{1}{21}\langle B(r)\rangle, \tag{6.48}$$

$$\langle u \mid V_{\mathrm{ax},z} \mid v\rangle = 0, \tag{6.49}$$

which give the energy eigenvalues,

$$\begin{aligned} E(A_{1g}) &= \langle u \mid V_{\mathrm{ax},z} \mid u\rangle, \\ E(B_{1g}) &= \langle v \mid V_{\mathrm{ax},z} \mid v\rangle. \end{aligned} \tag{6.50}$$

Thus, the splitting Q_2 of the ^2E_g term is given as

$$\begin{aligned} Q_2 &= E(A_{1g}) - E(B_{1g}) \\ &= \frac{4}{7}\langle A(r)\rangle + \frac{5}{21}\langle B(r)\rangle. \end{aligned} \tag{6.51}$$

When the term separation between $^2T_{2g}$ and ^2E_g is not much larger than the tetragonal splittings Q_1 and Q_2, it is important to take into account the nondiagonal matrix elements of $V_{\mathrm{ax}\,z}$ between the components of $^2T_{2g}$ and ^2E_g. However, in the present problem, one can show that all nondiagonal elements are zero: This is clear from the group-theoretical point of view since same irreducible representation of the D_{4h}-group does not appear in the reduction of both $^2T_{2g}$ and ^2E_g. The splittings of the $^2T_{2g}$ and ^2E_g terms are schematically illustrated in Fig. 6.3.

The next example is the calculation of the splittings of the $^2T_{2g}$ and ^2E_g terms in the one-electron system with the trigonal symmetry. The angular dependences of the wavefunctions referred to the trigonal axis are assumed to be those given in Table 6.2. Again by using (1.15) and

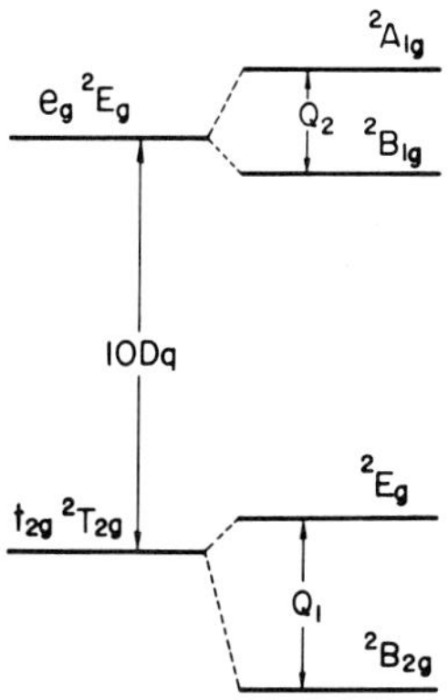

FIG. 6.3. Splittings of the $e_g\ ^2E_g$ and $t_{2g}\ ^2T_{2g}$ terms of O_h-symmetry in a tetragonal (D_{4h}) field ($Q_1 > 0$, $Q_2 > 0$).

(6.36) the matrix elements of $V_{\text{ax},Z}$ between the components of the $^2T_{2g}$ term are given as

$$\langle x_+ \mid V_{\text{ax},Z} \mid x_+\rangle = \langle x_- \mid V_{\text{ax},Z} \mid x_-\rangle = -\frac{1}{7}\langle A(r)\rangle - \frac{2}{63}\langle B(r)\rangle, \tag{6.52}$$

$$\langle x_0 \mid V_{\text{ax},Z} \mid x_0\rangle = \frac{2}{7}\langle A(r)\rangle + \frac{2}{7}\langle B(r)\rangle, \tag{6.53}$$

$$\langle x_+ \mid V_{\text{ax},Z} \mid x_-\rangle = \langle x_+ \mid V_{\text{ax},Z} \mid x_0\rangle = \langle x_- \mid V_{\text{ax},Z} \mid x_0\rangle = 0, \tag{6.54}$$

which give the energy eigenvalues,

$$E(E_g) = \langle x_+ \mid V_{\text{ax},Z} \mid x_+\rangle, \quad E(A_{1g}) = \langle x_0 \mid V_{\text{ax},Z} \mid x_0\rangle. \tag{6.55}$$

Thus, the splitting $3K$ of the $^2T_{2g}$ term is given as

$$3K = E(E_g) - E(A_{1g}) = -\frac{3}{7}\langle A(r)\rangle - \frac{20}{63}\langle B(r)\rangle. \tag{6.56}$$

Similarly, the matrix elements between the components of the ^2E_g term are given as

$$\langle u_+ \mid V_{\text{ax},Z} \mid u_+\rangle = \langle u_- \mid V_{\text{ax},Z} \mid u_-\rangle = \langle x_\pm \mid V_{\text{ax},Z} \mid x_\pm\rangle. \tag{6.57}$$

Therefore, no splitting but a shift of the ^2E_g term is predicted in agreement with the group-theoretical result in (6.22). Since the E_g trigonal representation appears in the reduction of both $^2T_{2g}$ and ^2E_g, one has nonvanishing matrix elements between the ^2E_g components of D_{3d} in $^2T_{2g}$ and ^2E_g as follows:

$$\begin{aligned}\langle x_+ \mid V_{\mathrm{ax},Z} \mid u_+\rangle &= \langle x_- \mid V_{\mathrm{ax},Z} \mid u_-\rangle \\ &= \frac{\sqrt{2}}{21}\left[-3\langle A(r)\rangle + \frac{5}{3}\langle B(r)\rangle\right] \\ &\equiv -\sqrt{2}\,K', \end{aligned} \tag{6.58}$$

which shift the two ^2E_g states in the opposite directions to increase their separation. The splitting and the shift of the $^2T_{2g}$ and ^2E_g terms without taking account of the nondiagonal elements (6.58) are schematically illustrated in Fig. 6.4. The inclusion of (6.58) decreases the splitting, $3K$, if K is positive.

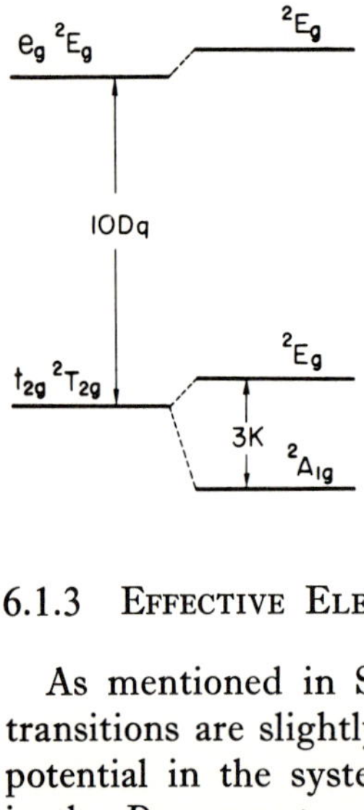

FIG. 6.4. Splitting and shift of the t_{2g} $^2T_{2g}$ and e_g ^2E_g terms of O_h-symmetry in a trigonal (D_{3d}) field ($K > 0$).

6.1.3 Effective Electric-Dipole Transition Moments

As mentioned in Section 5.2.1, the parity-forbidden electric-dipole transitions are slightly allowed by the presence of the static odd-parity potential in the system having no inversion symmetry. For example, in the D_3-symmetry system, the static odd-parity potential is given in (6.40). In this subsection the selection rules and relative intensities will be discussed by calculating the effective electric-dipole transition moments between the split components of the terms in the one-electron system with the D_3-symmetry.

It is easy to see that the use of the perturbation theory gives the effective electric-dipole transition moment $\mathbf{P}_{\text{eff}}(a - b)$ between even-parity states a and b as follows:

$$\mathbf{P}_{\text{eff}}(a - b) = \sum_i \frac{\langle a \mid V_{\text{odd}} \mid i \rangle \langle i \mid \mathbf{P} \mid b \rangle}{E_a - E_i} + \sum_i \frac{\langle a \mid \mathbf{P} \mid i \rangle \langle i \mid V_{\text{odd}} \mid b \rangle}{E_b - E_i} \tag{6.59}$$

where i is the odd-parity states whose energies are the E_i's, and $\mathbf{P}$ is the electric dipole moment, $-e \sum \mathbf{r}_i$. The first term of (6.59) is the matrix element of $\mathbf{P}$ between the small odd-parity component in state a and the even-parity state b, and the second term is that between the even-parity state a and the small odd-parity component in state b: The odd-parity components are brought into the even-parity states by perturbation V_{odd}. Equation (6.59) is often simplified by using the approximation in which the denominators $(E_a - E_i)$ and $(E_b - E_i)$ are replaced by a suitable average, ΔE. This approximation is called *closure approximation.* To this approximation, (6.59) may be expressed as

$$\mathbf{P}_{\text{eff}}(a - b) = \frac{2}{\Delta E} \langle a \mid V_{\text{odd}} \mathbf{P} \mid b \rangle. \tag{6.60}$$

Now, let us calculate the transition matrix element in (6.60) between the split components in the one-electron system of D_3-symmetry. We first neglect the mixing of the ${}^2E_g$ and ${}^2T_{2g}$ terms due to (6.58), assuming that the cubic-field splitting is much larger than the trigonal splitting. For calculating (6.60) for the left and right circular polarizations, σ_+ and σ_-, in the plane perpendicular to the trigonal axis and for the linear polarization π along the trigonal axis, it is convenient to express $\mathbf{P}$ as

$$\mathbf{P} = -P_- \hat{\mathbf{k}}^+ - P_+ \hat{\mathbf{k}}^- + P_0 \hat{\mathbf{k}}^0, \tag{6.61}$$

where

$$P_+ = -\frac{1}{\sqrt{2}} (P_X + iP_Y), \tag{6.62a}$$

$$P_- = \frac{1}{\sqrt{2}} (P_X - iP_Y), \tag{6.62b}$$

$$P_0 = P_Z, \tag{6.62c}$$

and

$$\hat{\mathbf{k}}^+ = -\frac{1}{\sqrt{2}} (\hat{\mathbf{i}} + i\hat{\mathbf{j}}), \tag{6.63a}$$

$$\hat{\mathbf{k}}^- = \frac{1}{\sqrt{2}} (\hat{\mathbf{i}} - i\hat{\mathbf{j}}), \tag{6.63b}$$

$$\hat{\mathbf{k}}^0 = \hat{\mathbf{k}}. \tag{6.63c}$$

In (6.63) $\hat{\mathbf{i}}$, $\hat{\mathbf{j}}$, and $\hat{\mathbf{k}}$ are the unit vectors in the directions of X, Y, and Z coordinate axes, respectively. In terms of spherical harmonics, (6.62) are expressed as

$$P_+ = -\frac{er(4\pi)^{1/2}}{\sqrt{3}} Y_{11}(\theta\varphi),$$

$$P_- = -\frac{er(4\pi)^{1/2}}{\sqrt{3}} Y_{1-1}(\theta\varphi), \tag{6.64}$$

$$P_0 = -\frac{er(4\pi)^{1/2}}{\sqrt{3}} Y_{10}(\theta\varphi).$$

We reduce the products of two spherical harmonics appearing in $\mathbf{P}V_{\text{odd}}$ into a linear combination of spherical harmonics: In the present problem V_{odd} is given by $V_{\text{trig}}^{\text{odd}}$ in (6.40). From (6.3) and (6.5) one may derive the relations,

$$Y_{1\pm1}(\theta\varphi)\, Y_{3\pm3}(\theta\varphi) = \frac{2\sqrt{3}}{4\pi}\, C_{\pm4}^{(4)}(\theta\varphi), \tag{6.65}$$

$$Y_{1\pm1}(\theta\varphi)\, Y_{3\mp3}(\theta\varphi) = -\frac{1}{4\pi}\left[3\left(\frac{5}{7}\right)^{1/2} C_{\mp2}^{(2)}(\theta\varphi) - \left(\frac{3}{7}\right)^{1/2} C_{\mp2}^{(4)}(\theta\varphi)\right], \tag{6.66}$$

$$Y_{10}(\theta\varphi)\, Y_{3\pm3}(\theta\varphi) = \frac{\sqrt{3}}{4\pi}\, C_{\pm3}^{(4)}(\theta\varphi). \tag{6.67}$$

By using these relations one obtains

$$P_\pm V_{\text{trig}}^{\text{odd}} = \frac{-erE(r)}{\sqrt{7}}\left[2C_{\pm4}^{(4)}(\theta\varphi) - \left(\frac{15}{7}\right)^{1/2} C_{\mp2}^{(2)}(\theta\varphi) + \frac{1}{\sqrt{7}}\, C_{\mp2}^{(4)}(\theta\varphi)\right], \tag{6.68}$$

$$P_0 V_{\text{trig}}^{\text{odd}} = \frac{-erE(r)}{\sqrt{7}}\,[C_{+3}^{(4)}(\theta\varphi) + C_{-3}^{(4)}(\theta\varphi)]. \tag{6.69}$$

Then, the transition matrix elements between the trigonal components for the π-polarization are calculated by using (6.69) and Table 6.2 as follows:

$$\langle x_0 \mid P_0 V_{\text{trig}}^{\text{odd}} \mid x_\pm\rangle = 0, \tag{6.70}$$

$$\langle x_0 \mid P_0 V_{\text{trig}}^{\text{odd}} \mid u_\pm\rangle = 0, \tag{6.71}$$

$$\langle x_\pm \mid P_0 V_{\text{trig}}^{\text{odd}} \mid u_\mp\rangle = 0, \tag{6.72}$$

$$\langle x_+ \mid P_0 V_{\text{trig}}^{\text{odd}} \mid u_+\rangle = -\langle x_- \mid P_0 V_{\text{trig}} \mid u_-\rangle = -\frac{\sqrt{5}\, e\langle rE(r)\rangle}{21} \tag{6.73}$$

Similarly, the transition matrix elements for the σ-polarizations are calculated as

$$\langle x_\pm \mid P_+ V_{\text{trig}}^{\text{odd}} \mid u_\pm \rangle = \langle x_\pm \mid P_- V_{\text{trig}}^{\text{odd}} \mid u_\pm \rangle = 0, \tag{6.74}$$

$$\langle x_\pm \mid P_\pm V_{\text{trig}}^{\text{odd}} \mid u_\mp \rangle = 0, \tag{6.75}$$

$$\langle x_\pm \mid P_\mp V_{\text{trig}}^{\text{odd}} \mid u_\mp \rangle = -\frac{2\sqrt{5}\, e\langle rE(r)\rangle}{21}, \tag{6.76}$$

$$\langle x_0 \mid P_\pm V_{\text{trig}}^{\text{odd}} \mid x_\pm \rangle = \langle x_0 \mid P_\pm V_{\text{trig}}^{\text{odd}} \mid u_\pm \rangle = 0. \tag{6.77}$$

$$\begin{aligned} \langle x_0 \mid P_+ V_{\text{trig}}^{\text{odd}} \mid x_- \rangle &= -\langle x_0 \mid P_- V_{\text{trig}}^{\text{odd}} \mid x_+ \rangle \\ &= -\frac{\sqrt{10}\, e\langle rE(r)\rangle}{21}, \end{aligned} \tag{6.78}$$

$$\begin{aligned} \langle x_0 \mid P_+ V_{\text{trig}}^{\text{odd}} \mid u_- \rangle &= -\langle x_0 \mid P_- V_{\text{trig}}^{\text{odd}} \mid u_+ \rangle \\ &= \frac{1}{\sqrt{2}} \langle x_0 \mid P_+ V_{\text{trig}}^{\text{odd}} \mid x_- \rangle. \end{aligned} \tag{6.79}$$

Since the transition probabilities are proportional to the absolute square of (6.60), one may calculate from (6.70)–(6.79) the selection rules and the relative intensities for the transitions between the split components in the D_3-symmetry system as shown in Table 6.3. In the table, σ_+ and σ_- are the constants having the same nonvanishing value only when light is left and right circularly polarized, respectively, and π is also the constant having the same nonvanishing value only when

TABLE 6.3

RELATIVE INTENSITIES FOR THE TRANSITIONS BETWEEN ^2E_g AND $^2T_{2g}$

Cubic term	Trigonal term		x_0	x_+	x_-	u_+	u_-
$^2T_{2g}$	$^2A_1$	x_0		$2\sigma_+$	$2\sigma_-$	σ_+	σ_-
	^2E	x_+				π	$4\sigma_+$
		x_-				$4\sigma_-$	π
^2E_g	^2E	u_+					
		u_-					

light is polarized parallel to the trigonal axis. For the light linearly polarized perpendicular to the trigonal axis, the relative intensities are given by the coefficients of $\hat{\mathbf{i}}^2$ and $\hat{\mathbf{j}}^2$ in the absolute square of (6.60) instead of those of $|\hat{\mathbf{k}}^+|^2$ and $|\hat{\mathbf{k}}^-|^2$ for the circularly polarized light. Noting the relations,

$$|\hat{\mathbf{k}}^+|^2 = |\hat{\mathbf{k}}^-|^2 = \tfrac{1}{2}(\hat{\mathbf{i}}^2 + \hat{\mathbf{j}}^2), \tag{6.80}$$

one may illustrate the transition diagram as shown in Fig. 6.5 for the light linearly polarized along (π) and perpendicular to (σ) the trigonal axis, in which σ is the same nonvanishing constant as π.

Problem 6.4. Derive both the even and odd-parity ligand-field potential in the D_2-symmetry system up to the term of $k = 4$: Here the rhombic distortion is along the x or y axis. Then, calculate the splittings and shifts of the terms in the one d-electron system of cubic symmetry. Also calculate the selection rules and relative intensities of the transitions between the split components. ◇

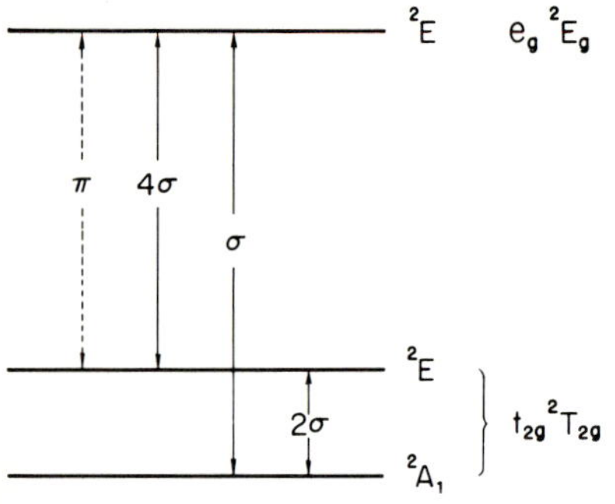

FIG. 6.5. Transition diagram for a single d-electron in a field of D_3 symmetry.

6.2 Wigner–Eckart Theorem

6.2.1 Low-Symmetry Field Potentials as Tensor Operators

Let us define irreducible tensor operators of type Γ, $X(\Gamma)$, in such a way that their components $X_\gamma(\Gamma)$ transform in the same way as the bases $\varphi(\Gamma\gamma)$ of the irreducible representation Γ under the symmetry operations of the O_h-group:

$$RX_\gamma(\Gamma)\,R^{-1} = \sum_{\gamma'} X_{\gamma'}(\Gamma)\,D^{(\Gamma)}_{\gamma'\gamma}(R), \tag{6.81}$$

where R is a symmetry operation of the O_h-group. As discussed in the previous section, low-symmetry field potentials are considered

TABLE 6.4

IRREDUCIBLE TENSOR OPERATORS IN TERMS OF CUBIC HARMONICS

Irred. tensor operators	Cubic harmonics
$V(A_{1g})$	$(x^4 + y^4 + z^4 - \frac{3}{5}r^4) + (\text{6th order}) + \cdots$
$V(A_{2g})$	$x^4(y^2 - z^2) + y^4(z^2 - x^2) + z^4(x^2 - y^2) + (\text{10th order}) + \cdots$
$V_u(E_g)$	$(3z^2 - r^2) + (\text{4th order}) + \cdots$
$V_v(E_g)$	$\sqrt{3}(x^2 - y^2) + (\text{4th order}) + \cdots$
$V_\gamma(T_{1g})$ [a]	$\sqrt{3}xy(x^2 - y^2) + (\text{6th order}) + \cdots$
$V_\zeta(T_{2g})$ [a]	$xy + (\text{4th order}) + \cdots$
$V(A_{1u})$	$xyz[x^4(y^2 - z^2) + y^4(z^2 - x^2) + z^4(x^2 - y^2)] + (\text{13th order}) + \cdots$
$V(A_{2u})$	$xyz + (\text{7th order}) + \cdots$
$V_u(E_u)$	$\sqrt{3}xyz(x^2 - y^2) + (\text{7th order}) + \cdots$
$V_v(E_u)$	$xyz(3z^2 - r^2) + (\text{7th order}) + \cdots$
$V_\gamma(T_{1u})$ [a]	$z + (\text{3rd order}) + \cdots$
$V_\zeta(T_{2u})$ [a]	$\sqrt{3}z(x^2 - y^2) + (\text{5th order}) + \cdots$

[a] The other components may be obtained from these by cyclic change of x, y, and z. For example, $V_\alpha(T_{1g}) \propto \sqrt{3}yz(y^2 - z^2)$, $V_\xi(T_{2u}) \propto \sqrt{3}x(y^2 - z^2)$, and so on.

to be some of these irreducible tensor operators, i.e., real irreducible tensor operators $V_\gamma(\Gamma)$. For example, the D_{4h}-symmetry field potential is $V_u(E_g)$, the D_{3d}-symmetry potential $[V_\xi(T_{2g}) + V_\eta(T_{2g}) + V_\zeta(T_{2g})]/\sqrt{3}$ in addition to the cubic field potential $V(A_{1g})$, and the odd-parity potential of the D_3-symmetry $[V_\xi(T_{2u}) + V_\eta(T_{2u}) + V_\zeta(T_{2u})]/\sqrt{3}$. Similarly, one can associate 18 real irreducible tensor operators $V_\gamma(\Gamma)$ with the ligand-field potentials of certain symmetries as follows:

cubic and tetrahedral: $V(A_{1g,u})$

tetragonal around the z-axis: $V_u(E_{g,u})$, $V_\gamma(T_{1g,u})$[‡]

trigonal around the [111]-axis:
$$\begin{cases} V(A_{2g,u})^{\S} \\ \dfrac{1}{\sqrt{3}}[V_\xi(T_{2g,u}) + V_\eta(T_{2g,u}) + V_\zeta(T_{2g,u})] \\ \dfrac{1}{\sqrt{3}}[V_\alpha(T_{1g,u}) + V_\beta(T_{1g,u}) + V_\gamma(T_{1g,u})]^{\S} \end{cases}$$

rhombic, the distortion along [100] or [010]: $V_v(E_{g,u})$

the distortion along [110] or $[1\bar{1}0]$: $V_\zeta(T_{2g,u})$.

‡ Here, $V_\gamma(T_{1g,u})$ are not associated with the D_{4h}- and D_4-potentials.

§ These are not associated with the D_{3d}- and D_3-potentials.

These irreducible tensor operators may be expanded in terms of suitable linear combinations of spherical harmonics: Such linear combinations of spherical harmonics, $\sum_m c_{km} Y_{km}(\theta\varphi)$, with a fixed value of k are called *cubic harmonics* of the kth order. For example, the cubic harmonic of the 4th order in $V(A_{1g})$ is given in (6.10) (except the first term with $k = 0$) and that in $V_u(E_g)$ is in (6.13) [the first term $C_0^{(2)}(\theta\varphi)$ is the cubic harmonic of $k = 2$ associated with $V_u(E_g)$]. The cubic harmonics of the kth order can be expressed in the form of polynomials $\sum c_{\alpha\beta\gamma} x^\alpha y^\beta z^\gamma$ $(\alpha + \beta + \gamma = k)$ if they are multiplied by r^k. To help intuitive understanding of the irreducible tensor operators, they are given in Table 6.4 in terms of cubic harmonics in the polynomial forms, although it is unnecessary in the following arguments to know these polynomial forms.

6.2.2 Factorization of Reduced Matrices

Now we consider the matrix elements of irreducible tensor operators,

$$\langle \alpha\Gamma\gamma \mid X_{\bar\gamma}(\bar\Gamma) \mid \alpha'\Gamma'\gamma' \rangle = \int d\tau \varphi^*(\alpha\Gamma\gamma)\, X_{\bar\gamma}(\bar\Gamma)\, \varphi(\alpha'\Gamma'\gamma'). \tag{6.82}$$

Since $X_{\bar\gamma}(\bar\Gamma)$ transforms like base $\varphi(\bar\Gamma\bar\gamma)$ of irreducible representation $\bar\Gamma$ of the O_h-group, it is clear from (2.37) that the linear combination of products $X_{\bar\gamma}(\bar\Gamma)\, \varphi(\alpha'\Gamma'\gamma')$,

$$\psi(\bar\Gamma\alpha'\Gamma'\Gamma''\gamma'') = \sum_{\bar\gamma\gamma'} X_{\bar\gamma}(\bar\Gamma)\, \varphi(\alpha'\Gamma'\gamma') \langle \Gamma'\gamma'\bar\Gamma\bar\gamma \mid \Gamma''\gamma'' \rangle, \tag{6.83}$$

transforms like base $\varphi(\Gamma''\gamma'')$ of irreducible representation Γ'' of the O_h-group. By using the orthogonality relation between C–G coefficients given in (2.34), (6.83) may be reexpressed as

$$X_{\bar\gamma}(\bar\Gamma)\, \varphi(\alpha'\Gamma'\gamma') = \sum_{\Gamma''\gamma''} \psi(\bar\Gamma\alpha'\Gamma'\Gamma''\gamma'') \langle \Gamma''\gamma'' \mid \Gamma'\gamma'\bar\Gamma\bar\gamma \rangle. \tag{6.84}$$

Inserting (6.84) into (6.82), one obtains

$$\langle \alpha\Gamma\gamma \mid X_{\bar\gamma}(\bar\Gamma) \mid \alpha'\Gamma'\gamma' \rangle = \sum_{\Gamma''\gamma''} \langle \alpha\Gamma\gamma \mid \bar\Gamma\alpha'\Gamma'\Gamma''\gamma'' \rangle \langle \Gamma''\gamma'' \mid \Gamma'\gamma'\bar\Gamma\bar\gamma \rangle, \tag{6.85}$$

where

$$\langle \alpha\Gamma\gamma \mid \bar\Gamma\alpha'\Gamma'\Gamma''\gamma'' \rangle = \int d\tau \varphi(\alpha\Gamma\gamma)^*\, \psi(\bar\Gamma\alpha'\Gamma'\Gamma''\gamma''). \tag{6.86}$$

Since (6.86) is nonvanishing only when the integrand transforms like the base of irreducible representation A_{1g}, we obtain (see Appendix II)

$$\langle \alpha\Gamma\gamma \mid \bar\Gamma\alpha'\Gamma'\Gamma''\gamma'' \rangle = I_R\, \delta_{\Gamma''\Gamma}\, \delta_{\gamma''\gamma} \tag{6.87}$$

where I_R does not depend upon γ and γ''. Therefore, we express I_R as

$$I_R = (\Gamma)^{-1/2}\langle\alpha\Gamma \| X(\bar{\Gamma}) \| \alpha'\Gamma''\rangle, \tag{6.88}$$

and call $\langle\alpha\Gamma \| X(\bar{\Gamma}) \| \alpha'\Gamma''\rangle$ *reduced matrix*. In (6.88) $(\Gamma)^{-1/2}$ is factorized for convenience of later use. Inserting (6.87) into (6.85) one finally obtains the formula,

$$\langle\alpha\Gamma\gamma \mid X_{\bar{\gamma}}(\bar{\Gamma}) \mid \alpha'\Gamma'\gamma'\rangle = (\Gamma)^{-1/2}\langle\alpha\Gamma \| X(\bar{\Gamma}) \| \alpha'\Gamma'\rangle\langle\Gamma\gamma \mid \Gamma'\gamma'\bar{\Gamma}\bar{\gamma}\rangle, \tag{6.89}$$

which is the mathematical expression of *Wigner–Eckart* theorem. Equation (6.89) shows that the matrix elements with the same $\alpha\Gamma$, $\alpha'\Gamma'$, and $\bar{\Gamma}$ but different sets of γ, γ', and $\bar{\gamma}$ are related to each other by C–G coefficients $\langle\Gamma\gamma \mid \Gamma'\gamma'\bar{\Gamma}\bar{\gamma}\rangle$. The usefullness of this theorem will be demonstrated in the subsequent subsections.

Since C–G coefficients have the properties,

$$(\Gamma)^{-1/2}\langle\Gamma\gamma \mid \Gamma'\gamma'\bar{\Gamma}\bar{\gamma}\rangle = \epsilon(\Gamma\bar{\Gamma}\Gamma')(\Gamma')^{-1/2}\langle\Gamma'\gamma' \mid \Gamma\gamma\bar{\Gamma}\bar{\gamma}\rangle, \tag{6.90}$$

where

$$\epsilon(\Gamma A_1\Gamma') = \epsilon(\Gamma T_2\Gamma') = 1, \tag{6.91a}$$

$$\epsilon(\Gamma A_2\Gamma') = \epsilon(\Gamma T_1\Gamma') = -1, \tag{6.91b}$$

and

$$\epsilon(\Gamma E\Gamma') = 1 \tag{6.91c}$$

except for

$$\epsilon(T_1ET_2) = \epsilon(T_2ET_1) = -1, \tag{6.91d}$$

one may easily prove the relation for real operators $X_{\bar{\gamma}}(\bar{\Gamma}) = V_{\bar{\gamma}}(\bar{\Gamma})$ as follows,

$$\langle\alpha\Gamma \| V(\bar{\Gamma}) \| \alpha'\Gamma'\rangle = \epsilon(\Gamma\bar{\Gamma}\Gamma')\langle\alpha'\Gamma' \| V(\bar{\Gamma}) \| \alpha\Gamma\rangle, \tag{6.92}$$

which shows that

$$\langle\alpha\Gamma \| V(\bar{\Gamma}) \| \alpha\Gamma\rangle = 0 \qquad \text{for} \quad \bar{\Gamma} = A_2 \quad \text{and} \quad T_1 . \tag{6.93}$$

For purely imaginary operators[‡] $X_{\bar{\gamma}}(\bar{\Gamma}) = T_{\bar{\gamma}}(\bar{\Gamma})$, a similar relation to (6.92) may be proved as

$$\langle\alpha\Gamma \| T(\bar{\Gamma}) \| \alpha'\Gamma'\rangle = -\epsilon(\Gamma\bar{\Gamma}\Gamma')\langle\alpha'\Gamma' \| T(\bar{\Gamma}) \| \alpha\Gamma\rangle, \tag{6.94}$$

which shows that

$$\langle\alpha\Gamma \| T(\bar{\Gamma}) \| \alpha\Gamma\rangle = 0 \qquad \text{for} \quad \bar{\Gamma} = A_1,\ E, \text{ and } T_2 . \tag{6.95}$$

‡ For example, the angular momentum operator $\mathbf{l} = -i\hbar\,\mathbf{r} \times \nabla$ is a purely imaginary operator $T(T_{1g})$ which will be discussed fully in the next chapter.

6.2.3 Simple Application of Wigner–Eckart Theorem

In Section 6.1 the term splittings and the relative intensities of the parity-forbidden transitions were calculated by the use of the explicit forms of low-symmetry potentials, to the approximation in which the t_{2g} and e_g orbitals are the d-functions. This assumption made it possible to cut off the higher order terms with $k > 4$ in the expansion of the potential in terms of cubic harmonics. However, this assumption is not necessarily valid in practice, and in what follows the same problem will be treated by using the Wigner–Eckart theorem without making such an assumption.

Let us first consider the splittings of the $t_{2g}\ {}^2T_{2g}$ and the $e_g\ {}^2E_g$ terms in the D_4-symmetry system. The even-parity D_4-potential may be represented by irreducible tensor operaotr $V_u(E_g)$.‡ The splitting of the $t_{2g}\ {}^2T_{2g}$ term is obtained by calculating the following matrix elements:

$$\langle t_{2g}\gamma \mid V_u(E_g) \mid t_{2g}\gamma' \rangle = \frac{1}{\sqrt{3}} \langle t_{2g} \| V(E_g) \| t_{2g} \rangle \langle T_2\gamma \mid T_2\gamma' Eu \rangle. \tag{6.96}$$

By using C–G coefficients in Appendix II, (6.96) with various γ and γ' are calculated as:

$$\begin{aligned} \langle t_{2g}\xi \mid V_u(E_g) \mid t_{2g}\xi \rangle &= \langle t_{2g}\eta \mid V_u(E_g) \mid t_{2g}\eta \rangle \\ &= -\frac{1}{2} \langle t_{2g}\zeta \mid V_u(E_g) \mid t_{2g}\zeta \rangle \\ &= -\frac{1}{2\sqrt{3}} \langle t_{2g} \| V(E_g) \| t_{2g} \rangle, \end{aligned} \tag{6.97}$$

the others are zero, which agree with (6.46) if the reduced matrix element is related to Q_1 as

$$\langle t_{2g} \| V(E_g) \| t_{2g} \rangle = -\frac{2}{\sqrt{3}} Q_1 . \tag{6.98}$$

In contrast to the present results in (6.97), the lack of the relation, $\langle \xi \mid V_{\mathrm{ax},z} \mid \xi \rangle = -\frac{1}{2} \langle \zeta \mid V_{\mathrm{ax},z} \mid \zeta \rangle$, in the previous results in (6.41) and (6.42) is due to the subtraction of the cubic potential part from V_{tet} in (6.14): tensor operator $V_u(E_g)$ corresponds to V_{tet}. Similarly the splitting of the $e_g\ {}^2E_g$ term is obtained by calculating

$$\langle e_g\gamma \mid V_u(E_g) \mid e_g\gamma' \rangle = \frac{1}{\sqrt{2}} \langle e_g \| V(E_g) \| e_g \rangle \langle E\gamma \mid E\gamma' Eu \rangle. \tag{6.99}$$

‡ Here, $V_\gamma(T_{1g})$ is not associated with this potential as it changes the sign under the operations in $2\hat{C}_2$ and $2\hat{C}_2'$ classes in Table 1.5 as easily seen from the expansion form of $V_\gamma(T_{1g})$ in Table 6.4.

By using C–G coefficients in Appendix II again, one obtains

$$\begin{aligned}\langle e_g u \mid V_u(E_g) \mid e_g u\rangle &= -\langle e_g v \mid V_u(E_g) \mid e_g v\rangle \\ &= -\tfrac{1}{2}\langle e_g \| V(E_g) \| e_g\rangle, \end{aligned} \tag{6.100}$$

the others are zero, which agree with (6.51) if

$$\langle e_g \| V(E_g) \| e_g\rangle = -Q_2 . \tag{6.101}$$

Again the lack of the relation, $\langle u \mid V_{\mathrm{ax},z} \mid u\rangle = -\langle v \mid V_{\mathrm{ax},z} \mid v\rangle$, in the previous results in (6.47) and (6.48) should be contrasted to (6.100). The absence of the nondiagonal elements of $V_u(E_g)$ between the ${}^2T_{2g}$ and ${}^2E_g$ terms is easily seen, as $\langle T_2\gamma \mid E\gamma' Eu\rangle$ are always zero: $E \times E = A_1 + A_2 + E$ involves no T_2.

The next example of the application of Wigner–Eckart theorem is the calculation of the relative intensities of the electric-dipole transitions as those discussed in Section 6.1.3. The odd-parity potential $V_{\mathrm{trig}}^{\mathrm{odd}}$ in the D_3-system may be represented by $V_{x_0}(T_{2u})$.‡ On the other hand, the electric-dipole moments, P_+, P_-, and P_0 are represented by irreducible tensor operators $U_{a_+}(T_{1u})$, $U_{a_-}(T_{1u})$, and $U_{a_0}(T_{1u})$, respectively. By using (6.84), the products of two tensor operators are reduced to linear combinations of tensor operators $W_M(\Gamma)$ as follows:

$$\begin{aligned}\tilde{P}_\pm &\equiv U_{a_\pm}(T_{1u})\, V_{x_0}(T_{2u}) \\ &= \mp \frac{i}{\sqrt{3}} W_{u_\pm}(E_g) - \frac{1}{\sqrt{6}} W_{a_\pm}(T_{1g}) + \frac{i}{\sqrt{2}} W_{x_\pm}(T_{2g}),\end{aligned} \tag{6.102}$$

$$\begin{aligned}\tilde{P}_0 &\equiv U_{a_0}(T_{1u})\, V_{x_0}(T_{2u}) \\ &= -\frac{1}{\sqrt{3}} W(A_{2g}) + \frac{\sqrt{2}}{\sqrt{3}} W_{a_0}(T_{1g}).\end{aligned} \tag{6.103}$$

The matrix elements of $\tilde{P}_0$ for the π-polarization are calculated as follows:

$$\langle t_{2g}x_0 \mid \tilde{P}_0 \mid t_{2g}x_\pm\rangle = 0 \tag{6.104}$$

$$\langle t_{2g}x_0 \mid \tilde{P}_0 \mid e_g u_\pm\rangle = \langle t_{2g}x_\pm \mid \tilde{P}_0 \mid e_g u_\mp\rangle = 0, \tag{6.105}$$

as all the C–G coefficients appearing in the matrix elements of $W(A_{2g})$

‡ Here, $V(A_{2u})$ and $V_{a_0}(T_{1u})$ cannot be the D_3-potential as they change their signs under the symmetry operations in the $3\hat{C}_2$ class as shown in Table 6.1 and (6.22).

and $W_{a_0}(T_{1g})$ between these trigonal components are zero. Nonvanishing matrix elements of $\tilde{P}_0$ are

$$\langle t_{2g}x_\pm \mid \tilde{P}_0 \mid e_g u_\pm \rangle = \left(\frac{2}{3}\right)^{1/2} \langle t_{2g}x_\pm \mid W_{a_0}(T_{1g}) \mid e_g u_\pm \rangle$$
$$= \mp \frac{i}{\sqrt{3}} \langle t_{2g} \| W(T_{1g}) \| e_g \rangle. \tag{6.106}$$

For the $\sigma_\pm$-polarization, one sees that

$$\langle t_{2g}x_\pm \mid \tilde{P}_+ \mid e_g u_\pm \rangle = \langle t_{2g}x_\pm \mid \tilde{P}_- \mid e_g u_\pm \rangle = \langle t_{2g}x_\pm \mid \tilde{P}_\pm \mid e_g u_\mp \rangle$$
$$= \langle t_{2g}x_0 \mid \tilde{P}_\pm \mid t_{2g}x_\pm \rangle = \langle t_{2g}x_0 \mid \tilde{P}_\pm \mid e_g u_\pm \rangle = 0, \tag{6.107}$$

as all the C–G coefficients appearing in the matrix elements of $W_{u_\pm}(E_g)$, $W_{a_\pm}(T_{1g})$, and $W_{x_\pm}(T_{2g})$ between these trigonal components are zero. Nonvanishing matrix elements of $\tilde{P}_\pm$ are

$$\langle t_{2g}x_\pm \mid \tilde{P}_\mp \mid e_g u_\mp \rangle$$
$$= \frac{i}{6} \langle t_{2g} \| W(T_{1g}) \| e_g \rangle - \frac{i}{2\sqrt{3}} \langle t_{2g} \| W(T_{2g}) \| e_g \rangle, \tag{6.108}$$

$$\langle t_{2g}x_0 \mid \tilde{P}_\pm \mid t_{2g}x_\mp \rangle$$
$$= \pm \frac{i}{3\sqrt{2}} \langle t_{2g} \| W(E_g) \| t_{2g} \rangle \mp \frac{i}{6} \langle t_{2g} \| W(T_{2g}) \| t_{2g} \rangle, \tag{6.109}$$

$$\langle t_{2g}x_0 \mid \tilde{P}_\pm \mid e_g u_\mp \rangle$$
$$= \pm \frac{i}{6} \langle t_{2g} \| W(T_{1g}) \| e_g \rangle \pm \frac{i}{2\sqrt{3}} \langle t_{2g} \| W(T_{2g}) \| e_g \rangle. \tag{6.110}$$

In (6.109) the reduced matrix $\langle t_{2g} \| W(T_{1g}) \| t_{2g} \rangle$ is set to be zero, as operators $U(T_{1u})$ and $V(T_{2u})$ may be chosen to be real, and one can apply (6.93) to this case.

In contrast to the relation derived from (6.73), (6.78), and (6.79) such as

$$\langle x_+ \mid \tilde{P}_0 \mid u_+ \rangle = \langle x_0 \mid \tilde{P}_+ \mid u_- \rangle$$
$$= -\tfrac{1}{2} \langle x_- \mid \tilde{P}_+ \mid u_+ \rangle, \tag{6.111}$$

which was obtained by approximating the t_{2g} and e_g orbitals as the *d*-functions, the present result in (6.106), (6.108), and (6.110) shows only the relation,

$$\langle x_0 \mid \tilde{P}_+ \mid u_- \rangle + \langle x_- \mid \tilde{P}_+ \mid u_+ \rangle = -\langle x_+ \mid \tilde{P}_0 \mid u_+ \rangle. \tag{6.112}$$

Relation (6.112), of course, is satisfied if (6.111) is assumed, but in the present calculation, (6.111) is satisfied only if we have the relation,

$$\langle t_{2g} \| W(T_{1g}) \| e_g \rangle = -\frac{1}{\sqrt{3}} \langle t_{2g} \| W(T_{2g}) \| e_g \rangle. \tag{6.113}$$

Similarly, the previously obtained relation (6.79) is valid only when the following relation holds in addition to (6.113):

$$\langle t_{2g} \| W(T_{1g}) \| e_g \rangle$$
$$= -\frac{1}{2} \langle t_{2g} \| W(E_g) \| t_{2g} \rangle + \frac{1}{2\sqrt{2}} \langle t_{2g} \| W(T_{2g}) \| t_{2g} \rangle. \tag{6.114}$$

Problem 6.5. By using the irreducible tensor operators for the ligand field and applying Wigner–Eckart theorem, derive the answers to Problem 6.4. In this case be free from the *d*-function approximation. ◇

6.3 Many Electrons in Fields of Low Symmetry

6.3.1 Calculation of the Matrix Elements

As discussed in Section 6.2, the matrix elements of the low-symmetry potential $V_{\bar{\gamma}}(\bar{\Gamma})$ between the

$$t_{2g}^n(S_1\Gamma_1)\, e_g^{\,m}(S_2\Gamma_2)S\Gamma M\gamma \qquad \text{and} \qquad t_{2g}^{n'}(S_1'\Gamma_1')\, e_g^{m'}(S_2'\Gamma_2')S'\Gamma'M'\gamma'$$

$(n + m = n' + m' = N)$ states are calculated by using Wigner–Eckart theorem as follows:

$$\langle t_{2g}^n(S_1\Gamma_1)\, e_g^{\,m}(S_2\Gamma_2)\, S\Gamma M\gamma \mid V_{\bar{\gamma}}(\bar{\Gamma}) \mid t_{2g}^{n'}(S_1'\Gamma_1')\, e_g^{m'}(S_2'\Gamma_2')\, S'\Gamma'M'\gamma' \rangle$$
$$= \delta(SS')\,\delta(MM')(\Gamma)^{-1/2} \langle \Gamma\gamma \mid \Gamma'\gamma'\bar{\Gamma}\bar{\gamma} \rangle$$
$$\times \langle t_{2g}^n(S_1\Gamma_1)\, e_g^{\,m}(S_2\Gamma_2)\, S\Gamma \| V(\bar{\Gamma}) \| t_{2g}^{n'}(S_1'\Gamma_1')\, e_g^{m'}(S_2'\Gamma_2')\, S'\Gamma' \rangle. \tag{6.115}$$

In deriving (6.115) we have used the fact that the ligand-field potential involves no spin operator. By using (6.115) it is possible to discuss the term splittings leaving reduced matrix

$$\langle t_{2g}^n(S_1\Gamma_1)\, e_g^{\,m}(S_2\Gamma_2)\, S\Gamma \| V(\bar{\Gamma}) \| t_{2g}^{n'}(S_1'\Gamma_1')\, e_g^{m'}(S_2'\Gamma_2')\, S\Gamma' \rangle$$

as adjustable parameters. However, to reduce the number of adjustable parameters, it is more convenient to express the reduced matrix for many electrons in terms of the reduced matrix for a single electron.

The ligand-field potential $V_{\bar{\gamma}}(\bar{\Gamma})$ in a many-electron system is given as the sum of the potentials for individual electrons i as follows:

$$V_{\bar{\gamma}}(\bar{\Gamma}) = \sum_i v_{i\bar{\gamma}}(\bar{\Gamma}), \tag{6.116}$$

where $v_{i\bar{\gamma}}(\bar{\Gamma})$ is the function of electron coordinate $\mathbf{r}_i$. Since the many-electron states $t_{2g}^n e_g{}^m S\Gamma M\gamma$ are given by the linear combination of Slater determinants, by using formulas (3.33)–(3.35) the matrix element in (6.115) may be expressed in terms of $\langle t_{2g} \| v(\bar{\Gamma}) \| t_{2g}\rangle$, $\langle e_g \| v(\bar{\Gamma}) \| e_g\rangle$, and $\langle t_{2g} \| v(\bar{\Gamma}) \| e_g\rangle$; in particular, from (3.35)

$$\langle t_{2g}^n e_g{}^m S\Gamma \| V(\bar{\Gamma}) \| t_{2g}^{n-k} e_g^{m+k} S\Gamma'\rangle = 0 \qquad \text{for} \quad |k| \geqslant 2, \tag{6.117}$$

from (3.34)

$$\langle t_{2g}^n e_g{}^m S\Gamma \| V(\bar{\Gamma}) \| t_{2g}^{n-1} e_g^{m+1} S\Gamma'\rangle = C_0 \langle t_{2g} \| v(\bar{\Gamma}) \| e_g\rangle, \tag{6.118}$$

and from (3.33)

$$\langle t_{2g}^n e_g{}^m S\Gamma \| V(\bar{\Gamma}) \| t_{2g}^n e_g{}^m S\Gamma'\rangle = C_1 \langle t_{2g} \| v(\bar{\Gamma}) \| t_{2g}\rangle + C_2 \langle e_g \| v(\bar{\Gamma}) \| e_g\rangle, \tag{6.119}$$

where C_0, C_1, and C_2 are numerical coefficients depending upon the states of interest and $\bar{\Gamma}$. From (6.93) and (6.119) one sees that

$$\langle t_{2g}^n e_g{}^m S\Gamma \| V(\bar{\Gamma}) \| t_{2g}^n e_g{}^m S\Gamma'\rangle = 0 \qquad \text{for} \quad \bar{\Gamma} = A_2 \quad \text{and} \quad T_1. \tag{6.120}$$

The calculation of C_0, C_1, and C_2 is straightforward as the wave-functions are already known. For example, C_0 for

$$\langle t_{2g}^3\ {}^4A_{2g} \| V(T_{2g}) \| t_{2g}^2({}^3T_{1g})e_g\ {}^4T_{1g}\rangle$$

may be calculated as follows: By using Tables 3.2 and 3.4, and also formula (3.34), one obtains

$$\left\langle t_2{}^3\ {}^4A_2 \frac{3}{2} e_2 \,|\, V_\zeta(T_2) \,|\, t_2{}^2({}^3T_1)e\ {}^4T_1 \frac{3}{2} \gamma \right\rangle$$

$$= -\int d\boldsymbol{\tau}\, |\, \xi\eta\zeta\, |^* V_\zeta(T_2) \,|\, \xi\eta u \,| = -\langle \zeta \,|\, v_\zeta(T_2) \,|\, u\rangle. \tag{6.121}$$

Noting the relations

$$\left\langle t_2{}^3\ {}^4A_2 \frac{3}{2} e_2 \,|\, V_\zeta(T_2) \,|\, t_2{}^2({}^3T_1)e\ {}^4T_1 \frac{3}{2} \gamma \right\rangle$$

$$= -\frac{1}{\sqrt{3}} \langle t_2{}^3\ {}^4A_2 \| V(T_2) \| t_2{}^2({}^3T_1)e\ {}^4T_1\rangle \tag{6.122}$$

and

$$\langle \zeta \mid v_\zeta(T_2) \mid u \rangle = \frac{1}{\sqrt{3}} \langle t_2 \| v(T_2) \| e \rangle, \tag{6.123}$$

one finally obtains

$$\langle t_2^3\, {}^4A_2 \| V(T_2) \| t_2^2({}^3T_1)e\, {}^4T_1 \rangle = \langle t_2 \| v(T_2) \| e \rangle. \tag{6.124}$$

Similarly C_1 and C_2 for $\langle t_2^2({}^3T_1)\, e\, {}^4T_2 \| V(T_2) \| t_2^2({}^3T_1)\, e\, {}^4T_2 \rangle$ may be obtained as follows: By using Table 3.4 and formula (3.33), one obtains

$$\left\langle t_2^2({}^3T_1)e\, {}^4T_2\, \tfrac{3}{2}\, \eta \mid V_\zeta(T_2) \mid t_2^2({}^3T_1)e\, {}^4T_2\, \tfrac{3}{2}\, \xi \right\rangle$$

$$= \tfrac{1}{4} \left[3 \int d\boldsymbol{\tau} \mid \xi\zeta u \mid^* V_\zeta(T_2) \mid \eta\zeta u \mid - \int d\boldsymbol{\tau} \mid \xi\zeta v \mid^* V_\zeta(T_2) \mid \eta\zeta v \mid \right]$$

$$= \tfrac{1}{2} \langle \xi \mid v_\zeta(T_2) \mid \eta \rangle. \tag{6.125}$$

Noting the relations

$$\left\langle t_2^2({}^3T_1)e\, {}^4T_2\, \frac{3}{2}\, \eta \mid V_\zeta(T_2) \mid t_2^2({}^3T_1)e\, {}^4T_2\, \frac{3}{2}\, \xi \right\rangle$$

$$= \frac{1}{\sqrt{6}} \langle t_2^2({}^3T_1)e\, {}^4T_2 \| V(T_2) \| t_2^2({}^3T_1)e\, {}^4T_2 \rangle \tag{6.126a}$$

and

$$\langle \xi \mid v_\zeta(T_2) \mid \eta \rangle = \frac{1}{\sqrt{6}} \langle t_2 \| v(T_2) \| t_2 \rangle, \tag{6.126b}$$

one finally proves that

$$\langle t_2^2({}^3T_1)e\, {}^4T_2 \| V(T_2) \| t_2^2({}^3T_1)e\, {}^4T_2 \rangle = \tfrac{1}{2} \langle t_2 \| v(T_2) \| t_2 \rangle. \tag{6.127}$$

In this case C_2 turns out to be zero.

Problem 6.6. Calculate C_1 and C_2 in (6.119) for

$$\langle t_2^2({}^3T_1)e\, {}^4T_1 \| V(T_2) \| t_2^2({}^3T_1)e\, {}^4T_1 \rangle$$

and

$$\langle t_2^2({}^3T_1)e\, {}^4\Gamma \| V(E) \| t_2^2({}^3T_1)e\, {}^4\Gamma \rangle$$

with $\Gamma = T_1$ and T_2. ◇

6.3.2 The Reduced Matrix in Complementary States

Since the matrix of the ligand-field potential is hermitian and real, the matrix elements of $V_{\bar{\gamma}}(\bar{\Gamma})$ in the complementary states are related to each other as

$$\langle t_2^n(S_1\Gamma_1)\, e^m(S_2\Gamma_2)\, S\Gamma M\gamma \mid V_{\bar{\gamma}}(\bar{\Gamma}) \mid t_2^{n'}(S_1'\Gamma_1')\, e^{m'}(S_2'\Gamma_2')\, S\Gamma' M\gamma' \rangle$$
$$= -\langle t_2^{6-n}(S_1\Gamma_1)\, e^{4-m}(S_2\Gamma_2)\, S\Gamma M\gamma \mid V_{\bar{\gamma}}(\bar{\Gamma}) \mid t_2^{6-n'}$$
$$\times (S_1'\Gamma_1')\, e^{4-m'}(S_2'\Gamma_2')\, S\Gamma' M\gamma' \rangle$$
$$(n + m = n' + m' \neq 5) \qquad (6.128)$$

as shown in (4.52). Here we neglect constant F_0 appearing in the diagonal elements. Equation (6.128) tells us that

$$\langle t_2^n(S_1\Gamma_1)\, e^m(S_2\Gamma_2)\, S\Gamma \,\|\, V(\bar{\Gamma}) \,\|\, t_2^{n'}(S_1'\Gamma_1')\, e^{m'}(S_2'\Gamma_2')\, S\Gamma' \rangle$$
$$= -\langle t_2^{6-n}(S_1\Gamma_1)\, e^{4-m}(S_2\Gamma_2)\, S\Gamma \,\|\, V(\bar{\Gamma}) \,\|\, t_2^{6-n'}(S_1'\Gamma_1')\, e^{4-m'}(S_2'\Gamma_2')\, S\Gamma' \rangle$$
$$(n + m = n' + m' \neq 5). \qquad (6.129)$$

In particular, if the states have a half-filled subshell configuration such as t_2^3, one may show from (6.129) and (4.22) that

$$\langle t_2^3 S\Gamma \,\|\, V(\bar{\Gamma}) \,\|\, t_2^3 S\Gamma' \rangle_L = -\langle t_2^3 S\Gamma \,\|\, V(\bar{\Gamma}) \,\|\, t_2^3 S\Gamma' \rangle_R$$
$$= -\mu_1\mu_1' \langle t_2^3 S\Gamma \,\|\, V(\bar{\Gamma}) \,\|\, t_2^3 S\Gamma' \rangle_L, \qquad (6.130)$$

where L and R refer to the L and R states, respectively, and μ_1 and μ_1' are the phase factors defined in (4.22) associated with $t_2^3 S\Gamma$ and $t_2^3 S\Gamma'$, respectively. From (6.130) one immediately sees that

$$\langle t_2^3 S\Gamma \,\|\, V(\bar{\Gamma}) \,\|\, t_2^3 S\Gamma \rangle = 0, \qquad (6.131)$$

and

$$\langle t_2^3 S\Gamma \,\|\, V(\bar{\Gamma}) \,\|\, t_2^3 S\Gamma' \rangle \neq 0 \qquad (\Gamma \neq \Gamma') \qquad (6.132a)$$

only for the combinations

$$S\Gamma = {}^2E,\ {}^2T_1 \quad \text{and} \quad S\Gamma' = {}^2T_2, \quad \text{and vice versa.} \qquad (6.132b)$$

Similarly,

$$\langle e^2 S\Gamma \,\|\, V(\bar{\Gamma}) \,\|\, e^2 S\Gamma \rangle = 0, \qquad (6.133)$$

and

$$\langle e^2 S\Gamma \,\|\, V(\bar{\Gamma}) \,\|\, e^2 S\Gamma' \rangle \neq 0 \qquad (\Gamma \neq \Gamma') \qquad (6.134a)$$

only for the combinations

$$S\Gamma = {}^1E \quad \text{and} \quad S\Gamma' = {}^1A_1, \quad \text{and vice versa.} \qquad (6.134b)$$

In the case of $n + m = n' + m' = 5$, one may derive by using (4.51) and (4.28) the relation,

$$\begin{aligned}
&\langle t_2^n(S_1\Gamma_1)\, e^m(S_2\Gamma_2)\, S\Gamma \,\|\, V(\bar{\Gamma}) \,\|\, t_2^{n'}(S_1'\Gamma_1')\, e^{m'}(S_2'\Gamma_2')\, S\Gamma'\rangle_L \\
&\quad = -\langle t_2^{6-n}(S_1\Gamma_1)\, e^{4-m}(S_2\Gamma_2)\, S\Gamma \,\|\, V(\bar{\Gamma}) \,\|\, t_2^{6-n'}(S_1'\Gamma_1')\, e^{4-m'}(S_2'\Gamma_2')\, S\Gamma'\rangle_R \\
&\quad = -\mu_1\mu_2\mu_1'\mu_2'\langle t_2^{6-n}(S_1\Gamma_1)\, e^{4-m}(S_2\Gamma_2)\, S\Gamma \,\|\, V(\bar{\Gamma}) \,\|\, t_2^{6-n'} \\
&\qquad \times (S_1'\Gamma_1')\, e^{4-m'}(S_2'\Gamma_2')\, S\Gamma'\rangle_L . \qquad (6.135)
\end{aligned}$$

In deriving (6.135) one has used the fact that $(-1)^{nm}(-1)^{n'm'} = 1$ for $n + m = n' + m' = 5$. Equation (6.135) shows that, if $n \neq 3$ and $n' \neq 3$, (6.128) holds even for $n + m = n' + m' = 5$ as $\mu_1\mu_2\mu_1'\mu_2' = 1$. If $n = 3$ and $n' \neq 3$, one has

$$\begin{aligned}
&\langle t_2^3(S_1\Gamma_1)\, e^2(S_2\Gamma_2)\, S\Gamma \,\|\, V(\bar{\Gamma}) \,\|\, t_2^{n'}(S_1'\Gamma_1')\, e^{m'}(S_2'\Gamma_2')\, S\Gamma'\rangle \\
&\quad = -\mu_1\mu_2\langle t_2^3(S_1\Gamma_1)\, e^2(S_2\Gamma_2)\, S\Gamma \,\|\, V(\bar{\Gamma}) \,\|\, t_2^{6-n'}(S_1'\Gamma_1')\, e^{4-m'}(S_2'\Gamma_2')\, S\Gamma'\rangle \\
&\qquad (n' \neq 3) \qquad (6.136)
\end{aligned}$$

and, if $n = 3$ and $n' = 3$,

$$\begin{aligned}
&\langle t_2^3(S_1\Gamma_1)\, e^2(S_2\Gamma_2)\, S\Gamma \,\|\, V(\bar{\Gamma}) \,\|\, t_2^3(S_1'\Gamma_1')\, e^2(S_2'\Gamma_2')\, S\Gamma'\rangle = 0 \\
&\qquad \text{for } S_1\Gamma_1,\ S_2\Gamma_2,\ S_1'\Gamma_1', \text{ and } S_2'\Gamma_2' \text{ giving } \mu_1\mu_2\mu_1'\mu_2' = 1, \qquad (6.137)
\end{aligned}$$

which shows

$$\langle t_2^3(S_1\Gamma_1)\, e^2(S_2\Gamma_2)\, S\Gamma \,\|\, V(\bar{\Gamma}) \,\|\, t_2^3(S_1\Gamma_1)\, e^2(S_2\Gamma_2)\, S\Gamma'\rangle = 0. \qquad (6.138)$$

As shown in (6.131), (6.133), and (6.138), the diagonal matrix elements of any low-symmetry potential in the states of the half-filled shell configurations, $t_2{}^3$, e^2, and $t_2{}^3e^2$ vanish. This tells us that the spectral lines due to the transitions between the terms of the same half-filled configuration are not broadened by the vibrational fluctuation of low-symmetry fields.

Problem 6.7. By using the argument given in Section 3.1.2, show that

$$\begin{aligned}
&\langle t_2^n(S_1\Gamma_1)\, e^m(S_2\Gamma_2)\, S\Gamma \,\|\, V(\Gamma) \,\|\, t_2^n(S_1'\Gamma_1')\, e^m(S_2'\Gamma_2')S\Gamma'\rangle \\
&\quad = \delta(S_1S_1')\,\delta(S_2S_2')\langle t_2^n(S_1\Gamma_1)\, e^m(S_2\Gamma_2)\, S\Gamma \,\|\, V(\Gamma) \,\|\, t_2^n(S_1\Gamma_1')\, e^m(S_2\Gamma_2')\, S\Gamma'\rangle. \diamond
\end{aligned}$$

Chapter VII SPIN-ORBIT INTERACTION

The spin-orbit interaction has its origin in the relativistic theory, and in the case of a single electron in atoms it is derived from the Dirac equation as

$$\mathscr{H}_{\mathrm{so}} = \xi(r)\,\mathbf{l}\cdot\mathbf{s}, \tag{7.1}$$

where $\xi(r)$ is

$$-\frac{e\hbar^2}{2m^2c^2}\frac{1}{r}\frac{dU(r)}{dr}$$

with spherically symmetric potential $U(r)$ for the electron. Classically, this interaction may be viewed as the interaction of the magnetic moment of an electron spin with the magnetic field induced by the motion of the nucleus around the electron: The nucleus is seen from the coordinate system fixed on the electron.

The strength of the spin-orbit interaction in iron-group ions is about one order of magnitude smaller than that of the Coulomb interaction, and is comparable to that of low-symmetry ligand fields. Therefore, the spin-orbit interaction as well as the low-symmetry ligand-field perturbation is responsible for the *fine structure* of multiplets. The spin-orbit interaction differs from the low-symmetry field perturbation in that the former involves spin operators, while the latter does not. The presence of spin operators makes it possible to connect the terms of different spin-multiplicities, which is important in discussing, for example, the intersystem combinations as briefly mentioned in Section 5.2.1.

7.1 The Problem of a Single d-Electron

7.1.1 Orbital Angular Momentum

Before begining the detailed discussion of the spin-orbit interaction, it is instructive to present the matrix elements of the orbital angular momentum in the cubic system with a single electron t_{2g} or e_g whose orbital is made from the d-function. From the well-known relations[‡]

$$\begin{aligned} l_z\varphi_{lm}(\mathbf{r}) &= m\varphi_{lm}(\mathbf{r}), \\ l_\pm\varphi_{lm}(\mathbf{r}) &= [l(l+1) - m(m\pm 1)]^{1/2}\,\varphi_{lm\pm 1}(\mathbf{r}), \end{aligned} \tag{7.2}$$

where

$$l_\pm = l_x \pm i l_y\,,$$

it is straightforward to calculate the following matrices of l_x, l_y, and l_z by using the explicit forms of the t_{2g} and e_g orbitals in (1.31) and (1.32):

$$l_x = \begin{array}{c} \begin{array}{ccccc} \xi & \eta & \zeta & u & v \end{array} \\ \left[\begin{array}{ccc:cc} 0 & 0 & 0 & -\sqrt{3}i & -i \\ 0 & 0 & i & 0 & 0 \\ 0 & -i & 0 & 0 & 0 \\ \hdashline \sqrt{3}i & 0 & 0 & 0 & 0 \\ i & 0 & 0 & 0 & 0 \end{array}\right] \end{array}, \tag{7.3}$$

$$l_y = \left[\begin{array}{ccc:cc} 0 & 0 & -i & 0 & 0 \\ 0 & 0 & 0 & \sqrt{3}i & -i \\ i & 0 & 0 & 0 & 0 \\ \hdashline 0 & -\sqrt{3}i & 0 & 0 & 0 \\ 0 & i & 0 & 0 & 0 \end{array}\right], \tag{7.4}$$

$$l_z = \left[\begin{array}{ccc:cc} 0 & i & 0 & 0 & 0 \\ -i & 0 & 0 & 0 & 0 \\ 0 & 0 & 0 & 0 & 2i \\ \hdashline 0 & 0 & 0 & 0 & 0 \\ 0 & 0 & -2i & 0 & 0 \end{array}\right]. \tag{7.5}$$

As seen above, the matrices of the angular momentum are hermitian and their elements are purely imaginary (note that the bases are real). It is

[‡] Throughout this book, the matrix elements of angular momenta are given in unit of $\hbar$.

also seen that all the matrix elements of l in the e_g state are zero. This means that the orbital angular momentum is completely *quenched* in the e_g state. Therefore, there is no first-order spin-orbit interaction in this state.

In the t_{2g} state the orbital angular momentum is not quenched. It is interesting to compare the matrix elements in the t_{2g} state with those in the p state in free atoms which are given as follows:

$$\begin{array}{c} \quad\quad p_x \quad p_y \quad p_z \end{array}$$

$$l_x = \begin{bmatrix} 0 & 0 & 0 \\ 0 & 0 & -i \\ 0 & i & 0 \end{bmatrix}, \tag{7.6}$$

$$l_y = \begin{bmatrix} 0 & 0 & i \\ 0 & 0 & 0 \\ -i & 0 & 0 \end{bmatrix}, \tag{7.7}$$

$$l_z = \begin{bmatrix} 0 & -i & 0 \\ i & 0 & 0 \\ 0 & 0 & 0 \end{bmatrix}. \tag{7.8}$$

Then, one immediately sees the relation

$$\mathbf{l}(t_{2g}) = -\mathbf{l}(p), \tag{7.8'}$$

which tells us that the expectation value of $\mathbf{l}^2 = l_x^2 + l_y^2 + l_z^2$ in the t_{2g} state made from the d-states is $l(l+1)$ not with $l = 2$ but with $l = 1$. This means that the orbital angular momentum is partially quenched in the t_{2g} state. Relation (7.8′) is called *T–P equivalence* which will be discussed in detail in Section 7.3.2. It should be noted that the T–P equivalence is only a formal matter as seen from the fact that $\mathbf{l}(t_{2g})$ does not satisfy the commutation relation which the angular momentum should satisfy: This is due to the neglect of the nondiagonal matrix elements between the t_{2g} and e_g states given in (7.3)–(7.5). However, if the cubic-field splitting is large, the neglect of the nondiagonal elements is justified and the T–P equivalence may conveniently be used for practical purposes.

7.1.2 Spin-Orbit Splitting

As the simplest example of the term splitting due to the spin-orbit interaction, let us calculate the splittings of the ^2E_g and $^2T_{2g}$ terms in a single-electron system assuming that the spin-orbit interaction is given by (7.1) and the t_{2g} and e_g orbitals are the d-functions. We further assume that the cubic splitting between the $^2T_{2g}$ and ^2E_g terms is much larger than the spin-orbit splitting of these terms. In this case the spin-orbit interaction is ineffective in the ^2E_g term as the orbital angular momentum is completely quenched in this term as mentioned in the previous subsection.

To calculate the matrix elements of the spin-orbit interaction in the $^2T_{2g}$ state, the following well-known matrices of the spin operator are used:

$$m_s = \tfrac{1}{2} \quad -\tfrac{1}{2}$$

$$s_x = \begin{bmatrix} 0 & 1 \\ 1 & 0 \end{bmatrix} \times \tfrac{1}{2},$$

$$s_y = \begin{bmatrix} 0 & -i \\ i & 0 \end{bmatrix} \times \tfrac{1}{2}, \tag{7.9}$$

$$s_z = \begin{bmatrix} 1 & 0 \\ 0 & -1 \end{bmatrix} \times \tfrac{1}{2}.$$

From (7.3)–(7.5) and (7.9), the matrix of the spin-orbit interaction,

$$\mathscr{H}_{so} = \xi(r)(l_x s_x + l_y s_y + l_z s_z), \tag{7.10}$$

in the $^2T_{2g}$ state is derived as

$$\begin{array}{c} \quad\quad \xi \quad\quad\quad\quad \eta \quad\quad\quad\quad \zeta \\ m_s = \tfrac{1}{2} \;\; -\tfrac{1}{2} \quad \tfrac{1}{2} \;\; -\tfrac{1}{2} \quad \tfrac{1}{2} \;\; -\tfrac{1}{2} \end{array}$$

$$\left[\begin{array}{cc:cc:cc} 0 & 0 & i & 0 & 0 & -1 \\ 0 & 0 & 0 & -i & 1 & 0 \\ \hdashline -i & 0 & 0 & 0 & 0 & i \\ 0 & i & 0 & 0 & i & 0 \\ \hdashline 0 & 1 & 0 & -i & 0 & 0 \\ -1 & 0 & -i & 0 & 0 & 0 \end{array}\right] \times \tfrac{1}{2}\zeta, \tag{7.11}$$

which can be reduced to two three-dimensional matrices as follows:

$$\begin{array}{ccc} \tfrac{1}{2}\xi & \tfrac{1}{2}\eta & -\tfrac{1}{2}\zeta \end{array} \qquad\qquad \begin{array}{ccc} -\tfrac{1}{2}\xi & -\tfrac{1}{2}\eta & \tfrac{1}{2}\zeta \end{array}$$

$$\begin{bmatrix} 0 & i & -1 \\ -i & 0 & i \\ -1 & -i & 0 \end{bmatrix} \times \tfrac{1}{2}\zeta, \qquad \begin{bmatrix} 0 & -i & 1 \\ i & 0 & i \\ 1 & -i & 0 \end{bmatrix} \times \tfrac{1}{2}\zeta. \tag{7.12}$$

In (7.11) and (7.12) ζ is given as

$$\zeta = \int_0^\infty dr\, r^2 R_d^2(r)\, \xi(r). \tag{7.13}$$

The matrices in (7.12) can be partially diagonalized as

$$\begin{array}{ccc} \tfrac{1}{2}t_+ & \tfrac{1}{2}t_- & -\tfrac{1}{2}\zeta \end{array} \qquad \begin{array}{ccc} -\tfrac{1}{2}t_+ & -\tfrac{1}{2}t_- & \tfrac{1}{2}\zeta \end{array}$$

$$\begin{bmatrix} -1 & 0 & 0 \\ 0 & 1 & -\sqrt{2} \\ 0 & -\sqrt{2} & 0 \end{bmatrix} \times \tfrac{1}{2}\zeta, \qquad \begin{bmatrix} 1 & 0 & -\sqrt{2} \\ 0 & -1 & 0 \\ -\sqrt{2} & 0 & 0 \end{bmatrix} \times \tfrac{1}{2}\zeta, \tag{7.14}$$

if one takes the following linear combination of ξ and η as new bases:

$$t_\pm = \mp \frac{1}{\sqrt{2}} (\xi \pm i\eta). \tag{7.15}$$

Then, one obtains the eigenvalues of (7.14) as

$$\epsilon_1 = -\tfrac{1}{2}\zeta, \qquad \epsilon_2 = \zeta, \tag{7.16}$$

in which ϵ_1 has fourfold degeneracy and ϵ_2 twofold degeneracy. The eigenfunctions associated with ϵ_1 are

$$\begin{aligned} \phi_1^{a,d} &= \varphi(t_\pm)\, \theta\left(\frac{1}{2} \pm \frac{1}{2}\right), \\ \phi_1^{b,c} &= \frac{1}{\sqrt{3}} \left[\varphi(t_\pm)\, \theta\left(\frac{1}{2} \mp \frac{1}{2}\right) + \sqrt{2}\, \varphi(\zeta)\, \theta\left(\frac{1}{2} \pm \frac{1}{2}\right)\right], \end{aligned} \tag{7.17}$$

and those associated with ϵ_2 are

$$\phi_2^{a,b} = \frac{1}{\sqrt{3}} \left[\sqrt{2}\, \varphi(t_\pm)\, \theta\left(\frac{1}{2} \mp \frac{1}{2}\right) - \varphi(\zeta)\, \theta\left(\frac{1}{2} \pm \frac{1}{2}\right)\right]. \tag{7.18}$$

It is interesting to compare these results with those of the ^2P state with a single p electron. We know that the ^2P level splits into $J = 3/2$ and $J = 1/2$ levels whose separation is given by Landé's interval rule as

$$\epsilon\,(J=\tfrac{3}{2}) - \epsilon\,(J=\tfrac{1}{2}) = \tfrac{3}{2}\zeta_p\,, \tag{7.19}$$

where ζ_p is given by an expression similar to (7.13) in which $R_d(r)$ is replaced by the radial part of the p function. Our result in (7.16) is identical to (7.19) if one makes the following replacement: $\epsilon_1 \to \epsilon(J = \tfrac{3}{2})$,

$\epsilon_2 \to \epsilon(J = \frac{1}{2})$, and $\zeta \to -\zeta_p$. The wave function associated with the $J = 3/2$ level are

$$\phi\left(J{=}\frac{3}{2}\,M_j{=}\pm\frac{3}{2}\right) = \varphi(p\,\pm 1)\,\theta\left(\frac{1}{2}\,\pm\frac{1}{2}\right),$$
$$\phi\left(J{=}\frac{3}{2}\,M_j{=}\pm\frac{1}{2}\right) = \frac{1}{\sqrt{3}}\left[\varphi(p\,\pm 1)\,\theta\left(\frac{1}{2}\,\mp\frac{1}{2}\right) + \sqrt{2}\,\varphi(p0)\,\theta\left(\frac{1}{2}\,\pm\frac{1}{2}\right)\right], \tag{7.20}$$

and those associated with $J = 1/2$ are

$$\phi\left(J{=}\frac{1}{2}\,M_j{=}\pm\frac{1}{2}\right) = \frac{1}{\sqrt{3}}\left[\sqrt{2}\,\varphi(p\,\pm 1)\,\theta\left(\frac{1}{2}\,\mp\frac{1}{2}\right) - \varphi(p0)\,\theta\left(\frac{1}{2}\,\pm\frac{1}{2}\right)\right]. \tag{7.21}$$

In (7.20) $\phi(\frac{3}{2}\pm\frac{1}{2})$ are obtained by operating $J_\mp = l_\mp + s_\mp$ on $\phi(\frac{3}{2}\pm\frac{3}{2})$ and $\phi(\frac{1}{2}\pm\frac{1}{2})$ are obtained by making them orthogonal to $\phi(\frac{3}{2}\pm\frac{1}{2})$. Wavefunctions (7.17) and (7.18) are, respectively, identical to (7.20) and (7.21) if one makes the replacement, $\varphi(t_\pm) \to \varphi(p\,\pm 1)$. A similarity of the spin-orbit splitting of the ${}^2T_{2g}$ term to that of the ${}^2P$ term comes from the T–P equivalence given in (7.9).

So far we have neglected the nondiagonal matrix elements of the spin-orbit interaction between the ${}^2T_{2g}$ and ${}^2E_g$ terms, assuming that the cubic field is much larger than the spin-orbit interaction. These nondiagonal elements are calculated from (7.3)–(7.5) and (7.9) as follows:

$$\begin{array}{cc} & \begin{array}{cccc} \multicolumn{2}{c}{u} & \multicolumn{2}{c}{v} \\ \frac{1}{2} & -\frac{1}{2} & \frac{1}{2} & -\frac{1}{2} \end{array} \\ \begin{array}{cc} \xi & \begin{array}{c} \frac{1}{2} \\ -\frac{1}{2} \end{array} \\ \eta & \begin{array}{c} \frac{1}{2} \\ -\frac{1}{2} \end{array} \\ \zeta & \begin{array}{c} \frac{1}{2} \\ -\frac{1}{2} \end{array} \end{array} & \left[\begin{array}{cc:cc} 0 & -\sqrt{3}i & 0 & -i \\ -\sqrt{3}i & 0 & -i & 0 \\ \hdashline 0 & \sqrt{3} & 0 & -1 \\ -\sqrt{3} & 0 & 1 & 0 \\ \hdashline 0 & 0 & 2i & 0 \\ 0 & 0 & 0 & -2i \end{array}\right] \times \tfrac{1}{2}\zeta. \end{array} \tag{7.22}$$

The nondiagonal elements can be shown to be zero between the ϵ_2 state and ${}^2E_g$ term as shown in the following problem:

Problem 7.1. Show that the spin-orbit interaction does not connect the ${}^2E_g$ and ϵ_2 states. ◇

Furthermore, they bring no splitting of the ϵ_1 and ${}^2E_g$ levels. These points will be discussed in the next section in the light of the group theory. The absence of the spin-orbit splitting of the ^2E_g term may simply be shown, in case the spin-orbit interaction is much smaller than the cubic field strength, by using the perturbation calculation: In the perturbation treatment the energy shifts of the $^2E_gM\gamma$ states are given by

$$\Delta E = \sum_{M'\gamma'} \frac{|\langle {}^2EM\gamma \mid \mathscr{H}_{\text{so}} \mid {}^2T_2M'\gamma'\rangle|^2}{10Dq} = \frac{3\zeta^2}{20Dq}, \tag{7.23}$$

which is independent of M and γ.

7.2 Double-Group

7.2.1 Rotation in Spin-Space

As is well-known, electron spins provide an additional freedom to electrons, a spin-space. Here, we consider how the wavefunctions involving spin coordinates are transformed by the rotation in the spin-space. For simplicity let us denote wave-functions $\Psi(\alpha S\Gamma M\gamma)$ as $\Psi(SM)$ omitting the orbital specification. Wavefunctions $\Psi(SM)$ were introduced as the eigenfunctions of spin operators $\mathbf{S}^2$ and S_z to satisfy

$$\begin{aligned} \mathbf{S}^2\Psi(SM) &= S(S+1)\,\Psi(SM), \\ S_z\Psi(SM) &= M\Psi(SM). \end{aligned} \tag{7.24}$$

In addition to these, the following relations can be derived from the commutation relations for spin operators:

$$S_\pm\Psi(SM) = [S(S+1) - M(M\pm1)]^{1/2}\,\Psi(S\,M\pm1), \tag{7.25}$$

where

$$S_\pm = S_x \pm iS_y\,.$$

Now, considering that S is transformed like a vector, one may show that

$$\begin{aligned} R_\alpha{}^{\text{s}}\mathbf{S}^2(R_\alpha{}^{\text{s}})^{-1} &= \mathbf{S}^2, \\ R_\alpha{}^{\text{s}}S_z(R_\alpha{}^{\text{s}})^{-1} &= S_z\,, \\ R_\alpha{}^{\text{s}}S_\pm(R_\alpha{}^{\text{s}})^{-1} &= e^{\mp i\alpha}S_\pm\,, \end{aligned} \tag{7.26}$$

where $R_\alpha{}^{\text{s}}$ is the spin rotation operator around the z-axis by angle α. Then, it follows from (7.24) that

$$R_\alpha{}^{\text{s}}\mathbf{S}^2(R_\alpha{}^{\text{s}})^{-1}R_\alpha{}^{\text{s}}\Psi(SM) = \mathbf{S}^2R_\alpha{}^{\text{s}}\Psi(SM) = S(S+1)\,R_\alpha{}^{\text{s}}\Psi(SM) \tag{7.27}$$

and

$$R_\alpha{}^{\text{s}}S_z(R_\alpha{}^{\text{s}})^{-1}R_\alpha{}^{\text{s}}\Psi(SM) = S_zR_\alpha{}^{\text{s}}\Psi(SM) = MR_\alpha{}^{\text{s}}\Psi(SM). \tag{7.28}$$

Equation (7.27) shows that $R_\alpha{}^s\Psi(SM)$ is also the eigenfunction of $\mathbf{S}^2$ with eigenvalue $S(S+1)$, consequently it is given by a linear combination $\sum_{M'} C_{MM'}\Psi(SM')$. Therefore, just as in the case of continuous rotations in the position-coordinate space, $\Psi(SM)$ ($M = S, S-1,..., -S$) form the bases of $(2S+1)$ -dimensional irreducible representation $D^{(S)}$. Combined with this fact, Eq. (7.28) shows that

$$R_\alpha{}^s\Psi(SM) = \mu(M\alpha)\,\Psi(SM), \tag{7.29}$$

where $\mu(M\alpha)$ is a constant depending upon M and α, and because of the normalization of $\Psi(SM)$ it should satisfy

$$|\,\mu(M\alpha)|^2 = 1 \qquad \text{or} \qquad \mu(M\alpha) = e^{-i\lambda(M\alpha)}. \tag{7.30}$$

In (7.30) $\lambda(M\alpha)$ is a real function of M and α. Since $\lambda(M0) = 0$, $\lambda(M\alpha)$ for very small α may be expressed as $\lambda(M\alpha) = \lambda_M\alpha$. Considering that the rotation by any angle α may be achieved by successive rotations by very small angles, one finally obtains

$$\mu(M\alpha) = \exp(-i\lambda_M\alpha) \tag{7.31}$$

for any angle α. On the other hand, it follows from (7.25) that

$$\begin{aligned} R_\alpha{}^s S_\pm (R_\alpha{}^s)^{-1} R_\alpha{}^s\Psi(SM) &= e^{\mp i\alpha} S_\pm R_\alpha{}^s\Psi(SM) \\ &= [S(S+1) - M(M\pm 1)]^{1/2}\, R_\alpha{}^s\Psi(S\,M\pm 1), \end{aligned} \tag{7.32}$$

which, by use of (7.29) and (7.31), is expressed as

$$e^{\mp i\alpha}\exp[-i(\lambda_M - \lambda_{M\pm 1})\alpha]\, S_\pm\Psi(SM) = [S(S+1) - M(M\pm 1)]^{1/2}\,\Psi(S\,M\pm 1). \tag{7.33}$$

Equation (7.33) shows that

$$\lambda_M - \lambda_{M\pm 1} = \mp 1, \tag{7.34}$$

whose general solution is

$$\lambda_M = C + M, \tag{7.35}$$

in which C is a real constant independent of M. We choose C to be zero so that the transformation property in the spin space is similar to that in the position-coordinate space as given in (1.74). Then one obtains the transformation

$$R_\alpha{}^s\Psi(SM) = e^{-iM\alpha}\Psi(SM). \tag{7.36}$$

Notice that this choice of C makes $R_\alpha{}^s$ commute with the time reversal operator $K = K_s K_0$ given in (4.30), in which K_0 is now considered as the complex conjugate operator acting also on the phase of the spin function: Readers will see from (7.31) and (4.32) that expressions

$$KR_\alpha^{s}\Psi(SM) = e^{i(C+M)\alpha}(-1)^{S-M}\Psi(S\ -M) \tag{7.37}$$

and

$$R_\alpha^{s}K\Psi(SM) = (-1)^{S-M}e^{-i(C-M)\alpha}\Psi(S\ -M) \tag{7.38}$$

are not identical to each other if $C \neq 0$. Just as in the case of calculating $\chi^{(l)}(\alpha)$ in (1.76), the character of $D^{(S)}(R_\alpha^{s})$ is obtained from (7.36) as

$$\chi^{(S)}(\alpha) = \frac{\sin(S+\frac{1}{2})\alpha}{\sin\frac{1}{2}\alpha}. \tag{7.39}$$

In contrast to the case of $\chi^{(l)}(\alpha)$, where l is always an integer, S in $\chi^{(S)}(\alpha)$ can be a half-integer. For half-integral S, one sees from (7.39) that

$$\chi^{(S)}(\alpha+2\pi) = -\chi^{(S)}(\alpha). \tag{7.40}$$

For example, for half-integral S one has

$$\begin{aligned}\chi^{(S)}(0) &= 2S+1,\\ \chi^{(S)}(2\pi) &= -(2S+1).\end{aligned} \tag{7.41}$$

Therefore, the representations for half-integral S are, in general, double-valued: The exceptional case is for $\alpha = \pi$, for which the representations are single-valued as $\chi^{(S)}(\pi) = \chi^{(S)}(3\pi) = 0$. Such a complexity of the double-valuedness may formally be avoided if the rotation period is considered as 4π instead of 2π in the spin space with half-integral S.

7.2.2 Cubic Double-Group

Let us consider the N-electron system with cubic symmetry whose electron Hamiltonian is given as

$$\mathscr{H} = \mathscr{H}_0 + \mathscr{H}_1 + \mathscr{H}_{SO}, \tag{7.42}$$

where

$$\mathscr{H}_0 = \sum_i f_i,$$

$$\mathscr{H}_1 = \sum_{i>j} g_{ij},$$

$$\mathscr{H}_{SO} = \sum \xi_i(r)\,\mathbf{l}_i\cdot\mathbf{s}_i.$$

In (7.42) f_i is the one-electron operator defined in (2.2) and is invariant to any symmetry operation of the O-group, and g_{ij} is the Coulomb interaction operator, which is also invariant to any symmetry operation of the O-group. Since $\mathscr{H}_{so}$ is the scalar product of $\mathbf{l}$ and $\mathbf{s}$, it is invariant if the same rotation is applied in both the spin-space and the position-coordinate space simultaneously: In this case $\mathbf{l}$ and $\mathbf{s}$ may be regarded as two vectors in a single space. The simultaneous and common rotation R^{os} in both the spin and position-coordinate spaces may be expressed as

$$R^{os} = R^{o}R^{s} = R^{s}R^{o}, \tag{7.43}$$

where R^{o} is the rotation in the position-coordinate space and R^{s} is the *same* rotation in the spin space. The result of our argument is

$$R^{os}\mathscr{H}(R^{os})^{-1} = \mathscr{H} \tag{7.44}$$

for R^{os} involving R^{o} of the O-group. It should be noted that $\xi(r)$ in $\mathscr{H}_{so}$ is not necessarily spherically symmetric but could be of cubic symmetry for $\mathscr{H}$ to satisfy (7.44).

Before discussing physical problems related to (7.44), we now study group theoretical problems associated with R^{os}. It is evident that the aggregate of R^{os} with R^{o} of the O-group forms a group. However, to avoid the double-valuedness of the spin-rotation group as mentioned in the previous subsection, we also assume the fiction that the period of the space rotation is 4π, i.e., the cubic system is not to go over into itself on rotation by 2π around an arbitrary axis but only on rotation by 4π. Then, the number of elements in this group is twice as many as in the O-group. This group is called *cubic double-group*. Although the number of elements is doubled, the number of classes is not necessarily so. The reason is explained as follows: In a double-group with the rotation period 4π, the inverse of a rotation by angle π around a symmetry axis denoted as R_π is equal to $R_\pi R$, where R is the rotation by angle 2π around the same axis. Then, if the direction of this axis can be inverted by a rotation in the group, R_π and $R_\pi R$ should be associated with the same class. For example, there are eight classes in the cubic double-group, while there are five classes in the O-group.

Since the cubic double-group has forty-eight elements and eight classes, the application of (1.71),

$$1^2 + 1^2 + 2^2 + 3^2 + 3^2 + 2^2 + 2^2 + 4^2 = 48,$$

shows that in this group we have two two-dimensional (denoted by E_1 and E_2) and one four-dimensional (denoted by G) irreducible representations in addition to those found in the O-group. These additional irreducible representations are called *double-valued representations*, whose characters are different in sign for rotations α and $\alpha + 2\pi$ ($\alpha \neq \pi$).

To construct the character table of the cubic double-group, we show the classes and the symmetry operations in each class in the tabulation.

Classes	Symmetry operations
$\hat{E}$	Rotation by 0 (or 4π)
$\hat{R}$	Rotation by 2π denoted by R
$\hat{C}_4'$	C_4, $C_4^3R = C_4^{-1}$
$\hat{C}_4''$	C_4^3, $C_4R = (C_4^3)^{-1}$
$\hat{C}_4^2$	C_4^2, $C_4^2R = (C_4^2)^{-1}$
$\hat{C}_3'$	C_3, $C_3^2R = C_3^{-1}$
$\hat{C}_3''$	C_3^2, $C_3R = (C_3^2)^{-1}$
$\hat{C}_2$	C_2, C_2R

The characters for single-valued representations A_1, A_2, E, T_1, and T_2 should be the same as those in Table 1.3 with the same values for the sets of E and R, $\hat{C}_4'$ and $\hat{C}_4''$, and $\hat{C}_3'$ and $\hat{C}_3''$. It is evident from (7.41) that the characters of double-valued representations, E_1, E_2, and G, for operation E are 2, 2, and 4, and for operation R are -2, -2, and -4, respectively. It is also evident that all the characters of the double-valued representations for $\hat{C}_4^2$ and $\hat{C}_2$ are zero as the rotation angles are π and 3π, and the characters for $\hat{C}_4''$ and $\hat{C}_3''$ are, respectively, just those for $\hat{C}_4'$ and $\hat{C}_3'$ with the signs changed. Then, with the help of the orthogonality relations (1.67) and (1.69b), the remaining unknown characters for the double-valued representations are calculated as shown in Table 7.1.

TABLE 7.1

CHARACTER TABLE FOR CUBIC DOUBLE-GROUP

Bethe		E	R	$6\hat{C}_4'$	$6\hat{C}_4''$	$6\hat{C}_4^2$	$8\hat{C}_3'$	$8\hat{C}_3''$	$12\hat{C}_2$
Γ_1	A_1	1	1	1	1	1	1	1	1
Γ_2	A_2	1	1	-1	-1	1	1	1	-1
Γ_3	E	2	2	0	0	2	-1	-1	0
Γ_4	T_1	3	3	1	1	-1	0	0	-1
Γ_5	T_2	3	3	-1	-1	-1	0	0	1
Γ_6	E_1	2	-2	$\sqrt{2}$	$-\sqrt{2}$	0	1	-1	0
Γ_7	E_2	2	-2	$-\sqrt{2}$	$\sqrt{2}$	0	1	-1	0
Γ_8	G	4	-4	0	0	0	-1	1	0

Calculating $\chi^{(S)}(\alpha)$ with half-integral S for particular α by use of (7.39) and comparing them with the characters in Table 7.1, one may derive the scheme of reducing $D^{(S)}$ with half-integral S into the double-valued representations of the cubic double group when rotations in the spin space are restricted to those of the cubic double-group. Such a reduction scheme is given in Table 7.2.

TABLE 7.2

REDUCTION OF $D^{(S)}$ INTO REPRESENTATIONS OF CUBIC DOUBLE-GROUP

S	Irred. reprs.
1/2	E_1
3/2	G
5/2	$E_2 + G$
7/2	$E_1 + E_2 + G$
9/2	$E_1 + 2G$
11/2	$E_1 + E_2 + 2G$

Problem 7.2. Derive the character table for the double D_4-group given in the tabulation, where

$$\hat{C}_4' : \ C_4, \ C_4^3R,$$

$$\hat{C}_4'' : \ C_4^3, \ C_4R,$$

$$\hat{C}_4^2 : \ C_4^2, \ C_4^2R,$$

$$\hat{C}_2 : \ C_2, \ C_2R,$$

$$\hat{C}_2' : \ C_2', \ C_2'R.$$

		$\hat{E}$	$\hat{R}$	$2\hat{C}_4'$	$2\hat{C}_4''$	$2\hat{C}_4^2$	$4\hat{C}_2$	$4\hat{C}_2'$
	A_1	1	1	1	1	1	1	1
	A_2	1	1	1	1	1	−1	−1
	B_1	1	1	−1	−1	1	−1	1
	B_2	1	1	−1	−1	1	1	−1
	E	2	2	0	0	−2	0	0
$(E_{1/2})$	E_1	2	−2	$\sqrt{2}$	$-\sqrt{2}$	0	0	0
$(E_{3/2})$	E_2	2	−2	$-\sqrt{2}$	$\sqrt{2}$	0	0	0

◇

Problem 7.3. Derive the character table for the double D_3-group as given in the tabulation. It should be noted that in this case C_2 and its inverse C_2R do not belong to the same class as the directions of the twofold symmetry axes perpendicular to the trigonal axis cannot be inverted by any symmetry operation in this group. To obtain pure imaginary characters, the orthogonality relations in the forms of (1.65a) and (1.69a) should be used. Two double-valued representations $\bar{A}_1$ and $\bar{A}_2$ are always combined together to assure the vanishing $\chi^{(S)}(\pi)$ and $\chi^{(S)}(3\pi)$ for half-integral S.

	$\hat{E}$	$\hat{R}$	$2\hat{C}_3'$	$2\hat{C}_3''$	$3\hat{C}_2'$	$3\hat{C}_2''$
A_1''	1	1	1	1	1	1
A_2''	1	1	1	1	-1	-1
E''	2	2	-1	-1	0	0
$\bar{A}_1$	1	-1	-1	1	i	$-i$
$\bar{A}_2$	1	-1	-1	1	$-i$	i
$\bar{E}$	2	-2	1	-1	0	0

$\hat{C}_3'$: C_3, $C_3{}^2R$, $\hat{C}_3''$: $C_3{}^2$, C_3R

$\hat{C}_2'$: C_2, $\hat{C}_2''$: C_2R

7.2.3 Labeling of the Spin-Orbit Split Components

As shown in (7.44), the electron Hamiltonian of a cubic system including the spin-orbit interaction is invariant to symmetry operation R^{os} defined in (7.43) in which R^o and R^s are rotations in the cubic double-group. Therefore, according to the argument given in Section 1.2.2, the energy levels of the system described by this Hamiltonian may be labeled with the irreducible representations of the cubic double-group. Then, a question arises: What irreducible representations of the cubic double group are derived from the $^{2S+1}\Gamma$ term?

For the purpose of answering this question, we examine the transformation property of wavefunction $\Psi(\alpha S\Gamma M\gamma)$ on the rotation R^{os}. From (2.19) and the argument given in Section 7.2.1, one sees

$$R^{os}\Psi(\alpha S\Gamma M\gamma) = \sum_{M'\gamma'} \Psi(\alpha S\Gamma M'\gamma')\, D^{(S)}_{M'M}(R^s)\, D^{(\Gamma)}_{\gamma'\gamma}(R^o), \tag{7.45}$$

where $D^{(S)}_{M'M}(R^s)$ and $D^{(\Gamma)}_{\gamma'\gamma}(R^o)$ are the matrix elements of the representations (not necessarily irreducible) of the cubic double-group. Now, just in the same way as done in (2.21) and (2.22), we look for the unitary transformation‡ U which reduces the product representation $D^{(S)} \times D^{(\Gamma)}$ into irreducible representations $D^{(\Gamma_J)}$ of the cubic double-group. Then, it follows that the wavefunctions

$$\Psi(\alpha S\Gamma\Gamma_J\gamma_J) = \sum_{M,\gamma} \Psi(\alpha S\Gamma M\gamma)U_{M\gamma,\Gamma_J\gamma_J} \tag{7.46}$$

are the bases of irreducible representation Γ_J of the cubic double-group as shown by

$$R^{os}\Psi(\alpha S\Gamma\Gamma_J\gamma_J) = \sum \Psi(\alpha S\Gamma\Gamma_J\gamma_J')D^{(\Gamma_J)}_{\gamma_J'\gamma_J}, \tag{7.47}$$

where the γ_J's are the degenerate components of Γ_J. However, for the purpose of merely knowing what irreducible representations are obtained by reducing $D^{(S)} \times D^{(\Gamma)}$, we may use the relations (2.23) and (1.78),

$$\begin{aligned}\chi(R^{os}) &= \chi^{(S)}(R^{s}) \times \chi^{(\Gamma)}(R^{o})\\ &= \sum_{\Gamma_J} \chi^{(\Gamma_J)}(R^{os}).\end{aligned} \tag{7.48}$$

Since we know from Table 7.2 the reduction scheme of $D^{(S)}$ with half-integral S into $D^{(\Gamma_J)}$ and from Table 1.4 that of $D^{(S)}$ with integral S, it is easy to show what irreducible representations of the cubic double-group are derived from the $^{2S+1}\Gamma$ term if one knows how the product representations are reduced. The scheme of reducing the product representations, which is partly given in Table 2.1, is supplemented by Table 7.3. The table is derived by using Table 7.1.

TABLE 7.3

PRODUCTS OF DOUBLE-VALUED REPRESENTATIONS OF CUBIC DOUBLE-GROUP

	A_1	A_2	E	T_1	T_2	E_1	E_2	G
E_1	E_1	E_2	G	$E_1 + G$	$E_2 + G$	$A_1 + T_1$	$A_2 + T_2$	$E + T_1 + T_2$
E_2	E_2	E_1	G	$E_2 + G$	$E_1 + G$	$A_2 + T_2$	$A_1 + T_1$	$E + T_1 + T_2$
G	G	G	$E_1 + E_2 + G$	$E_1 + E_2 + 2G$	$E_1 + E_2 + 2G$	$E + T_1 + T_2$	$E + T_1 + T_2$	$A_1 + A_2 + E + 2T_1 + 2T_2$

For example, $^4T_1$ is reduced to $G \times T_1 = E_1 + E_2 + 2G$, ^3E to $T_1 \times E = T_1 + T_2$, and $^6A_1$ to $(E_2 + G) \times A_1 = E_2 + G$. The results of Section 7.1.2 may be interpreted on the basis of the group theory as

‡ In the present problem, the same $D^{(\Gamma_J)}$ may appear more than once when $D^{(S)} \times D^{(\Gamma)}$ is reduced as we will see later. Then, the unitary transformation is not uniquely determined.

follows: Since the $^2T_2$ term is reduced to $E_1 \times T_2 = E_2 + G$, the term splits into two levels. One of them has twofold and another fourfold degeneracy in agreement with the result in Section 7.1.2 if the state described by ϕ_1's is G and that described by ϕ_2's is E_2. The ^2E term is reduced to $E_1 \times E = G$ so that no splitting is expected. The group theoretical consideration also explains the previous result that the spin-orbit interaction connects the ^2E term and the fourfold degenerate component of the $^2T_2$ term, as both these states are labeled with the same G irreducible representation.

7.3 Method of Operator Equivalent

7.3.1 Application of Wigner–Eckart Theorem

In this subsection, the factorization for the matrix elements

$$\langle \alpha S\Gamma M\gamma \mid \sum_i \xi(r_i)\,\mathbf{l}_i \cdot \mathbf{s}_i \mid \alpha' S'\Gamma' M'\gamma' \rangle \tag{7.49}$$

will be discussed. For this purpose, we rearrange the terms in $\mathscr{H}_{\text{so}}$ as follows:

$$\mathscr{H}_{\text{so}} = \frac{1}{\sqrt{2}}[-V_{+1\alpha}(1T_1) + iV_{+1\beta}(1T_1)] + \frac{1}{\sqrt{2}}[V_{-1\alpha}(1T_1) + iV_{-1\beta}(1T_1)] + V_{0\gamma}(1T_1), \tag{7.50}$$

where

$$V_{\pm 1\alpha}(1T_1) = \sum_i s_{i\pm 1} t_{i\alpha}\,, \tag{7.51a}$$

$$V_{\pm 1\beta}(1T_1) = \sum_i s_{i\pm 1} t_{i\beta}\,, \tag{7.51b}$$

$$V_{0\gamma}(1T_1) = \sum_i s_{i0} t_{i\gamma}\,, \tag{7.51c}$$

$$s_{i\pm 1} = \mp \frac{1}{\sqrt{2}}(s_{ix} \pm i s_{iy}), \tag{7.52a}$$

$$s_{i0} = s_{iz}\,, \tag{7.52b}$$

and

$$t_{i\alpha} = \xi(\mathbf{r}_i) l_{ix}\,, \tag{7.53a}$$

$$t_{i\beta} = \xi(\mathbf{r}_i) l_{iy}\,, \tag{7.53b}$$

$$t_{i\gamma} = \xi(\mathbf{r}_i) l_{iz}\,. \tag{7.53c}$$

Then, (7.49) is expressed as the linear combination of

$$\langle \alpha S\Gamma M\gamma \mid V_{q\bar{\gamma}}(1T_1) \mid \alpha' S'\Gamma' M'\gamma' \rangle\,, \tag{7.54}$$

where $q = 0, \pm 1$, and $\bar{\gamma} = \alpha, \beta, \gamma$.

First note that spin operator **S** transforms like a vector, consequently like the base of irreducible representation $D^{(S)}$ with $S = 1$ in the spin-rotation group, and operator **t** like the base of T_1 in the cubic-group if $\xi(r)$ is the function with spherical or cubic symmetry.‡ Then, it turns out that on rotation R^{os} operator $V_{q\bar{\gamma}}(1T_1)$ transforms like $\Psi(^3T_1 q\bar{\gamma})$. Therefore, the following linear combination of $V_{q\bar{\gamma}}(1T_1)\,\Psi(\alpha' S'\Gamma' M'\gamma')$,

$$\tilde{\Psi}(\alpha' S''\Gamma'' M''\gamma'') \equiv \sum_{\substack{M'q\\ \gamma'\bar{\gamma}}} V_{q\bar{\gamma}}(1T_1)\,\Psi(\alpha' S'\Gamma' M'\gamma')\langle S'M'1q \mid S''M''\rangle \times \langle \Gamma'\gamma' T_1\bar{\gamma} \mid \Gamma''\gamma''\rangle, \tag{7.55}$$

transforms on R^{os} like the γ'' base of irreducible representation Γ'' and is the eigenfunction of $\mathbf{S}^2$ and S_z with eigenvalues $S''(S''+1)$ and M''. By using this property, one may derive

$$\sum_{\sigma} \int d\boldsymbol{\tau}\Psi^*(\alpha S\Gamma M\gamma)\,\tilde{\Psi}(\alpha' S''\Gamma'' M''\gamma'') = [(2S+1)(\Gamma)]^{-1/2}\langle \alpha S\Gamma \,\|\, V(1T_1) \,\|\, \alpha' S'\Gamma' \rangle\, \delta_{\Gamma''\Gamma}\,\delta_{\gamma''\gamma}\,\delta_{S''S}\,\delta_{M''M}\,, \tag{7.56}$$

which is similar to (6.87). In (7.56) the reduced matrix

$$\langle \alpha S\Gamma \,\|\, V(1T_1) \,\|\, \alpha' S'\Gamma' \rangle$$

is independent of γ, γ', and $\bar{\gamma}$ as well as M, M', and q. From (7.55) and (7.56) one finally obtains the formula

$$\langle \alpha S\Gamma M\gamma \mid V_{q\bar{\gamma}}(1T_1) \mid \alpha' S'\Gamma' M'\gamma' \rangle = [(2S+1)(\Gamma)]^{-1/2}\langle \alpha S\Gamma \,\|\, V(1T_1) \,\|\, \alpha' S'\Gamma' \rangle \times \langle SM \mid S'M'1q\rangle\langle \Gamma\gamma \mid \Gamma'\gamma' T_1\bar{\gamma}\rangle, \tag{7.57}$$

which is the expression of the Wigner–Eckart theorem for the matrix of the spin-orbit interaction.

‡ In crystals and complex ions, the Hamiltonian of the spin-orbit interaction may differ from (7.1). However, in Chapter X we will show that, even in this case, the following argument is still applicable.

As seen from the example in Section 7.1.2, the matrix elements (7.54) are purely imaginary and have the property

$$\langle \alpha S\Gamma M\gamma \mid V_{q\bar{\gamma}}(1T_1) \mid \alpha' S'\Gamma' M'\gamma' \rangle$$
$$= -(-1)^q \langle \alpha' S'\Gamma' M'\gamma' \mid V_{-q\bar{\gamma}}(1T_1) \mid \alpha S\Gamma M\gamma \rangle, \qquad (7.58)$$

which assures the Hermitian property of $\mathscr{H}_{so}$,

$$\langle \alpha S\Gamma M\gamma \mid \mathscr{H}_{so} \mid \alpha' S'\Gamma' M'\gamma' \rangle = \langle \alpha' S'\Gamma' M'\gamma' \mid \mathscr{H}_{so} \mid \alpha S\Gamma M\gamma \rangle^*. \quad (7.59)$$

Applying the Wigner–Eckart theorem to (7.58), and using the relations

$$(\Gamma)^{-1/2} \langle \Gamma\gamma \mid \Gamma'\gamma' T_1\bar{\gamma} \rangle = -(\Gamma')^{-1/2} \langle \Gamma'\gamma' \mid \Gamma\gamma T_1\bar{\gamma} \rangle \qquad (7.60)$$

and

$$(2S+1)^{-1/2} \langle SM \mid S'M'1q \rangle = (-1)^q (-1)^{S-S'} (2S'+1)^{-1/2} \langle S'M' \mid SM1-q \rangle, \qquad (7.61)$$

one can show the relation

$$\langle \alpha S\Gamma \| V(1T_1) \| \alpha' S'\Gamma' \rangle = (-1)^{S-S'} \langle \alpha' S'\Gamma' \| V(1T_1) \| \alpha S\Gamma \rangle. \qquad (7.62)$$

Relation (7.60) is derived from (6.90) and (6.91). Since Wigner coefficients and C–G coefficients with the cubic bases in (7.57) are real, $\langle \alpha S\Gamma \| V(1T_1) \| \alpha' S'\Gamma' \rangle$ are purely imaginary.

Problem 7.4. Check (7.11) and (7.22) by using the Wigner–Eckart theorem. ◇

Problem 7.5. Show that with the trigonal bases the matrix elements of $\mathscr{H}_{so}$ are given as

$$\langle \alpha S\Gamma M_s M \mid \mathscr{H}_{so} \mid \alpha' S'\Gamma' M_s' M' \rangle$$
$$= (-1)^{M_s - M_s'} [(2S+1)(\Gamma)]^{-1/2} \langle \alpha S\Gamma \| V(1T_1) \| \alpha' S'\Gamma' \rangle$$
$$\times \langle SM_s \mid S'M_s' 1\; M_s - M_s' \rangle \langle \Gamma M \mid \Gamma' M' T_1\; M_s' - M_s \rangle,$$

where the M's indicate the trigonal components x_M, a_M, and u_M for T_2, T_1, and E, respectively, and they are zero for the components of A_1 and A_2. In deriving the above formula, note that

$$\mathscr{H}_{so} = \sum_{q=0,\pm 1} (-1)^q V_{q\,-q}(1T_1),$$

where

$$V_{\pm 1 \mp 1}(1T_1) = \pm \frac{1}{\sqrt{2}} [V_{\pm 1\alpha}(1T_1) \mp i V_{\pm 1\beta}(1T_1)],$$
$$V_{00}(1T_1) = V_{0\gamma}.$$

Here, all the components, α, β, γ of **l** and x, y, z of **s** in $V_{q\,-q}(1T_1)$, are referred to the coordinate system introduced for the trigonal system (see Fig. 6.2). ◇

7.3.2 First-Order Spin-Orbit Splittings of Cubic Terms

The extensive use of formula (7.57) derived in the previous subsection provides a simple method of calculating the first-order spin-orbit splittings of the cubic terms $^{2S+1}\Gamma$. It is clear that the terms with $\Gamma = A_1$, A_2, and E show no first-order splitting as the orbital angular momentum is quenched in these terms. Therefore, one may confine oneself to the splittings of terms $^{2S+1}T_1$ and $^{2S+1}T_2$.

For the purpose of obtaining the simple method to calculate the first-order spin-orbit splittings of the $^{2S+1}T_1$ and $^{2S+1}T_2$ terms, let us introduce purely imaginary operator $\mathbf{T}(T_1)$ whose component $T_{\bar{\gamma}}(T_1)$ transforms like the $\bar{\gamma}$ base of irreducible representation T_{1g} of the O_h-group and whose reduced matrix is given as

$$\langle \alpha T_1 \| T(T_1) \| \alpha T_1 \rangle = \langle \alpha T_2 \| T(T_1) \| \alpha T_2 \rangle = \sqrt{6}\, i. \tag{7.63}$$

Applying formula (6.89), one can show that, for example, the matrices of $T_\gamma(T_1)$ in the T_1 and T_2 states are given as

$$\begin{array}{cccc} & (\xi) & (\eta) & (\zeta) \\ & \alpha & \beta & \gamma \\ T_\gamma(T_1) = & \left[\begin{array}{c} 0 \\ \binom{+}{-}i \\ 0 \end{array}\right. & \begin{array}{c} \binom{-}{+}i \\ 0 \\ 0 \end{array} & \left.\begin{array}{c} 0 \\ 0 \\ 0 \end{array}\right], \end{array} \tag{7.64}$$

in which the signs and bases in brackets are for the T_2 state. Comparing (7.64) with (7.8), one sees that the matrix of $T_\gamma(T_1)$ in the T_1 state is identical to that of L_z in the P state and the matrix of $T_\gamma(T_1)$ in the T_2 state is identical to that of $-L_z$ in the P state. The same relation is found between the matrices of $T_\alpha(T_1)$ and L_x and between $T_\beta(T_1)$ and L_y.

On the other hand, if one considers the matrix element of $\lambda S_q T_{\bar{\gamma}}(T_1)$ in the $\alpha S\Gamma$ term as

$$\begin{aligned} &\langle \alpha S\Gamma M\gamma \mid \lambda S_q T_{\bar{\gamma}}(T_1) \mid \alpha S\Gamma M'\gamma' \rangle \\ &\quad = [(2S+1)(\Gamma)]^{-1/2} \lambda \langle S\Gamma \| \mathbf{ST}(T_1) \| S\Gamma \rangle \\ &\qquad \times \langle SM \mid SM'1q \rangle \langle \Gamma\gamma \mid \Gamma\gamma' T_1\bar{\gamma} \rangle, \end{aligned} \tag{7.65}$$

where λ is a constant and the S_q's ($q = 0, \pm 1$) are the components of resultant spin $\mathbf{S} = \sum \mathbf{s}_i$ defined in a way similar to (7.52) for s_{iq}, one notices that (7.65) is identical to (7.57) for $\alpha S\Gamma = \alpha' S' \Gamma'$ when λ is given as

$$\lambda(\alpha S\Gamma) = \frac{\langle \alpha S\Gamma \| V(1T_1) \| \alpha S\Gamma \rangle}{\langle S\Gamma \| \mathbf{ST}(T_1) \| S\Gamma \rangle}. \tag{7.66}$$

This means that, when the first-order spin-orbit splittings of the $\alpha S\Gamma$ terms are calculated, $\mathscr{H}_{\mathrm{so}}$ may be expressed as

$$\mathscr{H}_{\mathrm{so}} = \lambda(\alpha S\Gamma)\, \mathbf{S} \cdot \mathbf{T}(T_1), \tag{7.67}$$

in which $\lambda(\alpha S\Gamma)$ in general takes different values for different $\alpha S\Gamma$ terms. Expression (7.67) is called *operator equivalent* of the spin-orbit interaction (7.1), and is quite similar to $\lambda(\alpha SL)\, \mathbf{S} \cdot \mathbf{L}$ for calculating the spin-orbit splittings of the ^{2S+1}L terms in free atoms and ions.

Now, from the previously mentioned simple relation between the matrices of $\mathbf{T}(T_1)$ in the T_1 and T_2 states and $\mathbf{L}$ in the P state, it is evident that $\lambda\mathbf{S} \cdot \mathbf{T}$ makes the $^{2S+1}T_1$ term split just in the same way as $\lambda\mathbf{S} \cdot \mathbf{L}$ does for the ^{2S+1}P atomic term, and the splitting pattern of the $^{2S+1}T_2$ term is just inverted to that of the ^{2S+1}P term. Therefore, $^{2S+1}T_1$ and $^{2S+1}T_2$ terms split at most into three sublevels, leaving more degeneracies unlifted than those expected from the group theory. In this way, the patterns of the first-order splittings of the $^{2S+1}T_1$ and $^{2S+1}T_2$ terms are obtained without performing any calculation. The splittings of the terms with $S = 1/2, 1, 3/2$, and 2, are shown in Table 7.4,

TABLE 7.4

THE FIRST-ORDER SPIN-ORBIT SPLITTINGS

$^{2S+1}T_1(T_2)$	Eigenvalues of $\lambda\, \mathbf{S} \cdot \mathbf{T}$ [a]	Symmetries of the split components
$^2T_1(T_2)$	$\mp\lambda$	$E_1\ (E_2)$
	$\pm\lambda/2$	$G\ (G)$
$^3T_1(T_2)$	$\mp 2\lambda$	$A_1\ (A_2)$
	$\mp\lambda$	$T_1\ (T_2)$
	$\pm\lambda$	$E, T_2\ (E, T_1)$
$^4T_1(T_2)$	$\mp 5\lambda/2$	$E_1\ (E_2)$
	$\mp\lambda$	$G\ (G)$
	$\pm 3\lambda/2$	$E_2, G\ (E_1, G)$
$^5T_1(T_2)$	$\mp 3\lambda$	$T_1\ (T_2)$
	$\mp\lambda$	$E, T_2\ (E, T_1)$
	$\pm 2\lambda$	$A_2, T_1, T_2\ (A_1, T_2, T_1)$

[a] The upper signs are for $^{2S+1}T_1$ and the lower signs for $^{2S+1}T_2$.

in which labeling of the split components is performed by using the method discussed in Section 7.2.3.

As mentioned previously the values of λ are different for different terms and determined by (7.66), in which the denominator is

$$\langle S\Gamma \| \mathbf{ST}(T_1) \| S\Gamma \rangle = \langle S \| \mathbf{S} \| S \rangle \langle \Gamma \| \mathbf{T}(T_1) \| \Gamma \rangle$$
$$= i\sqrt{6}\,[S(S+1)(2S+1)]^{1/2}. \qquad (7.68)$$

The numerator of (7.66) may be expressed in terms of the reduced matrix of $V(1T_1)$ for a single electron. This will be discussed in the next section. The values of λ for some interesting terms are calculated in Problem 7.6.

7.4 Spin-Orbit Interaction in Many-Electron Systems

7.4.1 Calculation of $\langle \alpha S\Gamma \| V(1T_1) \| \alpha' S'\Gamma' \rangle$

In dealing with the spin-orbit interaction in many-electron systems, the reduced matrices $\langle t_2{}^n e^m S\Gamma \| V(1T_1) \| t_2^{n'} e^{m'} S'\Gamma' \rangle$ are left for the calculation as shown in (7.57). Since $\mathscr{H}_{\mathrm{so}}$ is given as the sum of one-electron operators, these reduced matrices can be expressed in terms of those for the one-electron system, $\langle t_2 \| v(1T_1) \| e \rangle$ and $\langle t_2 \| v(1T_1) \| t_2 \rangle$. The reduced matrix $\langle e \| v(1T_1) \| e \rangle$ does not appear as there is no spin-orbit interaction in the E_g state. By using an argument similar to that given in Section 6.3.1 for low-symmetry fields, one can show that

$$\langle t_2{}^n e^m S\Gamma \| V(1T_1) \| t_2^{n-k} e^{m+k} S'\Gamma' \rangle = 0 \qquad \text{for} \quad |k| \geqslant 2, \qquad (7.69)$$

$$\langle t_2{}^n e^m S\Gamma \| V(1T_1) \| t_2^{n-1} e^{m+1} S'\Gamma' \rangle = C_0 \langle t_2 \| v(1T_1) \| e \rangle, \qquad (7.70)$$

$$\langle t_2{}^n e^m S\Gamma \| V(1T_1) \| t_2{}^n e^m S'\Gamma' \rangle = C_1 \langle t_2 \| v(1T_1) \| t_2 \rangle, \qquad (7.71)$$

where C_0 and C_1 are numerical constants depending upon the terms in the matrix element of interest.

The calculation of C_0 and C_1 is straightforward as the wavefunctions are already known. For example, C_1 for

$$\langle t_2{}^2({}^3T_1)e\ {}^4T_2 \| V(1T_1) \| t_2{}^2({}^3T_1)e\ {}^4T_2 \rangle$$

can be calculated as follows: We had better choose a nonvanishing element involving the reduced matrix we want to calculate, which is as simple as possible for the calculation. In the present case, it is

$\langle t_2^2(^3T_1)\, e\; ^4T_2\tfrac{3}{2}\xi \mid V_{0\gamma}(1T_1)|\, t_2^2(^3T_1)\, e\; ^4T_2\tfrac{3}{2}\eta\rangle$. By using Table 3.4 and formula (3.34), one obtains

$$\left\langle t_2^2(^3T_1)e\; ^4T_2\, \tfrac{3}{2}\, \xi \,\middle|\, V_{0\gamma}(1T_1) \,\middle|\, t_2^2(^3T_1)e\; ^4T_2\, \tfrac{3}{2}\, \eta\right\rangle$$

$$= \sum_\sigma \int d\tau\, \tfrac{1}{2}\, [\sqrt{3} \mid \eta\zeta u \mid + \mid \eta\zeta v \mid]^* \sum_i s_{i0}t_{i\gamma}\, \tfrac{1}{2}\, [-\sqrt{3} \mid \zeta\xi u \mid + \mid \zeta\xi v \mid]$$

$$= \tfrac{1}{2} \left\langle t_2\, \tfrac{1}{2}\, \eta \,\middle|\, v_{0\gamma}(1T_1) \,\middle|\, t_2\, \tfrac{1}{2}\, \xi\right\rangle. \tag{7.72}$$

Noting the relations

$$\left\langle t_2^2(^3T_1)e\; ^4T_2\, \frac{3}{2}\, \xi \,\middle|\, V_{0\gamma}(1T_1) \,\middle|\, t_2^2(^3T_1)e\; ^4T_2\, \frac{3}{2}\, \eta\right\rangle$$

$$= \frac{1}{2\sqrt{10}} \langle t_2^2(^3T_1)e\; ^4T_2 \,\|\, V(1T_1) \,\|\, t_2^2(^3T_1)e\; ^4T_2\rangle \tag{7.73}$$

and

$$\langle t_2\tfrac{1}{2}\eta \mid v_{0\gamma}(1T_1) \mid t_2\tfrac{1}{2}\xi\rangle = -\tfrac{1}{6}\langle t_2 \,\|\, v(1T_1) \,\|\, t_2\rangle, \tag{7.74}$$

one finally obtains

$$\langle t_2^2(^3T_1)e\; ^4T_2 \,\|\, V(1T_1) \,\|\, t_2^2(^3T_1)e\; ^4T_2\rangle = -\frac{\sqrt{5}}{3\sqrt{2}} \langle t_2 \,\|\, v(1T_1) \,\|\, t_2\rangle. \tag{7.75}$$

If we assume the *d*-function approximation and use $\mathscr{H}_{\text{so}}$ in (7.1), (7.74) is given as $-i\zeta/2$ as seen in (7.11). Therefore, to this approximation, one has

$$\langle t_2 \,\|\, v(1T_1) \,\|\, t_2\rangle = 3i\zeta. \tag{7.76}$$

The next example is the calculation of C_0 for

$$\langle t_2^3\; ^2E \,\|\, V(1T_1) \,\|\, t_2^2(^3T_1)e\; ^4T_2\rangle.$$

We choose $\langle t_2^3\; ^2E\tfrac{1}{2}u \mid V_{-1\alpha}(1T_1)|\, t_2^2(^3T_1)\, e\; ^4T_2\tfrac{3}{2}\xi\rangle$ for this purpose. By using Tables 3.2 and 3.4 and formula (3.34), one can show that

$$\left\langle t_2^3\; ^2E\, \frac{1}{2}\, u \,\middle|\, V_{-1\alpha}(1T_1) \,\middle|\, t_2^2(^3T_1)e\; ^4T_2\, \frac{3}{2}\, \xi\right\rangle$$

$$= \sum_\sigma \int d\tau\, \frac{1}{\sqrt{2}} [\mid \xi\bar{\eta}\zeta \mid - \mid \bar{\xi}\eta\zeta \mid]^* \sum_i s_{i-1}t_{i\alpha}\, \frac{1}{2}\, [\sqrt{3} \mid \eta\zeta u \mid + \mid \eta\zeta v \mid]$$

$$= -\frac{1}{2\sqrt{2}} \left[\sqrt{3} \left\langle t_2 - \frac{1}{2}\, \xi \,\middle|\, v_{-1\alpha}(1T_1) \,\middle|\, e\, \frac{1}{2}\, u\right\rangle\right.$$

$$\left. + \left\langle t_2 - \frac{1}{2}\, \xi \,\middle|\, v_{-1\alpha}(1T_1) \,\middle|\, e\, \frac{1}{2}\, v\right\rangle\right]. \tag{7.77}$$

On the other hand, one has the relations

$$\left\langle t_2^3\ {}^2E\ \tfrac{1}{2}\ u \,\middle|\, V_{-1\alpha}(1T_1) \,\middle|\, t_2^2({}^3T_1)e\ {}^4T_2\ \tfrac{3}{2}\ \xi \right\rangle$$
$$= -\tfrac{1}{4}\langle t_2^3\ {}^2E \,\|\, V(1T_1) \,\|\, t_2^2({}^3T_1)e\ {}^4T_2\rangle, \tag{7.78}$$

$$\left\langle t_2 - \frac{1}{2}\ \xi \,\middle|\, v_{-1\alpha}(1T_1) \,\middle|\, e\ \frac{1}{2}\ u \right\rangle = \frac{1}{2\sqrt{3}}\langle t_2 \,\|\, v(1T_1) \,\|\, e\rangle, \tag{7.79}$$

and

$$\langle t_2 - \tfrac{1}{2}\xi \mid v_{-1\alpha}(1T_1) \mid e\tfrac{1}{2}v\rangle = \tfrac{1}{6}\langle t_2 \,\|\, v(1T_1) \,\|\, e\rangle. \tag{7.80}$$

Therefore, one finally obtains

$$\langle t_2^3\ {}^2E \,\|\, V(1T_1) \,\|\, t_2^2({}^3T_1)e\ {}^4T_2\rangle = \frac{2\sqrt{2}}{3}\langle t_2 \,\|\, v(1T_1) \,\|\, e\rangle. \tag{7.81}$$

If we use the d-function approximation and $\mathscr{H}_{so}$ in (7.1), (7.80) is given as $-i\zeta/\sqrt{2}$‡ as seen from (7.22). However, even if the d-function approximation is used, the e_g orbital may have the radial function which differs from that of t_{2g}. Therefore, we replace ζ by ζ'. Then, to this approximation, one has

$$\langle t_2 \,\|\, v(1T_1) \,\|\, e\rangle = -3\sqrt{2}\, i\zeta'. \tag{7.82}$$

All the nonvanishing reduced matrices of the spin-orbit interaction in N-electron systems ($N = 1, 2,\ldots, 5$) are given in Appendix VII. The reduced matrices in the remaining systems with $N = 6, 7, 8, 9$, are related to those already calculated. This point will be discussed in the next subsection.

7.4.2 The $\langle \alpha S\Gamma \,\|\, V(1T_1) \,\|\, \alpha' S'\Gamma'\rangle$ in Complementary States

Let us first consider $F_{kk'}^N$ in (4.41) and $F_{kk'}^{10-N}$ in (4.43) in the case in which one-electron operator F is $V_{q\bar{\gamma}}(1T_1)$. In this case it follows from (4.30) that for $\alpha_p = (\tfrac{1}{2}\Gamma_p m_p \gamma_p)$ with $m_p = \pm\tfrac{1}{2}$ one has

$$F_{kk'}^{10-N} = -\langle K\alpha_{p'} \mid v_{q\bar{\gamma}}(1T_1) \mid K\alpha_p\rangle$$
$$= (-1)^{m_p+m_{p'}}\langle -m_{p'}\gamma_{p'} \mid v_{q\bar{\gamma}}(1T_1) \mid -m_p\gamma_p\rangle, \tag{7.83}$$

where $p' = N + q$ and quantum numbers $s = 1/2$ and Γ_p are not

‡ Note factor $1/\sqrt{2}$ of $V_{-1\alpha}(1T_1)$ in $\mathscr{H}_{so}$ as shown in (7.50).

written for simplicity. In (7.83) we have used the fact that the orbital functions are real. By using (7.58), (7.83) may be reexpressed in the form

$$\begin{aligned} F_{kk'}^{10-N} &= -(-1)^{m_{p'}+m_p+q}\langle -m_p\gamma_p \mid v_{-q\bar{\gamma}}(1T_1) \mid -m_p'\gamma_p'\rangle \\ &= (-1)^{2m_p}\langle m_p\gamma_p \mid v_{q\bar{\gamma}}(1T_1) \mid m_p'\gamma_p'\rangle \\ &= -\langle \alpha_p \mid v_{q\bar{\gamma}}(1T_1) \mid \alpha_{N+q}\rangle \\ &= -F_{kk'}^{N}. \end{aligned} \tag{7.84}$$

In deriving (7.84) the relation[‡]

$$\langle m\gamma \mid v_{q\bar{\gamma}}(1T_1) \mid m'\gamma'\rangle = -\langle -m\gamma \mid v_{-q\bar{\gamma}}(1T_1) \mid -m'\gamma'\rangle \tag{7.85}$$

is used. Quite similarly, one can show that[§]

$$\begin{aligned} F_{kk}^{10-N} &= \sum_{i=N+1}^{10} \langle K\alpha_i \mid v_{q\bar{\gamma}}(1T_1) \mid K\alpha_i\rangle \\ &= \sum_{i=N+1}^{10} \langle \alpha_i \mid v_{q\bar{\gamma}}(1T_1) \mid \alpha_i\rangle. \end{aligned} \tag{7.86}$$

Furthermore, since F_0 is the matrix element of $V_{q\bar{\gamma}}(1T_1)$ in the $^1A_1$ state, it is zero. After all, (4.52) shows the relation

$$\begin{aligned} &\langle t_2^n(S_1\Gamma_1)\, e^m(S_2\Gamma_2)\, S\Gamma M\gamma \mid V_{q\bar{\gamma}}(1T_1) \mid t_2^{n'}(S_1'\Gamma_1')\, e^{m'}(S_2'\Gamma_2')\, S'\Gamma'M'\gamma'\rangle \\ &\quad = -\langle t_2^{6-n}(S_1\Gamma_1)\, e^{4-m}(S_2\Gamma_2)\, S\Gamma M\gamma \mid V_{q\bar{\gamma}}(1T_1) \mid t_2^{6-n'}(S_1'\Gamma_1')\, e^{4-m'} \\ &\qquad \times (S_2'\Gamma_2')\, S'\Gamma'M'\gamma'\rangle \qquad (n+m = n'+m' \neq 5), \end{aligned} \tag{7.87}$$

which leads to

$$\begin{aligned} &\langle t_2^n(S_1\Gamma_1)\, e^m(S_2\Gamma_2)\, S\Gamma \,\|\, V(1T_1) \,\|\, t_2^{n'}(S_1'\Gamma_1')\, e^{m'}(S_2'\Gamma_2')\, S'\Gamma'\rangle \\ &\quad = -\langle t_2^{6-n}(S_1\Gamma_1)\, e^{4-m}(S_2\Gamma_2)\, S\Gamma \,\|\, V(1T_1) \,\|\, t_2^{6-n'}(S_1'\Gamma_1')\, e^{4-m'}(S_2'\Gamma_2')\, S'\Gamma'\rangle \\ &\qquad\qquad (n+m = n'+m' \neq 5). \end{aligned} \tag{7.88}$$

Equation (7.88) is similar to (6.129) for the case of low-symmetry ligand fields. Because of this similarity, relations similar to all those derived for

‡ This can be proved by using the relation

$$\langle j_1m_1j_2m_2 \mid j_1j_2j_3m_3\rangle = (-1)^{j_1+j_2+j_3-2m_3}\langle j_1\ -m_1j_2\ -m_2 \mid j_1j_2j_3\ -m_3\rangle\text{t}$$

M. Rotenberg, R. Bivins, N. Metropolis, and J. K. Wooten, Jr. "The 3-*j* and 6-*j* Symbols." Technology Press, MIT, 1959.

§ In our problem of $\Gamma_i = t_2$ and e, F_{kk}^N and F_{kk}^{10-N} are zero.

low-symmetry fields in the systems of half-filled subshell configurations also hold in the present case: Corresponding to (6.131) and (6.132), one can show that

$$\langle t_2{}^3 S\Gamma \| V(1T_1) \| t_2{}^3 S\Gamma \rangle = 0, \tag{7.89}$$

$$\langle t_2{}^3 S\Gamma \| V(1T_1) \| t_2{}^3 S'\Gamma' \rangle \neq 0 \qquad (S\Gamma \neq S'\Gamma') \tag{7.90a}$$

only for the combinations

$$S\Gamma = {}^4A_2\,,\ {}^2E,\ {}^2T_1 \qquad \text{and} \qquad S'\Gamma' = {}^2T_2\,, \tag{7.90b}$$

and vice versa. Corresponding to (6.135) one can show that (7.88) holds even for $n + m = 5$ if $n \neq 3$ and $n' \neq 3$. Furthermore, corresponding to (6.136), (6.137), and (6.138), one has, respectively,

$$\begin{aligned}&\langle t_2{}^3(S_1\Gamma_1)\, e^2(S_2\Gamma_2)\, S\Gamma \| V(1T_1) \| t_2^{n'}(S_1'\Gamma_1')\, e^{m'}(S_2'\Gamma_2')\, S'\Gamma' \rangle \\ &\quad = -\mu_1\mu_2 \langle t_2{}^3(S_1\Gamma_1)\, e^2(S_2\Gamma_2)\, S\Gamma \| V(1T_1) \| t_2^{6-n'}(S_1'\Gamma_1')\, e^{4-m'}(S_2'\Gamma_2')\, S'\Gamma' \rangle \\ &\qquad\qquad (n' \neq 3, \quad m' \neq 2),\end{aligned} \tag{7.91}$$

$$\langle t_2{}^3(S_1\Gamma_1)\, e^2(S_2\Gamma_2)\, S\Gamma \| V(1T_1) \| t_2{}^3(S_1'\Gamma_1')\, e^2(S_2'\Gamma_2')\, S'\Gamma' \rangle = 0$$

$$\text{for } S_1\Gamma_1\,,\ S_2\Gamma_2\,,\ S_1'\Gamma_1', \text{ and } S_2'\Gamma_2' \text{ giving } \mu_1\mu_2\mu_1'\mu_2' = 1, \tag{7.92}$$

and

$$\langle t_2{}^3(S_1\Gamma_1)\, e^2(S_2\Gamma_2)\, S\Gamma \| V(1T_1) \| t_2{}^3(S_1\Gamma_1)\, e^2(S_2\Gamma_2)\, S'\Gamma' \rangle = 0. \tag{7.93}$$

Problem 7.6. Calculate $\lambda(\alpha S\Gamma)$ in terms of $\langle t_2 \| v(1T_1) \| t_2 \rangle$, and then confirm the following values of $\lambda(\alpha S\Gamma)$ obtained with the d-function approximation:

$$N = 1 \qquad \lambda(t_2\,{}^2T_2) = \zeta$$

$$N = 2 \qquad \begin{cases} \lambda(t_2{}^2\,{}^3T_1) = -\zeta/2, \\ \lambda(t_2e\,{}^3T_1) = \zeta/4, \\ \lambda(t_2e\,{}^3T_2) = -\zeta/4, \end{cases}$$

$$N = 3 \qquad \begin{cases} \lambda(t_2{}^2({}^3T_1)e\,{}^4T_1) = \zeta/6, \\ \lambda(t_2{}^2({}^3T_1)e\,{}^4T_2) = -\zeta/6, \\ \lambda(t_2e^2({}^3A_2)\,{}^4T_1) = -\zeta/3, \end{cases}$$

$$N = 4 \qquad \begin{cases} \lambda(t_2{}^2({}^3T_1)\,e^2({}^3A_2)\,{}^4T_2) = \zeta/4, \\ \lambda(t_2e^3\,{}^3T_1) = \zeta/4, \\ \lambda(t_2e^3\,{}^3T_2) = -\zeta/4, \end{cases}$$

$$N = 5 \qquad \begin{cases} \lambda(t_2{}^4({}^3T_1)e\,{}^4T_1) = -\zeta/6, \\ \lambda(t_2{}^4({}^3T_1)e\,{}^4T_2) = \zeta/6, \\ \lambda(t_2{}^4({}^3T_1)e\,{}^2T_1) = -\zeta/6, \\ \lambda(t_2{}^4({}^3T_1)e\,{}^2T_2) = \zeta/6. \end{cases}$$

The values of $\lambda(\alpha S\Gamma)$ for $N > 5$ are easily obtained from those for $N \leqslant 5$ by using (7.88). ◇

Problem 7.7. In ruby where Cr^{3+} ions are at the C_3-symmetry site, the effective electric-dipole transition moment $\mathbf{P}_{\text{eff}}$ for the intersystem combinations $t_2^3\,{}^4A_2M_s \leftrightarrow t_2^3\,{}^2EM_s'M'$ is approximately calculated from

$$\mathbf{P}_{\text{eff}}(t_2^3\,{}^4A_2M_s \leftrightarrow t_2^3\,{}^2EM_s'M')$$
$$= \sum_{\Gamma''M''} \langle t_2^3\,{}^4A_2M_s \mid \tilde{\mathbf{P}} \mid t_2^2e\,{}^4\Gamma''M_sM''\rangle$$
$$\times \langle t_2^2e\,{}^4\Gamma''M_sM'' \mid \mathscr{H}_{\text{so}} \mid t_2^3\,{}^2EM_s'M'\rangle[W({}^2E) - W({}^4\Gamma'')]^{-1},$$

where $\tilde{\mathbf{P}}$ is the effective electric-dipole moment combined with odd-parity ligand fields as introduced in Section 6.2.3, and M_s and M are those explained in Problem 7.5. Show that in this case intermediate states $t_2^2e\,{}^4\Gamma''$ are essentially restricted to only one term, $t_2^2({}^3T_1)e\,{}^4T_2$. Then, calculating $\mathbf{P}_{\text{eff}}$, show that the dipole strengths which are given as absolute squares of $\mathbf{P}_{\text{eff}}$ are calculated as follows:

${}^4A_2M_s$ \ ${}^2E$ $M_s'M'$	$\frac{1}{2}u_+$	$-\frac{1}{2}u_-$	$-\frac{1}{2}u_+$	$\frac{1}{2}u_-$
3/2	$\pi_\alpha/2$			$\sigma_{+\alpha}/2$
1/2	$\sigma_{+\alpha}/3$	$\sigma_{+\alpha}/6$	$\pi_\alpha/6$	$\sigma_{-\alpha}/3$
—1/2	$\sigma_{-\alpha}/6$	$\sigma_{-\alpha}/3$	$\sigma_{+\alpha}/3$	$\pi_\alpha/6$
—3/2		$\pi_\alpha/2$	$\sigma_{-\alpha}/2$	

where

$$\pi_\alpha = \frac{4}{3}\zeta'^2\left|\frac{P_\alpha^{0}}{W({}^2E) - W({}^4T_2)}\right|^2,$$

$$\sigma_{\pm\alpha} = \frac{4}{3}\zeta'^2\left|\frac{P_\alpha^{\pm}}{W({}^2E) - W({}^4T_2)}\right|^2,$$

and

$$P_\alpha^{0,\pm} = \langle {}^4A_2 \mid \tilde{P}_\alpha \mid {}^4T_2x_{0,\pm}\rangle.$$

Suffix α of $\tilde{P}_\alpha$ indicates α component of $\tilde{\mathbf{P}}$. Note that

$$\langle t_2^3\,{}^2E \,\|\, V(1T_1) \,\|\, t_2^2({}^3T_1)e\,{}^4T_2\rangle$$

has already been calculated in (7.81). The dipole strengths calculated here are used for the analysis of the Zeeman patterns of the R lines in ruby. ◇

Problem 7.8. By using the argument given in Section 3.1.2, show that

$$\langle t_2^n(S_1\Gamma_1)\,e^m(S_2\Gamma_2)\,S\Gamma \,\|\, V(1T_1) \,\|\, t_2^n(S_1'\Gamma_1')\,e^m(S_2'\Gamma_2')\,S'\Gamma'\rangle$$
$$= \delta(S_2S_2')\,\delta(\Gamma_2\Gamma_2')$$
$$\times \langle t_2^n(S_1\Gamma_1)\,e^m(S_2\Gamma_2)\,S\Gamma \,\|\, V(1T_1) \,\|\, t_2^n(S_1'\Gamma_1')\,e^m(S_2\Gamma_2)\,S'\Gamma'\rangle. \quad ◇$$

Chapter VIII FINE STRUCTURE OF MULTIPLETS

8.1 Kramers Degeneracy

8.1.1 Time-Reversal Operator

Operator K was introduced in (4.30) for the purpose of obtaining $\Psi(\alpha S\Gamma M\gamma)$ from $\Psi(\alpha S\Gamma -M\gamma)$ in a simple fashion, and was called a time-reversal operator. However, the physical implication of this operator was left unexplained. This operator was also used in Section 7.2.1, where K was shown to commute with rotation operator R^S in the spin space. In this subsection we will clarify the physical meaning of this operator, which helps one to understand Kramers degeneracy.

Let us first consider the time-dependent Schrödinger equation involving no spin,

$$i\hbar \frac{\partial\psi(t)}{\partial t} = \mathscr{H}\psi(t). \tag{8.1}$$

By denoting the time variable as t' instead of t, the complex conjugate to (8.1) is

$$-i\hbar \frac{\partial\psi^*(t')}{\partial t'} = \mathscr{H}^*\psi^*(t') = \mathscr{H}\psi^*(t'), \tag{8.2}$$

in which $\mathscr{H}^* = \mathscr{H}$ is assumed as it is the case for $\mathscr{H} = \mathscr{H}_0 + \mathscr{H}_1$ in (7.42). Replacing t' in (8.2) by $-t$, one obtains

$$i\hbar \frac{\partial\psi^*(-t)}{\partial t} = \mathscr{H}\psi^*(-t), \tag{8.3}$$

which shows that $\psi^*(-t)$ is also the solution of Eq. (8.1). Therefore, $\psi^*(-t)$ represents the states in which all velocities have opposite directions to those in $\psi(t)$. State $\theta\psi(t) \equiv \psi^*(-t)$ is called the time-reversed state of $\psi(t)$.

In time-independent problems, the time-reversed state of $\psi(r)$ is given by

$$\theta\psi(\mathbf{r}) = K_0\psi(\mathbf{r}) = \psi^*(\mathbf{r}) \tag{8.4}$$

In (8.4) operator K_O is the complex conjugation operator introduced in (4.30). Equation (8.4) indicates that K_O is a time reversal operator for orbital functions, and this reversal is called *Wigner's time reversal.* Using the well-known relation,

$$K_O Y_{lm}(\theta\varphi) = Y^*_{lm}(\theta\varphi) = (-1)^m Y_{l-m}(\theta\varphi), \tag{8.5}$$

one can show that the orbital angular momentum operator is transformed by K_O as

$$K_O \mathbf{l} K_O^{-1} = -\mathbf{l}, \tag{8.6}$$

which confirms the property of time reversal of K_O.

The proof of (8.6) is as follows: The relations

$$\begin{aligned} K_O l_z K_O^{-1} K_O Y_{lm}(\theta\varphi) &= (-1)^m K_O l_z K_O^{-1} Y_{l-m}(\theta\varphi) \\ &= m K_O Y_{lm}(\theta\varphi) = (-1)^m m Y_{l-m}(\theta\varphi) \end{aligned} \tag{8.7}$$

and

$$\begin{aligned} K_O l_\pm K_O^{-1} K_O Y_{lm}(\theta\varphi) &= (-1)^m K_O l_\pm K_O^{-1} Y_{l-m}(\theta\varphi) \\ &= [l(l+1) - m(m \pm 1)]^{1/2} K_O Y_{lm}(\theta\varphi) \\ &= (-1)^m [l(l+1) - (-m)(-m \mp 1)]^{1/2} Y_{l-m}(\theta\varphi) \end{aligned} \tag{8.8}$$

indicate that

$$K_O l_z K_O^{-1} = -l_z \tag{8.9}$$

and

$$K_O l_\pm K_O^{-1} = -l_\mp, \tag{8.10}$$

which means

$$K_O l_x K_O^{-1} \mp i K_O l_y K_O^{-1} = -(l_x \mp i l_y); \tag{8.11}$$

thus

$$K_O l_x K_O^{-1} = -l_x, \qquad K_O l_y K_O^{-1} = -l_y. \tag{8.12}$$

Quite similarly, using the relations (7.24), (7.25), (4.30), and (4.33), one can prove the transformation

$$K\mathbf{s}K^{-1} = -\mathbf{s}. \tag{8.13}$$

Equation (8.13) shows that K is a time reversal operator for the systems involving spins.

8.1.2 Kramers Theorem

Kramers theorem states that, in a system with an odd number of electrons whose electron Hamiltonian is invariant to time reversal, the electronic energy levels are at least twofold degenerate.

To prove the theorem it is convenient to express K in the following form:

$$K = \prod_{j=1}^{N} (-i\sigma_y)_j \, K_O, \tag{8.14}$$

where N is the total number of electrons and

$$\sigma_{yj} = \begin{matrix} \alpha & \beta \end{matrix} \atop \begin{bmatrix} 0 & -i \\ i & 0 \end{bmatrix}_j \tag{8.15}$$

is a transformation matrix (Pauli matrix) for spin functions α and β of electron j: the equivalence of $-i\sigma_y$ to K_S in (4.30) is seen by comparing

$$(\alpha, \beta) \begin{bmatrix} 0 & -1 \\ 1 & 0 \end{bmatrix} = (\beta, -\alpha) \tag{8.16}$$

with (4.31). Then, it follows from $(-i\sigma_y)^2 = -1$‡ that

$$K^2 = (-1)^N. \tag{8.17}$$

Now, since the eigenvalue equation

$$\mathscr{H}\Psi = E\Psi \tag{8.18}$$

is also satisfied by $K\Psi$ if $K\mathscr{H}K^{-1} = \mathscr{H}$, one has two possibilities:

(i) $K\Psi = e^{i\delta}\Psi$,

or

(ii) $K\Psi \neq e^{i\delta}\Psi$,

where δ is a real number. In the case of (i), one sees that

$$K^2\Psi = e^{-i\delta}K\Psi = \Psi \tag{8.19}$$

contradicts (8.17) if N is odd. Therefore, in the system with an odd

‡ The number 1 is the unit matrix.

number of electrons, $K\Psi$ and Ψ represent different states with the same energy, which proves the theorem.

Kramers theorem does not exclude the possibility that some electronic levels in a system with an even number of electrons are doubly degenerate due to time reversal. Regardless of whether N is odd or even, if $K\Psi$ is different from Ψ, the pair of these states are degenerate. This degeneracy is called a *time-reversal degeneracy*, and $K\Psi$ and Ψ are called *time-reversal pair states*. The time-reversal degeneracy is lifted if an external magnetic field is applied, as the interaction energy of the magnetic moment proportional to the angular momentum with the magnetic field changes its sign when the direction of the magnetic moment is reversed by time reversal.

8.2 Higher-Order Splittings of Cubic Terms

As shown in (6.131), (6.133), (6.138), (7.89), (7.91), and (7.95), the terms of half-filled subshell configurations undergo no first-order splitting due to low-symmetry fields and the spin-orbit interaction. These terms, however, may split due to the higher-order action of these interactions. In this section, we shall show several examples of the higher-order splitting of the cubic terms by using the perturbation theory. A general treatment of the splittings of the cubic terms including the higher-order perturbations will be found in the next section.

8.2.1 Splitting of the $t_2^3\ {}^2E$ Term

Let us first consider the splitting of the $t_2^3\ {}^2E$ term in a system of the D_3-symmetry. As shown in (6.131) and (7.89), this term undergoes no first-order splitting due to either the trigonal field or the spin-orbit interaction, so that we have to consider the second-order splitting.

As discussed in Section 6.2, the even-parity field in this system is expressed by the tensor operator $V_{x_0}(T_{2g}) = V_0(T_{2g})$. If we ignore the $t_2^2e\ {}^2\Gamma$ terms whose energies are much higher than that of $t_2^3\ {}^2E$ in the range of $Dq/B \sim 2$ to 3 as seen in Fig. 5.3, the trigonal field may connect $t_2^3\ {}^2E$ only with the spin doublets within the t_2^3 electron configuration. Furthermore, (6.132) shows that the trigonal field may connect $t_2^3\ {}^2E$ only with $t_2^3\ {}^2T_2$. The nondiagonal matrix elements of $V_0(T_{2g})$ between the $t_2^3\ {}^2E$ and $t_2^3\ {}^2T_2$ terms with the trigonal bases, $\langle t_2^3\ {}^2EM_sM \mid V_0(T_{2g}) \mid t_2^3\ {}^2T_2M_sM'\rangle$, are calculated by using (6.89) and the method given in Section 6.31 as follows:

$$\langle t_2^3\ {}^2EM_sM \mid V_0(T_{2g}) \mid t_2^3\ {}^2T_2M_sM'\rangle = -\sqrt{6}\, K\, \delta(MM'), \tag{8.20}$$

where the relation

$$\langle t_2^3\ {}^2E \,\|\, V(T_{2g}) \,\|\, t_2^3\ {}^2T_2 \rangle = \sqrt{2}\,\langle t_2 \,\|\, v(T_{2g}) \,\|\, t_2 \rangle \tag{8.21}$$

is used and K is defined as

$$\langle t_2 \,\|\, v(T_{2g}) \,\|\, t_2 \rangle = -3\sqrt{2}\,K. \tag{8.22}$$

The definition of K in (8.22) is identical to that of K in (6.56). From(8.20) one sees that the second-order perturbation energy

$$\begin{aligned} &\sum_{M'} \langle t_2^3\ {}^2EM_sM \mid V_0(T_{2g}) \mid t_2^3\ {}^2T_2M_sM' \rangle \\ &\qquad \times \langle t_2^3\ {}^2T_2M_sM' \mid V_0(T_{2g}) \mid t_2^3\ {}^2EM_sM'' \rangle / [W({}^2E) - W({}^2T_2)] \\ &\quad = \delta(MM'')6K^2/[W({}^2E) - W({}^2T_2)] \end{aligned} \tag{8.23}$$

is independent of M, giving no splitting of $t_2^3\ {}^2E$. In this calculation the spin-orbit interaction is not considered, and this result is evident from the group theoretical view as irreducible representation E of the O-group goes to E of the D_3-group without any decomposition.

Next, we consider the nondiagonal matrix elements of $\mathscr{H}_{so}$ which connect $t_2^3\ {}^2E$ with the other terms. We ignore the $t_2^2e\ {}^2\Gamma$ terms also in this case. Equation (7.90) shows that the spin-orbit interaction connects $t_2^3\ {}^2E$ only with $t_2^3\ {}^2T_2$ if we confine ourselves to the terms of the t_2^3 configuration. The nondiagonal matrix elements

$$\langle t_2^3\ {}^2EM_sM \mid \mathscr{H}_{so} \mid t_2^3\ {}^2T_2M_s'M' \rangle$$

are calculated by the use of the formula in Problem 7.5 and the method in Section 7.4.1 as follows:

$$\begin{array}{cc|cc:cc:cc} {}^2E \diagdown {}^2T_2 & M' & \multicolumn{2}{c}{x_+} & \multicolumn{2}{c}{x_-} & \multicolumn{2}{c}{x_0} \\ M & M_s \diagdown M_s' & \tfrac{1}{2} & -\tfrac{1}{2} & \tfrac{1}{2} & -\tfrac{1}{2} & \tfrac{1}{2} & -\tfrac{1}{2} \\ \hline u_+ & \tfrac{1}{2} & -1 & 0 & 0 & \sqrt{2} & 0 & 0 \\ & -\tfrac{1}{2} & 0 & 1 & 0 & 0 & \sqrt{2} & 0 \\ \hdashline u_- & \tfrac{1}{2} & 0 & 0 & 1 & 0 & 0 & \sqrt{2} \\ & -\tfrac{1}{2} & -\sqrt{2} & 0 & 0 & -1 & 0 & 0 \end{array} \times \frac{\zeta}{\sqrt{6}}, \tag{8.24}$$

where the relation

$$\langle t_2^3\ {}^2E \,\|\, V(1T_1) \,\|\, t_2^3\ {}^2T_2 \rangle = -(2/3)^{1/2} \langle t_2 \,\|\, v(1T_1) \,\|\, t_2 \rangle \tag{8.25}$$

is used together with (7.76). If the $t_2^2e\ {}^4\Gamma$ terms are considered in addition to the $t_2^3S\Gamma$ terms, it will be found that the spin-orbit interaction

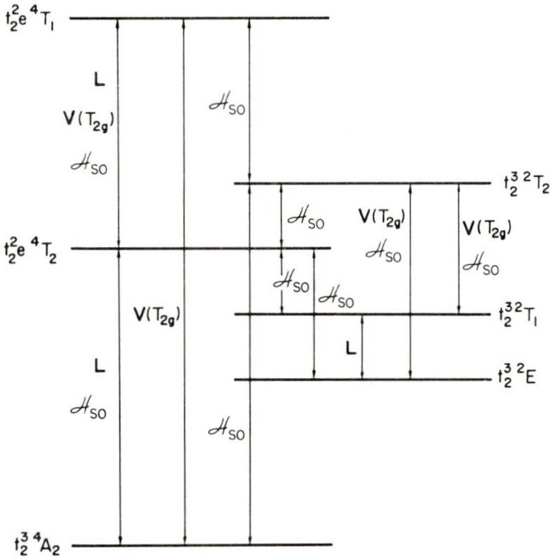

FIG. 8.1. Coupling scheme among terms of d^3 through $\mathbf{V}(T_{2g})$, $\mathscr{H}_{so}$, and $\mathbf{L}$.

also connects $t_2^3\ ^2E$ with $t_2^2(^3T_1)e\ ^4T_2$ but not with $t_2^2(^3T_1)e\ ^4T_1$. The coupling through $V(T_{2g})$ and $\mathscr{H}_{so}$ among the terms of the t_2^3 configuration ($^4A_2$, ^2E, $^2T_1$, $^2T_2$) and the $t_2^2e\ ^4\Gamma$ terms ($^4T_2$, $^4T_1$) is shown in Fig. 8.1, where the coupling through $\mathbf{L}$ is also indicated for later use.

Problem 8.1. Show that there is no nondiagonal matrix element of $\mathscr{H}_{so}$ between the $t_2^3\ ^2E$ and $t_2^2(^3T_1)e\ ^4T_1$ terms. ◇

From the group theoretical consideration, it is evident that the second-order perturbation involving $\mathscr{H}_{so}$ twice does not split the ^2E term: Such a perturbation still exists in a cubic system, where ^2E is reduced to $E_1 \times E = G$. Therefore, we have to consider the second-order pertrubation involving $V_0(T_{2g})$ and $\mathscr{H}_{so}$. The perturbation energies of this type are calculated from (8.20) and (8.24) as

$$\Big[\sum_{M'} \langle t_2^3\ ^2EM_sM \mid V_0(T_{2g}) \mid t_2^3\ ^2T_2M_sM'\rangle \times \langle t_2^3\ ^2T_2M_sM' \mid \mathscr{H}_{so} \mid t_2^3\ ^2EM_s''M''\rangle + \sum_{M'} \langle t_2^3\ ^2EM_sM \mid \mathscr{H}_{so} \mid t_2^3\ ^2T_2M_s''M'\rangle \times \langle t_2^3\ ^2T_2M_s''M' \mid V_0(T_{2g}) \mid t_2^3\ ^2EM_s''M''\rangle\Big]/[W(^2E) - W(^2T_2)]$$

$$= \begin{cases} 2K\zeta\, \delta(M_sM_s'')\, \delta(MM'')/[W(^2E) - W(^2T_2)] & \text{for} \quad M_sM = \pm\tfrac{1}{2}u_\pm, \\ -2K\zeta\, \delta(M_sM_s'')\, \delta(MM'')/[W(^2E) - W(^2T_2)] & \text{for} \quad M_sM = \pm\tfrac{1}{2}u_\mp, \end{cases} \tag{8.26}$$

which gives the splitting

$$W(^2E \pm\tfrac{1}{2}u_\pm) - W(^2E \pm\tfrac{1}{2}u_\mp) = \frac{4K\zeta}{W(^2E) - W(^2T_2)}\,. \tag{8.27}$$

The doubly degenerate split components, $\pm\frac{1}{2}u_\pm$ and $\pm\frac{1}{2}u_\mp$, cannot be split further because of Kramers theorem. The splitting of the $t_2{}^3\,{}^2E$ term calculated here may be observed in the spectrum of ruby as the separation of the R_1 and R_2 lines, and it has been found that the second-order perturbation calculation described here explains well the observed separation.

Problem 8.2. Examining the transformation properties of the wavefunctions associated with the split components, $\pm\frac{1}{2}u_\pm$ and $\pm\frac{1}{2}u_\mp$, show that they are, respectively, labeled with double-valued irreducible representation $\bar{A}_1 + \bar{A}_2$ and $\bar{E}$ of the double D_3-group given in Problem 7.3. ◇

Problem 8.3. Calculate the splitting of the $t_2{}^3\,{}^2E$ term in the D_2-symmetry system. ◇

8.2.2 Splitting of the $t_2{}^3\,{}^4A_2$ Term

Here we will discuss the splitting of the $t_2{}^3\,{}^4A_2$ term in the D_3-symmetry system. In the cubic system the ${}^4A_2$ term shows no spin-orbit splitting as $G \times A_2 = G$, but in the D_3-symmetry system G is decomposed into $\bar{A}_1 + \bar{A}_2 + \bar{E}$ so that the ${}^4A_2$ term may split: The decomposition of G in the D_3-symmetry system is achieved by comparing Table 7.1 and the table in Problem 7.3. It follows from Kramers theorem that the ${}^4A_2$ term may split into two Kramers doublets labeled as $\bar{A}_1 + \bar{A}_2$ and $\bar{E}$.

From the above-mentioned argument, it is evident that the perturbations involving only $\mathscr{H}_{\text{so}}$ give rise to no splitting of ${}^4A_2$. It is also evident that the perturbations involving only $V_0(T_{2g})$ give no splitting as ${}^4A_2$ is an orbital singlet. Therefore, to calculate the splitting, the combined action of $\mathscr{H}_{\text{so}}$ and $V_0(T_{2g})$ has to be taken into account. A detailed examination shows that the lowest-order perturbations giving rise to the splitting are the third-order ones involving $\mathscr{H}_{\text{so}}$ twice and $V_0(T_{2g})$ once. If the $t_2{}^2\,e^2\Gamma$ terms are ignored, the coupling scheme of Fig. 8.1 shows that the third-order perturbations of this type are given by the following processes:

$$\text{(I)}\quad t_2{}^3\,{}^4A_2 \xrightarrow{\mathscr{H}_{\text{so}}} t_2{}^2e\,{}^4T_2 \xrightarrow{V_0(T_{2g})} t_2{}^2e\,{}^4T_2 \xrightarrow{\mathscr{H}_{\text{so}}} t_2{}^3\,{}^4A_2\,, \tag{8.28a}$$

$$\text{(II)}\begin{cases} t_2^3\,{}^4A_2 \xrightarrow{V_0(T_{2g})} t_2^2e\,{}^4T_1 \xrightarrow{\mathscr{H}_{\rm SO}} t_2^2e\,{}^4T_2 \xrightarrow{\mathscr{H}_{\rm SO}} t_2^3\,{}^4A_2\,, \\ t_2^3\,{}^4A_2 \xrightarrow{\mathscr{H}_{\rm SO}} t_2^2e\,{}^4T_2 \xrightarrow{\mathscr{H}_{\rm SO}} t_2^2e\,{}^4T_1 \xrightarrow{V_0(T_{2g})} t_2^3\,{}^4A_2\,, \end{cases} \tag{8.28b}$$

$$\text{(III)}\begin{cases} t_2^3\,{}^4A_2 \xrightarrow{V_0(T_{2g})} t_2^2e\,{}^4T_1 \xrightarrow{\mathscr{H}_{\rm SO}} t_2^3\,{}^2T_2 \xrightarrow{\mathscr{H}_{\rm SO}} t_2^3\,{}^4A_2\,, \\ t_2^3\,{}^4A_2 \xrightarrow{\mathscr{H}_{\rm SO}} t_2^3\,{}^2T_2 \xrightarrow{\mathscr{H}_{\rm SO}} t_2^2e\,{}^4T_1 \xrightarrow{V_0(T_{2g})} t_2^3\,{}^4A_2\,. \end{cases} \tag{8.28c}$$

All these perturbation processes give comparable contributions to the splitting of the $t_2^3\,{}^4A_2$ term. For example, process (I) is calculated as follows: It follows from (6.89), (6.127), and (8.22) that

$$\langle t_2^2e\,{}^4T_2M_sM \mid V_0(T_{2g}) \mid t_2^2e\,{}^4T_2M_s'M'\rangle = -\tfrac{1}{2}K\,\delta(M_sM_s')\,\delta(MM')\,\alpha(M), \tag{8.29a}$$

where

$$\alpha(M=\pm 1) = -1, \qquad \alpha(M=0) = 2. \tag{8.29b}$$

From the formula in Problem 7.5, (7.82), and

$$\langle t_2^3\,{}^4A_2 \,\|\, V(1T_1) \,\|\, t_2^2e\,{}^4T_2\rangle = \frac{\sqrt{10}}{3}\langle t_2 \,\|\, v(1T_1) \,\|\, e\rangle, \tag{8.30}$$

one obtains $\langle t_2^3\,{}^4A_2M_se_2 \mid \mathscr{H}_{\rm SO} \mid t_2^2e\,{}^4T_2M_s'M'\rangle$ as

${}^4A_2$ M_s \ T_2 M' / M_s'	x_+ $\frac{3}{2}$	x_+ $\frac{1}{2}$	x_+ $-\frac{1}{2}$	x_+ $-\frac{3}{2}$	x_- $\frac{3}{2}$	x_- $\frac{1}{2}$	x_- $-\frac{1}{2}$	x_- $-\frac{3}{2}$	x_0 $\frac{3}{2}$	x_0 $\frac{1}{2}$	x_0 $-\frac{1}{2}$	x_0 $-\frac{3}{2}$
$\frac{3}{2}$	0	0	0	0	0	$\sqrt{6}$	0	0	-3	0	0	0
$\frac{1}{2}$	$-\sqrt{6}$	0	0	0	0	0	$2\sqrt{2}$	0	0	-1	0	0
$-\frac{1}{2}$	0	$-2\sqrt{2}$	0	0	0	0	0	$\sqrt{6}$	0	0	1	0
$-\frac{3}{2}$	0	0	$-\sqrt{6}$	0	0	0	0	0	0	0	0	3

$$\times\; -\frac{i\zeta'}{3}\,. \tag{8.31}$$

Then, by using (8.29) and (8.31), one knows that ${}^4A_2$ splits into two Kramers doublets, $M_s = \pm\frac{3}{2}$ and $\pm\frac{1}{2}$, and their separation is

$$W\left(M_s = \pm\frac{1}{2}\right) - W\left(M_s = \pm\frac{3}{2}\right) = \frac{4}{3}\,\frac{\zeta'^2K}{[W({}^4T_2) - W({}^4A_2)]^2}\,. \tag{8.32}$$

This separation could be enhanced or canceled by comparable contributions from the other processes, even from those not mentioned in (8.28), so that the calculation of the small splitting of the orbital singlet such as $t_2^3\,{}^4A_2$ requires a very careful treatment.

Problem 8.4. Calculate contributions to the splitting of the $t_2^3\,{}^4A_2$ term from processes (II) and (III) in (8.28). ◇

8.3 Effective Hamiltonian

8.3.1 Spin Hamiltonian

Let us consider an orbital singlet with the $2S + 1$ spin multiplicity, ${}^{2S+1}A$ ($A = A_1$ or A_2), which is well separated from the other terms. This state, in general, is mixed with the other terms through low-symmetry fields and the spin-orbit coupling and shows small splittings. Therefore, the $(2S + 1)$ wavefunctions of this state are given by a linear combination of the wavefunctions of various terms. We denote such $(2S + 1)$ functions by $\psi_1, \psi_2, \ldots, \psi_{2S+1}$, and suppose that, with these functions as the bases, the matrix of the total Hamiltonian involving low-symmetry fields and the spin-orbit interactions is partially diagonal, i.e., the nondiagonal matrix elements connecting the orbital singlet with all the other terms are zero as shown below:

$$
\begin{array}{cc|cccc|ccc}
 & & \multicolumn{4}{c}{{}^{2S+1}A} & & & \\
 & & \psi_1 & \psi_2 & \cdots & \psi_{2S+1} & & & \\
\hline
 & \psi_1 & \times & \times & \cdots & \times & & & \\
{}^{2S+1}A & \psi_2 & \times & & & \vdots & & 0 & \\
 & \vdots & \vdots & & & \vdots & & & \\
 & \psi_{2S+1} & \times & \cdots & \cdots & \times & & & \\
\hline
 & & & & & & \times & \times & \cdots \\
 & & & 0 & & & \times & & \\
 & & & & & & \vdots & &
\end{array} .
$$

In principle it is possible to obtain functions ψ_i by applying a unitary transformation to the wavefunctions of the cubic terms, but here without trying to look for the explicit forms of ψ_i's we assume that these functions have already been obtained. Then, our problem of calculating the splitting of the orbital singlet is to diagonalize the small $(2S + 1)$-dimensional secular matrix. This secular matrix is, of course, hermitian and reflects symmetry properties, including time-reversal symmetry, of the system. Our purpose in this subsection is, without knowing the ψ_i's and the small secular matrix, to find an *effective Hamiltonian* which involves only spin operators and which reproduces the same small secular matrix with spin functions $\Theta(SM)$ as the bases. This effective Hamiltonian is called *spin Hamiltonian.* In doing this it is inevitable to leave unknown parameters in the spin Hamiltonian.

For this purpose, we first study an example with $S = 1/2$, ${}^2A_1$. In this case the Hermitian secular matrix is given in the form

$$\begin{bmatrix} a & c - id \\ c + id & b \end{bmatrix}, \tag{8.33}$$

where a, b, c, and d are real numbers. Considering the matrices of $\mathbf{1}$, S_x, S_y, and S_z with bases $\Theta(\frac{1}{2}M)$ as follows,

$$M = \tfrac{1}{2} \quad -\tfrac{1}{2}$$

$$\mathbf{1} = \begin{bmatrix} 1 & 0 \\ 0 & 1 \end{bmatrix}, \tag{8.34a}$$

$$S_x = \begin{bmatrix} 0 & 1 \\ 1 & 0 \end{bmatrix} \times \tfrac{1}{2}, \tag{8.34b}$$

$$S_y = \begin{bmatrix} 0 & -i \\ i & 0 \end{bmatrix} \times \tfrac{1}{2}, \tag{8.34c}$$

$$S_z = \begin{bmatrix} 1 & 0 \\ 0 & -1 \end{bmatrix} \times \tfrac{1}{2}, \tag{8.34d}$$

one sees that the matrix of the spin Hamiltonian,

$$\mathscr{H}_S = \tfrac{1}{2}(a + b)\mathbf{1} + (a - b)S_z + 2cS_x + 2dS_y, \tag{8.35}$$

with bases $\Theta(\frac{1}{2}\,\frac{1}{2})$ and $\Theta(\frac{1}{2} - \frac{1}{2})$ is identical to (8.33). The spin Hamiltonian in the form of (8.35) involves as many unknown parameters as in (8.33), thus providing no advantage. However, if one considers that bases $\Theta(\frac{1}{2}\,\frac{1}{2})$ and $\Theta(\frac{1}{2} - \frac{1}{2})$ have, respectively, the same transformation properties‡ as those of ψ_1 and ψ_2 with respect to the symmetry operation which bring the system into itself, one notices that the spin Hamiltonian should be invariant to those symmetry operations as the

‡ The wavefunctions of the ${}^{2S+1}A_1$ cubic terms may be the bases of irreducible representations Γ's of the double group of a certain symmetry. Since low-symmetry fields and the spin-orbit interaction connect the states with the same Γ, ψ_i's (or the linear combinations of ψ_i) may also be the bases of the Γ's. Furthermore, since the transformation properties of wavefunctions $\Psi({}^{2S+1}A_1Me_1)$ are the same as those of spin functions $\Theta(SM)$, we may conclude that the transformation properties of $\Theta(SM)$ and the ψ_i's (or the linear combinations of ψ_i) should be the same. A similar argument may also be applied to the case of ${}^{2S+1}A_2$: In this case $\Theta(SM)f(A_2)$ transforms in the same way as $\Psi({}^{2S+1}A_2Me_2)$ does, where $f(A_2)$ is a function which transforms as the base of irreducible representation A_2 of the O-group. The presence of factor $f(A_2)$ in the bases of the spin hamiltonian brings no essential difference in the argument. Afterall, it may be said that, as long as orbital singlets are concerned, the spin Hamiltonian should be invariant to the symme tryoperations which bring the system into itself. An example of ${}^1E$ for which this is not the case will be discussed later.

real Hamiltonian is. Now imposing the condition of time-reversal invariance on (8.35), we know that the linear terms in S_x, S_y, and S_z should be zero, i.e.,

$$\mathscr{H}_S = \tfrac{1}{2}(a + b)\mathbf{1}. \tag{8.36}$$

Spin Hamiltonian (8.36) indicates the absence of the splitting of the $^2A_1$ (or $^2A_2$) term in any symmetry system in agreement with Kramers theorem: spin doublets ($S = \frac{1}{2}$) always appear in the system with an odd number of electrons.

In the case of $^3A_{1,2}$, the secular matrix corresponding to (8.33) is three-dimensional and, in general, involves nine unknown parameters. By using spin functions $\Theta(11)$, $\Theta(10)$, and $\Theta(1-1)$ as the bases, the diagonal elements involving three parameters may be reproduced by a suitable linear combination of operators $\mathbf{1}$, S_z, and S_z^2 whose matrices are independent of each other and are given as

$$\mathbf{1} = \begin{bmatrix} 1 & 0 & 0 \\ 0 & 1 & 0 \\ 0 & 0 & 1 \end{bmatrix}, \tag{8.37a}$$

$$S_z = \begin{bmatrix} 1 & 0 & 0 \\ 0 & 0 & 0 \\ 0 & 0 & -1 \end{bmatrix}, \tag{8.37b}$$

$$S_z^2 = \begin{bmatrix} 1 & 0 & 0 \\ 0 & 0 & 0 \\ 0 & 0 & 1 \end{bmatrix}. \tag{8.37c}$$

The independence means that none of the matrices can be expressed by a linear combination of the others. The elements one-off the diagonal involving four parameters may be reproduced by a suitable linear combination of the following Hermitian matrices:

$$S_x = \begin{bmatrix} 0 & 1 & 0 \\ 1 & 0 & 1 \\ 0 & 1 & 0 \end{bmatrix} \times \frac{1}{\sqrt{2}}, \tag{8.38a}$$

$$S_y = \begin{bmatrix} 0 & -i & 0 \\ i & 0 & -i \\ 0 & i & 0 \end{bmatrix} \times \frac{1}{\sqrt{2}}, \tag{8.38b}$$

$$S_xS_z + S_zS_x = \begin{bmatrix} 0 & 1 & 0 \\ 1 & 0 & -1 \\ 0 & -1 & 0 \end{bmatrix} \times \frac{1}{\sqrt{2}}, \tag{8.38c}$$

$$S_yS_z + S_zS_y = \begin{bmatrix} 0 & -i & 0 \\ i & 0 & i \\ 0 & -i & 0 \end{bmatrix} \times \frac{1}{\sqrt{2}}, \tag{8.38d}$$

which are independent of each other. Finally, the elements two-off the diagonal involving two parameters may be reproduced by a suitable linear combination of the following Hermitian matrices:

$$S_x^2 - S_y^2 = \begin{bmatrix} 0 & 0 & 1 \\ 0 & 0 & 0 \\ 1 & 0 & 0 \end{bmatrix},$$
$$S_xS_y + S_yS_x = \begin{bmatrix} 0 & 0 & -i \\ 0 & 0 & 0 \\ i & 0 & 0 \end{bmatrix}, \tag{8.39}$$

which are also independent of each other. From all these it follows that the spin Hamiltonian for $^3A_{1,2}$ is given by a linear combination of **1**, S_x, S_y, S_z, $S_yS_z + S_zS_y$, $S_zS_x + S_xS_z$, $S_xS_y + S_yS_x$, S_z^2, and $S_x^2 - S_y^2$. Imposing the time-reversal invariance on $\mathscr{H}_s$, one sees that the terms linear to S_x, S_y, and S_z should be zero. Furthermore, for example, if the system has the D_2-symmetry whose twofold symmetry axes are x, y, and z, only the terms proportional to **1**, S_z^2 and $S_x^2 - S_y^2$ should be nonvanishing as they are invariant to the rotations of the D_2-group. Thus, one finally obtains the spin Hamiltonian

$$\mathscr{H}_s = D[S_z^2 - \tfrac{1}{3}S(S+1)] + E(S_x^2 - S_y^2). \tag{8.40}$$

In (8.40) the constant term is included in $-\frac{1}{3}DS(S+1)$ so that the trace of the matrix be zero. The trace gives the shift of energy levels as a whole, which we are not interested in. In (8.40) D and E are undetermined parameters. The secular matrix of (8.40) with bases $\Theta(SM)$ $(S = 1)$ is easily obtained by using (8.37) and (8.39) as

$$M = \quad 1 \qquad 0 \qquad -1$$

$$\begin{bmatrix} D/3 & 0 & E \\ 0 & -2D/3 & 0 \\ E & 0 & D/3 \end{bmatrix}, \tag{8.41}$$

whose eigenvalues are

$$\epsilon_1 = \frac{D}{3} + E, \qquad \epsilon_2 = \frac{D}{3} - E, \qquad \epsilon_3 = -\frac{2D}{3}. \tag{8.42}$$

Equation (8.42) shows that $^3A_{1,2}$ split into three nondegenerate components in the D_2-symmetry system.

Considering these examples, we can find the general method of constructing the spin Hamiltonian with spin S for the $^{2S+1}A_{1,2}$ terms.

For this purpose we introduce irreducible spin tensor operators $\mathbf{S}^{(k)}$ whose components are denoted by $S_q^{(k)}$. The $S_q^{(k)}$'s are obtained first by symmetrizing the polynomial expansion, $\Sigma C_{\alpha\beta\gamma}x^\alpha y^\beta z^\gamma$ ($\alpha + \beta + \gamma = k$), of the cubic harmonics of the kth order, and then by substituting S_x, S_y, and S_z for x, y, and z, respectively. For example, $S_xS_y + S_yS_x$ in (8.39) is a component of $\mathbf{S}^{(2)}$ and is derived by symmetrizing the polynomial expansion of the second-order cubic harmonic $V_\zeta(T_{2g}) = xy$ as $xy + yx$, and then by substituting S_x and S_y for x and y, respectively. The symmetrization is necessary to make the spin operators hermitian.

As seen from such a process of constructing the irreducible spin-tensor operators, $\mathbf{S}^{(k)}$ has $2k + 1$ components, which are transformed like the γ base of irreducible representation Γ of the O-group. Therefore, q of $S_q^{(k)}$ may be denoted by $\Gamma\gamma$. Furthermore, as seen in examples of the matrices of $S_q^{(2)}$ given in (8.37)–(8.39), the matrices of all the components of $\mathbf{S}^{(k)}$'s ($k = 0, 1, 2, \ldots, 2S$) with bases $\Theta(SM)$ ($M = S, S - 1, \ldots, -S$) are independent of each other.

Now considering that the $(2S + 1)$-dimensional hermitian matrix generally involves the $(2S + 1)^2$ parameters, one sees that this matrix may be equivalent to the matrix of a suitable linear combination of $\mathbf{1}$, $S_q^{(1)}, \ldots, S_q^{(2S)}$ with bases $\Theta(SM)$: In this case the total number of $S_q^{(k)}$'s is $\sum_{k=0}^{2S} (2k + 1) = (2S + 1)^2$. Imposing the symmetry requirements on this linear combination, one finally obtains the spin Hamiltonian for the state with spin S.

For example, let us construct the spin Hamiltonian for the $^4A_2$ state in the D_3-symmetry system. We first make a linear combination of the components of irreducible spin tensor operators, $\mathbf{1}$, $\mathbf{S}^{(1)}$, $\mathbf{S}^{(2)}$, and $\mathbf{S}^{(3)}$. Then, requesting the time-reversal invariance and the invariance to any rotation in the D_3-group, one obtains

$$\mathscr{H}_S = D[S_Z^2 - \tfrac{1}{3}S(S + 1)], \tag{8.43}$$

where the Z-axis is along the trigonal axis. The eigenvalues of (8.43) for the state with the fictitious spin $S = 3/2$ are obtained as

$$\begin{aligned} \epsilon_1 &= D \quad \text{(twofold degeneracy)}, \\ \epsilon_2 &= -D \quad \text{(twofold degeneracy)}, \end{aligned} \tag{8.44}$$

in agreement with the result in Section 8.2.2.

Parameters are left undetermined in the spin Hamiltonian. These parameters are determined from experiments or by the theoretical calculation as described in the previous section. The method of the spin Hamiltonian demonstrates its great power when the splitting due to an external magnetic field is treated. This will be discussed in the next section.

So far we have confined ourselves to orbital singlets. However, it is generally true that the $(2S + 1)$-dimensional Hermitian matrix may be replaced by the matrix of a suitable linear combination of $S_q^{(k)}$ $(0 \leqslant k \leqslant 2S)$ with bases $\Theta(SM)$. Therefore, in principle, for any cubic term with a $(2S + 1)$-fold degeneracy including both spin and orbital degeneracies, one may construct a "spin Hamiltonian" with the *fictitious* spin S. For example, the ^1E term may be dealt with as though it has a fictitious spin of $\frac{1}{2}$ and the "spin Hamiltonian" in the form of (8.35) is applicable. The most serious drawback in this case comes from the fact that the transformation properties of ψ_1 and ψ_2 may not necessarily be the same as those of $\Theta(\frac{1}{2}\,\frac{1}{2})$ and $\Theta(\frac{1}{2}\,-\frac{1}{2})$. This makes us unable to simplify the linear combination of spin-tensor operators by using symmetry arguments. To show this, let us consider an example of the $t_2^2\,{}^1E$ term in the D_4-symmetry system. The group theory shows that this term splits into A_1 and B_1, while $\Theta(\frac{1}{2}\,\frac{1}{2})$ and $\Theta(\frac{1}{2}\,-\frac{1}{2})$ transform as the bases of E_1 of the double D_4-group. Therefore, the "spin Hamiltonian" cannot be expected to satisfy all the symmetry requirements which are satisfied in the real Hamiltonian. Actually, if one calculates the diagonal matrix element of the D_4-symmetry potential, $V_u(E_g)$, as

$$\langle t_2^2\,{}^1E\gamma \mid V_u(E_g) \mid t_2^2\,{}^1E\gamma' \rangle = -(2/3)^{1/2}\,\delta_{\gamma\gamma'}\beta_\gamma \langle t_2 \| v(E_g) \| t_2 \rangle, \tag{8.45}$$

where

$$\beta_u = -1/\sqrt{2}, \qquad \text{and} \qquad \beta_v = 1/\sqrt{2},$$

one sees that the splitting of the $t_2^2\,{}^1E$ term really occurs. This contradicts the result obtained from (8.36). This means that we can not impose the time-reversal invariance on the "spin Hamiltonian" in (8.35). This example shows that great caution is necessary when the terms with orbital degeneracy are treated with the "spin Hamiltonian." However, in the treatment of a Kramers doublet, we may always assume the time-reversal invariance of the "spin Hamiltonian," since ψ_1 and ψ_2 as well as $\Theta(\frac{1}{2}\,\frac{1}{2})$ and $\Theta(\frac{1}{2}\,-\frac{1}{2})$ are the time-reversal pair states.

Problem 8.5. By using the method of spin Hamiltonian show that six-fold degeneracy of the $^6A_1$ term is partially lifted in a cubic system. ◇

Problem 8.6. Treat the ^2E term in the D_3-symmetry system as if it has a fictitious spin of 3/2. ◇

8.3.2 General Effective Hamiltonian

In the previous subsection, we showed that the method of spin Hamiltonian is simple and powerful in predicting qualitative features of the splitting of orbital singlets but not necessarily so for the terms with orbital degeneracies. This defect may be eliminated if one chooses in place of spin functions bases which transform in the same way as the n functions ψ_i, associated with the term of n-fold degeneracy including both orbital and spin. With these functions ψ_i the nondiagonal matrix elements of the total Hamiltonian connecting the term of interest with the others are assumed to be zero. This choice of base functions makes one unable to construct simple effective Hamiltonian by using spin operators only. However, we will show in this subsection that a relatively simple effective Hamiltonian may be constructed by using both spin operators and irreducible tensor operators $X(\bar{\Gamma})$, introduced in Section 6.2.

Let us start with the simple case of obtaining the effective Hamiltonian for spin singlets having orbital degeneracies, $^1\Gamma$ ($\Gamma = E, T_1, T_2$). Note that the (Γ)-dimensional secular matrix with bases ψ_i's involves $(\Gamma)^2$ parameters. This matrix may be replaced by the matrix of a linear combination of irreducible tensor operators $X_{\bar{\gamma}}(\bar{\Gamma})$ $(\Gamma \times \Gamma = \Sigma\bar{\Gamma})$ with bases $\tilde{\varphi}(\Gamma\gamma)$. Here $\tilde{\varphi}(\Gamma\gamma)$ is any given function which transforms like the γ base of the irreducible representation Γ of the O-group. The reason why this replacement is possible is clear if one considers the fact that the matrices of $X_{\bar{\gamma}}(\bar{\Gamma})$ with different $\bar{\Gamma}\bar{\gamma}$ are independent of each other and the number of operators $X_{\bar{\gamma}}(\bar{\Gamma})$ with $\bar{\Gamma}$ appearing in $\Gamma \times \Gamma$ is $(\Gamma)^2$. Furthermore, since $\tilde{\varphi}(\Gamma\gamma)$'s transform in the same way as ψ_i's do, the linear combination of $X_{\bar{\gamma}}(\bar{\Gamma})$ should satisfy the symmetry requirements which are satisfied by the real Hamiltonian.

For example, the primitive form of the effective Hamiltonian for the ^1E term is given by

$$\mathscr{H}_{\text{eff}} = aV(A_{1g}) + bT(A_{2g}) + cV_u(E_g) + dV_v(E_g), \tag{8.46}$$

where a, b, c, and d are real unknown parameters. In (8.46) $X(A_{1g})$, $X_u(E_g)$, and $X_v(E_g)$ are given by real operators $V_{\bar{\gamma}}(\bar{\Gamma})$ and $X(A_{2g})$ by purely imaginary operator $T(A_{2g})$, as we are dealing with diagonal matrices [see (6.93) and (6.95)]. Requiring $K\mathscr{H}_{\text{eff}}K^{-1} = \mathscr{H}_{\text{eff}}$ for (8.46), one knows that all the irreducible tensor operators should be real so that parameter b should be zero. Furthermore, since $\mathscr{H}_{\text{eff}}$ should be invariant to any rotation of the D_3-group, all the terms except $V(A_{1g})$ should be zero. Thus, the final form of the effective Hamiltonian for ^1E in the D_3-symmetry potential is obtained as

$$\mathscr{H}_{\text{eff}} = aV(A_{1g}), \tag{8.47}$$

which shows that no splitting is expected in this case.

For the ^1E term in the D_2-symmetry whose twofold symmetry axes are x, y, and z, a similar consideration leads us to

$$\mathscr{H}_{\text{eff}} = cV_u(E_g) + dV_v(E_g), \tag{8.48}$$

where constant term $aV(A_{1g})$ is omitted for simplicity. For this effective Hamiltonian, bases $\tilde{\varphi}(E\gamma)$ may be chosen so that

$$\langle\tilde{\varphi}(E) \| V(E_g) \| \tilde{\varphi}(E)\rangle = 2, \tag{8.49}$$

which gives

$$V_u(E_g) = \begin{bmatrix} -1 & 0 \\ 0 & 1 \end{bmatrix}, \quad (\text{columns } u, v)$$
$$V_v(E_g) = \begin{bmatrix} 0 & 1 \\ 1 & 0 \end{bmatrix}. \tag{8.50}$$

Then, the use of (8.48) gives the splitting,

$$\epsilon_{1,2} = \pm(c^2 + d^2)^{1/2}. \tag{8.51}$$

Now, let us discuss a general method of constructing effective Hamiltonian for the terms with both orbital and spin degeneracies, $^{2S+1}\Gamma$. We look for an effective Hamiltonian whose matrix with given bases $\tilde{\Psi}(S\Gamma M\gamma)$ is identical to the $(2S + 1)(\Gamma)$-dimensional secular matrix with bases ψ_i. Here, $\tilde{\Psi}(S\Gamma M\gamma)$ transforms in the same way as unperturbed wavefunction $\Psi(\alpha S\Gamma M\gamma)$ as well as ψ_i does. Considering the arguments previously given for the spin Hamiltonian for spin singlets, we immediately notice that the effective Hamiltonian we are looking for is given by a suitable linear combination of the products $\bar{S}_q^{(k)}X_\gamma(\bar{\Gamma})$, where $0 \leqslant k \leqslant 2S$ and $\Gamma \times \Gamma = \Sigma\bar{\Gamma}$. The number of these products is $(2S + 1)^2(\Gamma)^2$ which is equal to that of parameters in the $(2S + 1)(\Gamma)$-dimensional secular matrix. It is evident that the matrices of these products with bases $\tilde{\Psi}(S\Gamma M\gamma)$ are independent of each other. The final form of the effective Hamiltonian is obtained by making the linear combination invariant to any symmetry operation which brings the system into itself.

For example, the effective Hamiltonian for the ^2E term in the D_3-symmetry is obtained as follows: We first make a linear combination of the products of an element in $(\mathbf{1}, S_x, S_y, S_z)$ and an element in $(V(A_{1g}), T(A_{2g}), V_u(E_g), V_v(E_g))$ as

$$\begin{aligned}\mathscr{H}_{\text{eff}} = {} & V(A_{1g})(a_1\mathbf{1} + a_2S_x + a_3S_y + a_4S_z) \\ & + T(A_{2g})(b_1\mathbf{1} + b_2S_x + b_3S_y + b_4S_z) \\ & + V_u(E_g)(c_1\mathbf{1} + c_2S_x + c_3S_y + c_4S_z) \\ & + V_v(E_g)(d_1\mathbf{1} + d_2S_x + d_3S_y + d_4S_z).\end{aligned} \tag{8.52}$$

Then, imposing time-reversal invariance on (8.52), one may simplify (8.52) as

$$\mathscr{H}_{\text{eff}} = a_1 V(A_{1g}) + T(A_{2g})(b_2 S_x + b_3 S_y + b_4 S_z) + c_1 V_u(E_g) + d_1 V_v(E_g). \quad (8.53)$$

Furthermore, imposing the D_3-rotation invariance on (8.53), one finally obtains

$$\mathscr{H}_{\text{eff}} = \Delta T(A_{2g}) S_Z\,, \quad (8.54)$$

where $S_Z = (S_x + S_y + S_z)/\sqrt{3}$,‡ and Δ is an undetermined parameter. In (8.54) the constant term $a_1 V(A_{1g})$ is omitted. It should be remarked that $T(A_{2g})$ alone is not invariant to rotations in $3\hat{C}_2$ as shown in Table 6.1, but $T(A_{2g})S_Z$ is the trigonal invariant. And $\tilde{\Psi}(^2EM_sM)$ may be chosen so that

$$\langle \tilde{\Psi}(^2E) \,\|\, T(A_{2g}) \,\|\, \tilde{\Psi}(^2E) \rangle = -i\sqrt{2}, \quad (8.55)$$

which gives the matrix of $T(A_{2g})$ as follows:

$$\begin{array}{rcc} M = & u_+ & u_- \end{array}$$

$$T(A_{2g}) = \begin{bmatrix} 1 & 0 \\ 0 & -1 \end{bmatrix}. \quad (8.56)$$

Then, the secular matrix of $\mathscr{H}_{\text{eff}}$ in (8.54) with bases $\tilde{\Psi}(^2EM_sM)$ is calculated as

$$\begin{array}{cccc} \multicolumn{2}{c}{u_+} & \multicolumn{2}{c}{u_-} \\ \tfrac{1}{2} & -\tfrac{1}{2} & \tfrac{1}{2} & -\tfrac{1}{2} \end{array}$$

$$\left[\begin{array}{cc:cc} 1 & 0 & \multicolumn{2}{c}{\multirow{2}{*}{0}} \\ 0 & -1 & & \\ \hdashline \multicolumn{2}{c:}{\multirow{2}{*}{0}} & -1 & 0 \\ & & 0 & 1 \end{array}\right] \times \tfrac{1}{2}\Delta, \quad (8.57)$$

which gives the splitting of the ^2E term as

$$\epsilon(\pm\tfrac{1}{2}u_\pm) - \epsilon(\pm\tfrac{1}{2}u_\mp) = \Delta. \quad (8.58)$$

Although $\tilde{\Psi}(^2EM_sM)$ are not wavefunctions $\Psi(\alpha\,^2EM_sM)$ of the $\alpha\,^2E$ term, the transformation properties of $\tilde{\Psi}(^2EM_sM)$ and $\Psi(\alpha\,^2EM_sM)$ are the same. Therefore, we may conclude from (8.58) that the $\alpha\,^2E$ term splits into two components, $\alpha\,^2E \pm\tfrac{1}{2}u_\pm$ and $\alpha\,^2E \pm\tfrac{1}{2}u_\mp$.

‡ Here Z is the trigonal coordinated axis introduced in Fig. 6.2.

In the method of effective Hamiltonian several parameters are left undetermined. They should be determined by experiments or by other kinds of theory. To calculate the values of these parameters one may use the perturbation theory as described in Section 8.2. For example, $\Delta(t_2^3\ {}^2E)$ was already calculated in (8.27).

Problem 8.7. Calculate the splitting of the $\alpha\,{}^2T_2$ term in the D_3-symmetry by the use of the effective Hamiltonian method. ◇

8.4 Zeeman Effects

8.4.1 Treatment by the Effective Hamiltonian Method

With the aid of the relativistic theory, it is known that the total magnetic moment $\mathbf{M}$ of electrons in a many electron system is given by

$$\mathbf{M} = -\mu_B \sum_i (\mathbf{l}_i + 2\mathbf{s}_i) = -\mu_B(\mathbf{L} + 2\mathbf{S}), \tag{8.59}$$

where

$$\mu_B = e\hbar/2mc, \tag{8.60}$$

is called Bohr magneton and e is positive for an electron. When an external magnetic field $\mathbf{H}$ is applied, this magnetic moment interacts with the magnetic field giving the interaction energy as

$$\mathscr{H}_Z = \mu_B \mathbf{H} \cdot (\mathbf{L} + 2\mathbf{S}), \tag{8.61}$$

which is called a *Zeeman term.* The purpose of this section is to discuss the additional term-splitting induced by the Zeeman term, which is called *Zeeman splitting.*

To study the effects of the Zeeman term by the effective Hamiltonian method, it should first be noticed that (8.61) changes its sign under time-reversal K acting on the electron system, but it is invariant to the time-reversal acting on both the electron system and the source of the magnetic field: The magnetic field is induced by a current which changes its direction by time-reversal. This time reversal will be denoted by $\bar{K}$. It follows that the total Hamiltonian including the Zeeman term is invariant to operation $\bar{K}$.

A similar generalization of symmetry operations may also be done for rotations. We first note that the magnetic field transforms like an axial vector if a rotation is applied to the source of the magnetic field. Then, it follows that the Zeeman term (8.61) is invariant to any simultaneous rotation of the electron system and the magnetic field source by the same angle, as **L** and **S** also transform like axial vectors by rotations, and (8.61) is the scaler product of two axial vectors. Such a simultaneous rotation will be denoted as $\bar{R}^{os}$, where R^{os} are elements of the group to which the rotational symmetry of the system belongs. It is clear that the total Hamiltonian including the Zeeman term is invariant to rotations $\bar{R}^{os}$.

Now we are ready to construct the effective Hamiltonian for the ${}^{2S+1}\Gamma$ term in a magnetic field. Since the $(2S+1)(\Gamma)$-dimensional secular matrix is still hermitian when the Zeeman term is included, it has to be expressed by the matrix of a suitable linear combination of the products, $S_q^{(k)}X_{\bar{\gamma}}(\bar{\Gamma})$ $(0 \leqslant k \leqslant 2S;\ \Gamma \times \Gamma = \Sigma\bar{\Gamma})$, with bases $\tilde{\Psi}(S\Gamma M\gamma)$. However in the presence of a magnetic field, the coefficients of the linear combination should be functions of the field. If these coefficients are expanded in powers of the magnetic field, the effective Hamiltonian will be given in general as

$$\begin{aligned} \mathscr{H}_{\text{eff}} &= \sum_{\substack{k,\bar{\Gamma} \\ q,\bar{\gamma}}} \alpha_{q\bar{\gamma}}^{k\bar{\Gamma}}(\mathbf{H})\, S_q^{(k)} X_{\bar{\gamma}}(\bar{\Gamma}), \\ \alpha_{q\bar{\gamma}}^{k\bar{\Gamma}}(\mathbf{H}) &= \sum_{n=0} \sum_{\substack{\alpha,\beta,\gamma \\ \alpha+\beta+\gamma=n}} A_{\alpha\beta\gamma}^{n}(k\bar{\Gamma}, q\bar{\gamma})\, H_x^{\alpha} H_y^{\beta} H_z^{\gamma}. \end{aligned} \tag{8.62}$$

Since the higher order terms in the expansion in powers of **H** are usually small, we retain only the terms with $n = 0$ and 1 in the following argrument.

To see how Zeeman splitting is calculated by the effective Hamiltonian method, let us consider an example of the ${}^2E$ term in the D_3-symmetry. In this case the final form of the field-independent terms in (8.62) are already obtained in (8.54). Imposing the K invariance on the field-dependent terms, one sees that

$$\begin{aligned} A_{\alpha\beta\gamma}^{1}(0\, A_{1g}\,,\, 0\, e_{1g}) &= A_{\alpha\beta\gamma}^{1}(1\, A_{2g}\,,\, q\, e_{2g}) \\ &= A_{\alpha\beta\gamma}^{1}(0\, E_g\,,\, 0\, \gamma) = 0 \end{aligned} \tag{8.63}$$

for all possible combinations of $\alpha\beta\gamma$. At this stage, the form of the effective Hamiltonian is

$$\begin{aligned} \mathscr{H}_{\text{eff}} &= \Delta T(A_{2g})(S_x + S_y + S_z)/\sqrt{3} + [a_2(\mathbf{H})S_x + a_3(\mathbf{H})S_y + a_4(\mathbf{H})S_z] \\ &\quad + b_1(\mathbf{H})\, T(A_{2g}) + V_u(E_g)[c_2(\mathbf{H})S_x + c_3(\mathbf{H})S_y + c_4(\mathbf{H})S_z] \\ &\quad + V_v(E_g)[d_2(\mathbf{H})S_x + d_3(\mathbf{H})S_y + d_4(\mathbf{H})S_z], \end{aligned} \tag{8.64}$$

in which $a_i(\mathbf{H})$, $b_i(\mathbf{H})$, $c_i(\mathbf{H})$, and $d_i(\mathbf{H})$ are linear functions of **H** (without terms independent of $\mathbf{H}$), and $V(A_{1g})$ is replaced by unity as it is equivalent to a unit matrix.

Now let us further impose the rotation invariance for $\bar{R}^{\text{os}}$ in the D_3-group on the field-dependent terms in (8.64). For this purpose it is convenient to introduce $H_{\pm 1}$ and H_0 defined by

$$\begin{aligned} H_{\pm 1} &= \mp \frac{1}{\sqrt{2}}(H_X \pm iH_Y), \\ H_0 &= H_Z \end{aligned} \tag{8.65}$$

and $S_{\pm 1}$ and S_0 defined by

$$\begin{aligned} S_{\pm 1} &= \mp \frac{1}{\sqrt{2}}(S_X \pm iS_Y), \\ S_0 &= S_Z , \end{aligned} \tag{8.66}$$

in which X, Y, and Z are the coordinate axes introduced in Fig. 6.2 for the trigonal system. Clearly H_M and S_M $(M = \pm 1, 0)$ transform like the M trigonal base of irreducible representation T_1 under the rotation $\bar{R}^{\text{os}}$ in the cubic group. Then, by using a method similar to that used for deriving low-symmetry ligand field potentials, the trigonal invariants given by linear combinations of the products of H_M and $S_{M'}$ are obtained as follows:

$$-H_{+1}S_{-1} - H_{-1}S_{+1} + H_0S_0 = H_XS_X + H_YS_Y + H_ZS_Z , \tag{8.67}$$

$$H_{+1}S_{-1} + H_{-1}S_{+1} + 2H_0S_0 = -(H_XS_X + H_YS_Y) + 2H_ZS_Z . \tag{8.68}$$

Equation (8.67) transforms like the base of A_1 and (8.68) like the x_0 base of T_2, which are both trigonal invariants. These invariants have to be obtained from the second term in (8.64) by assuming appropriate relations among nine parameters. By making suitable linear combinations of (8.67) and (8.68), the trigonal-invariant form of the second term in (8.64) may be given by

$$g_{\parallel}\mu_B H_Z S_Z + g_{\perp}\mu_B(H_XS_X + H_YS_Y), \tag{8.69}$$

where $g_{\parallel}$ and $g_{\perp}$ are undetermined parameters. The trigonal invariant form of the third term in (8.64) is easily obtained as

$$g_{\parallel}'\mu_B H_Z T(A_{2g}), \tag{8.70}$$

where $g_{\parallel}'$ is an undetermined parameter. It would not be difficult to see that the trigonal-invariant forms of the fourth plus the fifth terms in (8.64) are given as

$$\sum_{\substack{MM' \\ M''M'''}} H_M S_{M'} V_{M''} \langle T_1 M' E M'' \mid T_1 M''' \rangle \langle T_1 M T_1 M''' \mid A_1 e_1 \rangle$$
$$\propto H_{+1}(S_{+1}V_+ + S_0 V_-) + H_{-1}(S_0 V_+ - S_{-1}V_-) + H_0(S_{-1}V_+ + S_{+1}V_-), \tag{8.71}$$

$$\sum_{\substack{MM' \\ M''M'''}} H_M S_{M'} V_{M''} \langle T_1 M' E M'' \mid T_1 M''' \rangle \langle T_1 M T_1 M''' \mid T_2 0 \rangle$$
$$\propto H_{+1}(S_{+1}V_+ + S_0 V_-) + H_{-1}(S_0 V_+ - S_{-1}V_-) - 2H_0(S_{-1}V_+ + S_{+1}V_-), \tag{8.72}$$

$$\sum_{\substack{MM' \\ M''M'''}} H_M S_{M'} V_{M''} \langle T_1 M' E M'' \mid T_2 M''' \rangle \langle T_1 M T_2 M''' \mid T_2 0 \rangle$$
$$\propto H_{+1}(S_{+1}V_+ - S_0 V_-) - H_{-1}(S_0 V_+ + S_{-1}V_-), \tag{8.73}$$

where the $V_\pm$'s are $V_M(E_g)$ with $M = \pm 1$. Making suitable linear combinations of (8.71), (8.72), and (8.73), one may finally obtain the trigonal invariants of the fourth and fifth terms in (8.64) in the following form:

$$g''_\parallel \mu_B H_0[S_{-1}V_+(E_g) + S_{+1}V_-(E_g)]/\sqrt{2}$$
$$-g_\perp{}' \mu_B[H_{+1}S_{+1}V_+(E_g) - H_{-1}S_{-1}V_-(E_g)]$$
$$-\sqrt{2}\, g''_\perp \mu_B[H_{+1}S_0 V_-(E_g) + H_{-1}S_0 V_+(E_g)], \tag{8.74}$$

where $g''_\parallel$, $g_\perp{}'$ and $g''_\perp$ are unknown parameters. Summing up all the trigonal invariants, we obtain the final form of the effective Hamiltonian as follows:

$$\mathscr{H}_{\text{eff}} = \Delta S_Z T(A_{2g}) + g_\parallel \mu_B H_Z S_Z + g_\parallel{}' \mu_B H_Z T(A_{2g})$$
$$+ g''_\parallel \mu_B H_Z[S_{-1}V_+(E_g) + S_{+1}V_-(E_g)]/\sqrt{2} \tag{8.75}$$

for $\mathbf{H} \parallel Z$, and

$$\mathscr{H}_{\text{eff}} = \Delta S_Z T(A_{2g}) + g_\perp \mu_B H_X S_X$$
$$+ g_\perp{}' \mu_B H_X[S_{+1}V_+(E_g) + S_{-1}V_-(E_g)]/\sqrt{2}$$
$$+ g''_\perp \mu_B H_X S_Z[V_+(E_g) - V_-(E_g)] \tag{8.76}$$

for **H** $\parallel X$. Base functions $\tilde{\Psi}(^2EM_SM)$ are chosen so that the matrices of $V_{\pm}(E_g)$ are given by

$$\begin{array}{l} \qquad\qquad\quad\; u_+ \quad u_- \\ V_+(E_g) = \begin{bmatrix} 0 & 0 \\ -1 & 0 \end{bmatrix}, \\ V_-(E_g) = \begin{bmatrix} 0 & 1 \\ 0 & 0 \end{bmatrix}. \end{array} \tag{8.77}$$

The matrices of $S_{\pm 1}$ are easily obtained from (8.34). Then, the matrix of $\mathscr{H}_{\text{eff}}$ for **H** $\parallel Z$ is calculated as

$$\begin{array}{cccc} \tfrac{1}{2}u_+ & -\tfrac{1}{2}u_- & \tfrac{1}{2}u_- & -\tfrac{1}{2}u_+ \end{array}$$

$$\left[\begin{array}{cc:cc} \Delta + (g_\parallel + g_\parallel')\mu_B H & -g_\parallel''\mu_B H & & \\ -g_\parallel''\mu_B H & \Delta - (g_\parallel + g_\parallel')\mu_B H & \multicolumn{2}{c}{0} \\ \hdashline & & -\Delta + (g_\parallel - g_\parallel')\mu_B H & 0 \\ \multicolumn{2}{c:}{0} & 0 & -\Delta - (g_\parallel - g_\parallel')\mu_B H \end{array}\right] \times \tfrac{1}{2}, \tag{8.78}$$

and the matrix of $\mathscr{H}_{\text{eff}}$ for **H** $\parallel X$ is calculated as

$$\begin{array}{cccc} \tfrac{1}{2}u_+ & -\tfrac{1}{2}u_- & \tfrac{1}{2}u_- & -\tfrac{1}{2}u_+ \end{array}$$

$$\left[\begin{array}{cc:cc} \Delta & 0 & -g_\perp''\mu_B H & g_\perp\mu_B H \\ 0 & \Delta & g_\perp\mu_B H & g_\perp''\mu_B H \\ \hdashline -g_\perp''\mu_B H & g_\perp\mu_B H & -\Delta & g_\perp'\mu_B H \\ g_\perp\mu_B H & g_\perp''\mu_B H & g_\perp'\mu_B H & -\Delta \end{array}\right] \times \tfrac{1}{2}. \tag{8.79}$$

The eigenvalues of (8.78) are easily obtained. The result shows the Zeeman splitting with **H** $\parallel Z$ as illustrated in Fig. 8.2. Such a splitting has been observed in ruby,‡ and the linearity of the splitting to a magnetic

‡ In ruby the site symmetry of Cr^{3+} ions is C_3, so that we have additional terms in (8.75) and (8.76). However, the final result for C_3 does not differ from that for D_3.

field has been confirmed by using pulsed magnetic fields up to 200 kOe. The proportionality factors to $\mu_B H$, called *g-values*, have been found to be 1.48 ± 0.08 for the $\pm\frac{1}{2}u_{\pm}$ component and 2.44 ± 0.08 for the $\pm\frac{1}{2}u_{\mp}$ component. The observed relation, $1.48 + 2.44 \sim 4$, shows that $g''_{\|}$ in the effective Hamiltonian is small, as the sum of the g-values is $2g_{\|} \sim 4$ if $g''_{\|}$ is neglected: $g_{\|}$ is expected to be close to the spin-only value, $g_{\|} \sim 2$.

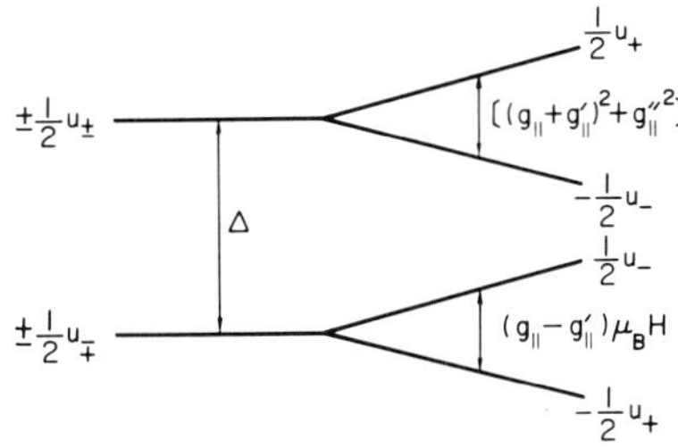

FIG. 8.2. Zeeman splitting of the ^2E term in the D_3-symmetry with $\mathbf{H} \parallel Z$.

By assuming that $\frac{1}{2}g_{\perp}\mu_B H$ and $\frac{1}{2}g''_{\perp}\mu_B H$ are small as compared with Δ, the eigenvalues of (8.79) are obtained as shown in Fig. 8.3. In this case no Zeeman splitting has been found in optical experiments for the two components of $t_2{}^3\,{}^2E$ in ruby. This shows that $g_{\perp}{}'$ is very small.

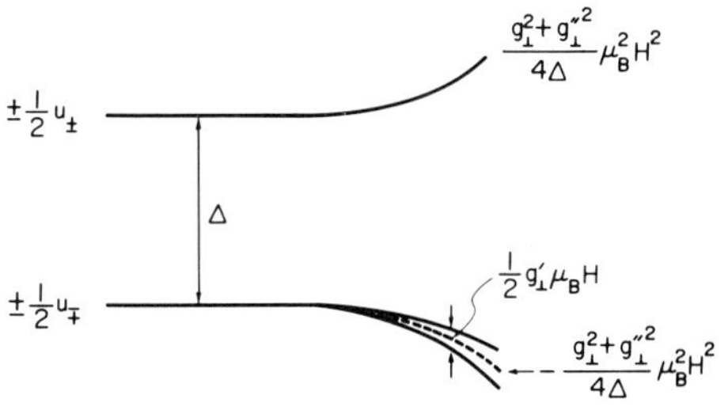

FIG. 8.3. Zeeman splitting of the ^2E term in the D_3-symmetry with $\mathbf{H} \parallel X$. Relations, $g_{\perp}\mu_B H \ll 2\Delta$ and $g''_{\perp}\mu_B H \ll 2\Delta$, are assumed.

However, a more accurate experiment by the use of an optical detection of electron spin resonance in the $\pm\frac{1}{2}u_{\pm}$ component‡ has confirmed the presence of nonvanishing $g_{\perp}{}'$, although this parameter has been found to be very small, $0 < g_{\perp}{}' < 0.06$.

‡ S. Geschwind, G. E. Devlin, R. L. Cohen, and S. R. Chinn, *Phys. Rev.* **137**, A1087 (1965).

Problem 8.8. Derive the spin Hamiltonian linear to $\mathbf{H}$ for the ${}^4A_2$ term in the D_3-symmetry in a magnetic field $\mathbf{H}$. ◇

Problem 8.9. Derive the effective Hamiltonian linear to $\mathbf{H}$ for the ${}^2T_2$ term in the D_3-symmetry in a magnetic field $\mathbf{H}$. ◇

8.4.2 Calculation of g-Values

Undetermined parameters in effective Hamiltonian may be calculated theoretically by the use of either the perturbation theory or more elaborate methods. Here, we present two examples of calculating g-values in effective Hamiltonian. The methods of calculating g-values in the other cases are inferable from these examples.

The first example is the calculation of g in the spin Hamiltonian

$$\mathscr{H}_S = g\mu_B \mathbf{H} \cdot \mathbf{S}. \tag{8.80}$$

for the $t_2{}^3\,{}^4A_2$ term in a cubic system. For simplicity, the magnetic field may be assumed to be along the z-direction without any loss of generality, so that the Zeeman term in the original Hamiltonian is given by

$$\mathscr{H}_z = \mu_B H(L_z + 2S_z). \tag{8.81}$$

The main contribution to g comes from the diagonal matrix elements of $-2\mu_B HS_z$ in the $t_2{}^3\,{}^4A_2$ term, and it gives the spin-only value, $g_0 = 2$ (more exactly $g_0 = 2.0023$). What we are interested in is the deviation of g from g_0, $g - g_0$, which is called a *g-shift*. The g-shift may be interpreted as coming from the contribution of the orbital angular momentum which is brought into the orbital singlet because of the spin-orbit interaction: Otherwise, the orbital angular momentum is completely quenched in the orbital singlet.

In the ${}^4A_2$ term, g in (8.80) is calculated from

$$g = \frac{1}{3\mu_B H}\left[\epsilon\left(M_s = \frac{3}{2}\right) - \epsilon\left(M_s = -\frac{3}{2}\right)\right]. \tag{8.82}$$

As seen from the physical argument given above, the g-shift comes from the second-order contribution to (8.82) involving $\mathscr{H}_{SO}$ and $\mu_B HL_z$ in the perturbation calculation. By using the coupling scheme in Fig. 8.1, the second-order contribution to (8.82) is found to be

$$\begin{aligned} \Delta g = \tfrac{1}{3}[&\langle {}^4A_2\,\tfrac{3}{2} \mid L_z \mid {}^4T_2\,\tfrac{3}{2}\zeta\rangle\langle {}^4T_2\,\tfrac{3}{2}\zeta \mid \mathscr{H}_{SO} \mid {}^4A_2\,\tfrac{3}{2}\rangle \\ &+ \langle {}^4A_2\,\tfrac{3}{2} \mid \mathscr{H}_{SO} \mid {}^4T_2\,\tfrac{3}{2}\zeta\rangle\langle {}^4T_2\,\tfrac{3}{2}\zeta \mid L_z \mid {}^4A_2\,\tfrac{3}{2}\rangle \\ &- \langle {}^4A_2 - \tfrac{3}{2} \mid L_z \mid {}^4T_2 - \tfrac{3}{2}\zeta\rangle\langle {}^4T_2 - \tfrac{3}{2}\zeta \mid \mathscr{H}_{SO} \mid {}^4A_2\,\tfrac{3}{2}\rangle \\ &- \langle {}^4A_2 - \tfrac{3}{2} \mid \mathscr{H}_{SO} \mid {}^4T_2 - \tfrac{3}{2}\zeta\rangle\langle {}^4T_2 - \tfrac{3}{2}\zeta \mid L_z \mid {}^4A_2 - \tfrac{3}{2}\rangle] \\ &\times [W(t_2{}^3\,{}^4A_2) - W(t_2{}^2e\,{}^4T_2)]^{-1}, \end{aligned} \tag{8.83}$$

in which t_2^2e for ${}^4T_2$ is suppressed in the matrix elements. Since the matrices of L_z and $\mathscr{H}_{SO}$ are hermitian and each term in (8.83) is real, (8.83) may be simplified as

$$\begin{aligned}
\Delta g &= -\frac{2}{3}\frac{\langle {}^4A_2 \mid L_z \mid {}^4T_2\rangle}{10Dq} \\
&\quad \times \left[\left\langle {}^4T_2\frac{3}{2}\zeta \middle| \mathscr{H}_{SO} \middle| {}^4A_2\frac{3}{2}\right\rangle - \left\langle {}^4T_2-\frac{3}{2}\zeta \middle| \mathscr{H}_{SO} \middle| {}^4A_2-\frac{3}{2}\right\rangle\right] \\
&= -\frac{2}{3}\frac{\langle t_2\zeta \mid l_z \mid ev\rangle}{10Dq}\left[\left\langle e\frac{1}{2}v \middle| \mathscr{H}_{SO} \middle| t_2\frac{1}{2}\zeta\right\rangle - \left\langle e-\frac{1}{2}v \middle| \mathscr{H}_{SO} \middle| t_2-\frac{1}{2}\zeta\right\rangle\right] \\
&= -\frac{4}{3}\frac{\langle t_2\zeta \mid l_z \mid ev\rangle\langle e\frac{1}{2}v \mid \mathscr{H}_{SO} \mid t_2\frac{1}{2}\zeta\rangle}{10Dq} \\
&= -\frac{8}{3}\frac{k'\zeta'}{10Dq},
\end{aligned} \tag{8.84}$$

where $10Dq$ is the cubic field splitting parameter equal to

$$W(t_2^2e\ {}^4T_2) - W(t_2^3\ {}^4A_2)$$

and k' is defined as

$$\langle t_2\zeta \mid l_z \mid ev\rangle = 2ik'. \tag{8.85}$$

Here k' is unity in the d-function approximation as seen from (7.5). In deriving (8.84), use was made of the relations

$$\langle {}^4T_2 \pm \tfrac{3}{2}\zeta \mid \mathscr{H}_{SO} \mid {}^4A_2 \pm \tfrac{3}{2}\rangle = \langle e \pm \tfrac{1}{2}v \mid \mathscr{H}_{SO} \mid t_2 \pm \tfrac{1}{2}\zeta\rangle, \tag{8.86}$$

$$\langle {}^4A_2 \mid L_z \mid {}^4T_2\zeta\rangle = \langle t_2\zeta \mid l_z \mid ev\rangle, \tag{8.87}$$

which are easily derived by using the explicit forms of the wavefunctions. The method of calculating the matrix elements of the orbital angular momentum in general is explained in detail in Appendix VIII. Since $10Dq \sim 10^4\ \text{cm}^{-1}$, $\zeta' \sim 10^2\ \text{cm}^{-1}$ and $k' \sim 1$ for the systems involving iron-group transition metal ions, (8.84) shows that the g-shift is of the order of 10^{-2}.

The next example is the calculation of $g_{\parallel}'$ in (8.75). Since the term $g_{\parallel}'\mu_B HT(A_{2g})$ involves no spin operator, the perturbation processes giving $g_{\parallel}'$ should involve spin operators an even number times. Furthermore, the orbital operator of type $T(A_{2g})$ has to be found in the reduction of the products of orbital operators appearing in the perturbation processes into irreducible representations. Of course, the perturbation processes should involve operators L_z or S_z once as the term of interest is linear to H_z. From these considerations and the coupling scheme in Fig. 8.1, it follows that none of the first- and second-order processes may contribute to a nonvanishing $g_{\parallel}'$;

$$\text{(I)}\ \begin{cases} t_2^3\,{}^2E\tfrac{1}{2}u_\pm \xrightarrow{L_z} t_2^3\,{}^2T_1\tfrac{1}{2}a_\pm \xrightarrow{V_0(T_{2g})} t_2^3\,{}^2T_2\tfrac{1}{2}x_\pm \xrightarrow{V_0(T_{2g})} t_2^3\,{}^2E\tfrac{1}{2}u_\pm, \\ \text{conjugate complex of the above}\quad [V_0(T_{2g})V_0(T_{2g})L_z], \end{cases} \tag{8.88a}$$

$$\text{(II)}\ t_2^3\,{}^2E\tfrac{1}{2}u_\pm \xrightarrow{V_0(T_{2g})} t_2^3\,{}^2T_2\tfrac{1}{2}x_\pm \xrightarrow{L_z} t_2^3\,{}^2T_2\tfrac{1}{2}x_\pm \xrightarrow{V_0(T_{2g})} t_2^3\,{}^2E\tfrac{1}{2}u_\pm, \tag{8.88b}$$

$$\text{(III)}\ \begin{cases} t_2^3\,{}^2E\tfrac{1}{2}u_\pm \xrightarrow{V_0(T_{2g})} t_2^3\,{}^2T_2\tfrac{1}{2}x_\pm \xrightarrow{S_z} t_2^3\,{}^2T_2\tfrac{1}{2}x_\pm \xrightarrow{\mathscr{H}_{\rm SO}} t_2^3\,{}^2E\tfrac{1}{2}u_\pm, \\ \text{conjugate complex of the above}\quad [\mathscr{H}_{\rm SO}S_zV_0(T_{2g})], \end{cases} \tag{8.88c}$$

$$\text{(IV)}\ \begin{cases} t_2^3\,{}^2E\tfrac{1}{2}u_\pm \xrightarrow{V_0(T_{2g})} t_2^3\,{}^2T_2\tfrac{1}{2}x_\pm \xrightarrow{\mathscr{H}_{\rm SO}} t_2^3\,{}^2E\tfrac{1}{2}u_\pm \xrightarrow{S_z} t_2^3\,{}^2E\tfrac{1}{2}u_\pm, \\ t_2^3\,{}^2E\tfrac{1}{2}u_\pm \xrightarrow{\mathscr{H}_{\rm SO}} t_2^3\,{}^2T_2\tfrac{1}{2}x_\pm \xrightarrow{V_0(T_{2g})} t_2^3\,{}^2E\tfrac{1}{2}u_\pm \xrightarrow{S_z} t_2^3\,{}^2E\tfrac{1}{2}u_\pm, \\ t_2^3\,{}^2E\tfrac{1}{2}u_\pm \xrightarrow{S_z} t_2^3\,{}^2E\tfrac{1}{2}u_\pm \xrightarrow{V_0(T_{2g})} t_2^3\,{}^2T_2\tfrac{1}{2}x_\pm \xrightarrow{\mathscr{H}_{\rm SO}} t_2^3\,{}^2E\tfrac{1}{2}u_\pm, \\ t_2^3\,{}^2E\tfrac{1}{2}u_\pm \xrightarrow{S_z} t_2^3\,{}^2E\tfrac{1}{2}u_\pm \xrightarrow{\mathscr{H}_{\rm SO}} t_2^3\,{}^2T_2\tfrac{1}{2}x_\pm \xrightarrow{V_0(T_{2g})} t_2^3\,{}^2E\tfrac{1}{2}u_\pm. \end{cases} \tag{8.88d}$$

The detailed examination of these processes shows that processes (III) and (IV) cancel out, and one obtains

$$g_{\parallel}' = -\frac{12K^2}{[W(t_2^3\,{}^2E) - W(t_2^3\,{}^2T_1)][W(t_2^3\,{}^2E) - W(t_2^3\,{}^2T_2)]} - \frac{6K^2}{[W(t_2^3\,{}^2E) - W(t_2^3\,{}^2T_2)]^2}, \tag{8.89}$$

where K is the trigonal splitting parameter defined in (8.22). In ruby $K \sim -350\ \text{cm}^{-1}$, $W(^2E) - W(^2T_1) \sim -700\ \text{cm}^{-1}$, and $W(^2E) - W(^2T_2) \sim -6500\ \text{cm}^{-1}$, so that (8.89) gives $g_{\parallel}' \sim -0.34$ which almost explains the observed g-values, $g(\pm\tfrac{1}{2}u_\pm) = 1.48 \pm 0.08$ and $g(\pm\tfrac{1}{2}u_\mp) = 2.44 \pm 0.08$, if $g_{\parallel}$ is assumed to be the spin-only value.

As seen in these examples, the form of effective Hamiltonian gives us some insight into the perturbation processes effective in calculating undetermined parameters in the effective Hamiltonian. This is also one of the merits in using the effective Hamiltonian method.

8.4.3 Electron Spin Resonance and Optical Zeeman Patterns

As discussed so far, the application of a magnetic field induces Zeeman splitting of terms, and transitions between the Zeeman levels within a term (or within a split component of a term) are observed in electron spin resonance experiments, and transitions between the Zeeman levels of two different terms (or of two different split components of a term in far-infrared spectroscopy) are observed as Zeeman patterns in optical experiments.

In electron-spin resonance experiments, the transitions are mainly of magnetic-dipole type and are allowed between the two Zeeman levels with quantum numbers M_s and M_s' which are related by $| M_s - M_s' | = 1$. Here M_s and M_s' are to be associated with the base functions $\Theta(SM_s)$ for the spin Hamiltonian as there is a one-to-one correspondence between these base functions and the real wavefunctions ψ_i. Therefore, it is straightforward to calculate transition intensities once the eigenvectors which diagonalize the matrix of spin Hamiltonian are obtained.

The calculations of transition intensities in optical Zeeman patterns are more complicated than those of transition intensities in electron-spin resonance experiments. Here we mention only one example of calculating the Zeeman pattern for the electric-dipole transitions between the $t_2^3\ {}^4A_2$ and $t_2^3\ {}^2E$ terms in the C_3-symmetry system, ruby. The caltulations of Zeeman patterns in many other cases may easily be inferred from this example. In this example with a magnetic field parallel to the trigonal axis, the calculations by using the effective Hamiltonian method show that the ${}^4A_2$ term splits into four Zeeman levels specified by $M_s = 3/2$, $1/2$, $-1/2$, $-3/2$, and the ${}^2E$ term also into four specified by $M_sM = \frac{1}{2}u_+$, $-\frac{1}{2}u_-$, $\frac{1}{2}u_-$, $-\frac{1}{2}u_+$.‡ The electric-dipole transition intensities between these Zeeman levels were already calculated in Problem 7.7. In Problem 7.7, one can show that

$$P_\alpha^{\ 0} = \delta_{\alpha 0} P_0^{\ 0},$$
$$P_\alpha^{\ \pm} = \delta_{\alpha \mp 1} P_{\mp 1}^{\pm}, \tag{8.90}$$

if one calculates transition moments $P_\alpha^{0,\pm}$ by using the method described in Section 6.2.3. Then, the Zeeman pattern in this case is given in terms of $\pi = \pi_0$, $\sigma_+ = \sigma_{+,\alpha=-1}$, and $\sigma_- = \sigma_{-,\alpha=1}$.

Besides performing the calculation as described in Problem 7.7, it is also important to examine the selection rule governing the transitions between the Zeeman levels. The selection rule may be found by considering the transformation properties of the initial and final states as well as those of the effective transition moment operator. In the present problem

‡ We assume that $g''_{\parallel} = 0$.

the selection rule may be found as follows: We first note that $\mathbf{P}_{\text{eff}}$ in Problem 7.7 can be reexpressed as

$$\begin{aligned}\mathbf{P}_{\text{eff}} = \sum_{\substack{\Gamma''M''\\ M_s''}} &\langle t_2^3\,{}^4A_2M_s \mid (R^{\text{os}})^{-1}R^{\text{os}}\tilde{\mathbf{P}}(R^{\text{os}})^{-1}R^{\text{os}} \mid t_2^2e\,{}^4\Gamma''M_s''M''\rangle \\ &\times \langle t_2^2e\,{}^4\Gamma''M_s''M'' \mid (R^{\text{os}})^{-1}R^{\text{os}}\mathscr{H}_{\text{so}}(R^{\text{os}})^{-1}R^{\text{os}} \mid t_2^3\,{}^2EM_s'M'\rangle \\ &\times [W({}^2E) - W({}^4\Gamma'')]^{-1} \\ = \sum_{\substack{\Gamma''M''\\ M_s''}} &\langle t_2^3\,{}^4A_2M_s \mid (R^{\text{os}})^{-1}R^{\text{os}}\mathbf{P}(R^{\text{os}})^{-1}V_{\text{odd}} \mid t_2^2e\,{}^4\Gamma''M_s''M''\rangle \\ &\times \langle t_2^2e\,{}^4\Gamma''M_s''M'' \mid \mathscr{H}_{\text{so}}R^{\text{os}} \mid t_2^3\,{}^2EM_s'M'\rangle \\ &\times [W({}^2E) - W({}^4\Gamma'')]^{-1},\end{aligned} \tag{8.91}$$

where R^{os} is a rotation in the double C_3-group. In deriving (8.91) we have used the relation

$$\begin{aligned}&\sum_{MM_s} R^{\text{os}} \mid S\Gamma M_sM\rangle\langle S\Gamma M_sM \mid (R^{\text{os}})^{-1} \\ &= \sum_{\substack{M_sM,M_s'M'\\ M_s''M''}} \mid S\Gamma M_s'M'\rangle\, U_{M_s'M',M_sM}U^*_{M_s''M'',M_sM}\langle S\Gamma M_s''M'' \mid \\ &= \sum_{M_s'M',M_s''M''} \mid S\Gamma M_s'M'\rangle\langle S\Gamma M_s''M'' \mid \sum_{M_sM} U_{M_s'M',M_sM}U^{-1}_{M_sM,M_s''M''} \\ &= \sum_{M_s'M'} \mid S\Gamma M_s'M'\rangle\langle S\Gamma M_s'M' \mid,\end{aligned} \tag{8.92}$$

where U is the unitary representation matrix for operation of R^{os} on the $S\Gamma$ term. Since all the elements in the double C_3-group are generated from a single element C_3, it is sufficient in our argument to consider only the case of $R^{\text{os}} = C_3$. Now, by using Table 6.2 and (7.36), one can show that

$$C_3\Psi({}^2EM_sM) = \exp\left[-\frac{2\pi i}{3}(M_s + M)\right]\Psi({}^2EM_sM), \tag{8.93a}$$

$$C_3\Psi({}^4A_2M_s) = \exp\left[-\frac{2\pi i}{3}M_s\right]\Psi({}^4A_2M_s), \tag{8.93b}$$

$$C_3P_MC_3^{-1} = \exp\left[-\frac{2\pi i}{3}M\right]P_M, \tag{8.93c}$$

where P_M are P_+, P_-, and P_0 defined in (6.64) for $M = +1, -1$, and 0, respectively. Then, Eqs. (8.93) enable us to reexpress (8.91) as

$$P_{\mathrm{eff},\bar{M}} = \exp\left\{-\frac{2\pi i}{3}[(M_s' + M') + \bar{M} - M_s]\right\} P_{\mathrm{eff},\bar{M}}, \qquad (8.94)$$

which shows that for $P_{\mathrm{eff},\bar{M}} \neq 0$

$$M_s - (M_s' + M') + 3n = \bar{M} \qquad (8.95)$$

(n = positive and negative integers including zero). The relation in (8.95) gives the selection rule we wanted to obtain. This selection rule, of course, is consistent with the result given in Problem 7.7, but cannot predict some of the zero-intensity transitions calculated there: For example, according to the selection rule, the π-transition ($M = 0$) is possible between ${}^4A_2\frac{3}{2}$ and ${}^2E - \frac{1}{2}u_-$ as $M_s - (M_s' + M') = 3$, but its intensity was found to be zero in Problem 7.7. In the trigonal system, $(M_s + M)$ of the $S\Gamma M_s M$ state are called *crystal quantum numbers*. It is possible to establish a more elaborate theory on this line for more complicated symmetry systems.[‡]

Next let us consider the case in which a magnetic field is applied perpendicular to the trigonal axis ($\mathbf{H} \parallel X$) and the magnetic field strength is in the range of $g\mu_B H \gg 2D$ and $g_\perp \mu_B H \ll 2\Delta$. Here $2D$ (0.38 cm^{-1} in ruby) is the initial splitting of the $t_2^3\,{}^4A_2$ term defined in (8.43), $g_\perp$ and Δ (29 cm^{-1} in ruby) are defined in (8.76), and $g_\perp'$ and $g_\perp''$ in (8.76) are assumed to be zero. This situation often occurs, as magnetic field strengths easily available are of the order of 10 to 100 kOe ($2\mu_B H \sim 1 - 10$ cm^{-1}), while in many cases the initial splittings of orbital-singlet ground states are $\sim 0.1 - 1$ cm^{-1}, and those of the excited states are larger than 10 cm^{-1}. In this case the ${}^4A_2$ term splits into four Zeeman levels specified by $M_s^{(X)} = 3/2, 1/2, -1/2, -3/2$, in which the quantization axis is the X-axis instead of Z. On the other hand, the Zeeman levels of the ${}^2E$ term may still be specified by $M_s M$ as in the case of $\mathbf{H} \parallel Z$. Then the effective transition moments are calculated by

$$\mathbf{P}_{\mathrm{eff}}(t_2^3\,{}^4A_2 M_s^{(X)} - t_2^3\,{}^2E M_s' M')$$
$$= \sum_{M_s^{(Z)}} \langle S = \tfrac{3}{2}\, M_s^{(X)} \mid S = \tfrac{3}{2}\, M_s^{(Z)}\rangle\, \mathbf{P}_{\mathrm{eff}}(t_2^3\,{}^4A_2 M_s^{(Z)} - t_2^3\,{}^2E M_s' M'), \qquad (8.96)$$

[‡] T. Murao, F. H. Spedding, and R. H. Good, Jr., *J. Chem. Phys.* **42**, 993 (1965); T. Murao, W. J. Haas, R. W. G. Syme, F. H. Spedding, and R. H. Good, Jr., *J. Chem. Phys.* **47**, 1572 (1967).

where $\langle S = \frac{3}{2} M_s^{(X)} \mid S = \frac{3}{2} M_s^{(Z)} \rangle$ is the matrix elements associated with the transformation of the quantization axis from Z to X. It is known‡ that, when the quantization axis is rotated by angle (θ, φ), the wave-functions $\Psi(jm')$ associated with the jm' state (j: angular momentum) quantized along the rotated axis are given in terms of $\Psi(jm)$ associated with the jm state quantized along the unrotated axis as follows:

$$\Psi(jm') = \sum_m \Psi(jm)\, U_{mm'}^{(j)}(\theta\varphi), \tag{8.97}$$

where $U_{mm'}^{(j)}(\theta\varphi) = e^{-im\varphi} U_{mm'}^{(j)}(\theta 0)$ and

$$\begin{aligned} U_{mm'}^{(j)}(\theta 0) &= (-1)^{j-m'} \left[\frac{(j+m)!}{(j-m)!} \frac{1}{(j+m')!\,(j-m')!}\right]^{1/2} \\ &\times \left[\left(\sin\frac{\theta}{2}\right)^{-m+m'} \left(\cos\frac{\theta}{2}\right)^{-m-m'}\right] \\ &\times \left[\left(\frac{d}{dt}\right)^{j-m} t^{j+m'}(1-t)^{j-m'}\right]_{t=(\cos\theta/2)^2}. \end{aligned} \tag{8.98}$$

For example,

$$U_{mm'}^{(1/2)}(\theta 0) = \begin{matrix} \frac{1}{2} & -\frac{1}{2} \end{matrix} \atop \begin{bmatrix} a & -b \\ b & a \end{bmatrix}, \tag{8.99}$$

$$U_{mm'}^{(3/2)}(\theta 0) = \begin{bmatrix} a^3 & -\sqrt{3}\,a^2 b & \sqrt{3}\,ab^2 & -b^3 \\ \sqrt{3}\,a^2 b & a(1-3b^2) & b(1-3a^2) & \sqrt{3}\,ab^2 \\ \sqrt{3}\,ab^2 & -b(1-3a^2) & a(1-3b^2) & -\sqrt{3}\,a^2 b \\ b^3 & \sqrt{3}\,ab^2 & \sqrt{3}\,a^2 b & a^3 \end{bmatrix}, \tag{8.100}$$

(columns labelled $\frac{3}{2}$, $\frac{1}{2}$, $-\frac{1}{2}$, $-\frac{3}{2}$)

where $a = \cos(\theta/2)$ and $b = \sin(\theta/2)$. From (8.96) and (8.100) with $\theta = \pi/2$, one may easily calculate the dipole strengths in terms of π, σ_+, and σ_-.

Problem 8.10. Calculate the dipole strengths of the

$$t_2^{3}\ {}^4A_2 M_s^{(X)} \to t_2^{3}\ {}^2E M_s' M'$$

transitions. ◇

‡ J. Schwinger, "Quantum Theory of Angular Momentum" (L. C. B. Biedenharn and H. Van Dam, eds.), p. 229. Academic Press, New York, 1965.

8.5 Linear Stark Effects

In the theory of atomic spectra it is well known that an external electric field $\mathbf{E}$ interacts with electrons and induces shifts and splittings of energy levels. This effect is called the *Stark effect.* The interaction Hamiltonian in this case is given by

$$\mathscr{H}_S = e\mathbf{E} \cdot \sum_i \mathbf{r}_i . \tag{8.101}$$

Since $\mathscr{H}_S$ changes its sign by inversion at the origin, it has no matrix element between the states with the same parity. Therefore, if the system has inversion symmetry, we can expect no Stark effect linear to $\mathbf{E}$. In this subsection, using the effective Hamiltonian method, we treat the Stark effect linear to $\mathbf{E}$ in the system without inversion symmetry. For this purpose we cite a particular example of the ^2E term in the C_3-symmetry system. It will also be shown that no linear Stark effect is expected for this term if the symmetry of the system is D_3 where no inversion symmetry exists. In the course of the argument we should keep in mind that Kramers degeneracies are not lifted by an electric field as $\mathscr{H}_S$ is invariant to time reversal.

If one reexamines the arguments given for obtaining (8.62), it is evident that the primitive form of the effective Hamiltonian in the presence of an electric field is also given by a form similar to (8.62) in which $\mathbf{H}$ is replaced by $\mathbf{E}$. Therefore, in our specific problem, after imposing the time-reversal invariance condition, the effective Hamiltonian is given by

$$\begin{aligned}\mathscr{H}_{\text{eff}} &= \Delta S_Z T(A_{2g}) + a(\mathbf{E})\, V(A_{1g}) + b(\mathbf{E})\, V_u(E_g) + c(\mathbf{E})\, V_v\, (E_g) \\ &\quad + T(A_{2g})[d_1(\mathbf{E})S_x + d_2(\mathbf{E})S_y + d_3(\mathbf{E})S_z],\end{aligned} \tag{8.102}$$

where $a(\mathbf{E})$, $b(\mathbf{E})$, $c(\mathbf{E})$, and $d_i(\mathbf{E})$ are linear functions of $\mathbf{E}$ without terms independent of $\mathbf{E}$. Now we notice that, if $\bar{\bar{R}}^{\text{os}}$ is defined as a rotation acting on both the electron system and the source of an electric field, the Stark term in (8.101) is invariant to the operation of any rotation $\bar{\bar{R}}^{\text{os}}$ as the electric field transforms like a polar vector and the Stark term is the scalar product of two polar vectors. Therefore, the total Hamiltonian is invariant to the operation of $\bar{\bar{R}}^{\text{os}}$ in which R^{os} is an element of the group to which symmetry of the system belongs. In the present problem R^{os} is an element of the C_3-group.

The trigonal (C_3)-invariants should transform as the bases of A_1, A_2, $T_1M = 0$, and $T_2M = 0$ of the O-group, and they are obtained from the terms of (8.102) by assuming suitable relations among the parameters as done in (8.71)–(8.73). To obtain the trigonal-invariants, we define

$$E_{\pm 1} = \mp \frac{1}{\sqrt{2}}(E_X \pm iE_Y), \qquad E_0 = E_Z . \tag{8.103}$$

The trigonal-invariant of the second term in (8.102) is

$$E_0 V(A_{1g}), \tag{8.104}$$

which transforms as the $T_1 M = 0$ base. The trigonal-invariant forms of the third plus fourth terms are

$$i(E_{-1}V_+ - E_{+1}V_-) \qquad (T_2\, M{=}0), \tag{8.105}$$

and

$$E_{-1}V_+ + E_{+1}V_- \qquad (T_1\, M{=}0). \tag{8.106}$$

The trigonal-invariant forms of the fifth term are

$$(E_{+1}S_{-1} + E_{-1}S_{+1} - E_0S_0)\, T(A_{2g}) \qquad (A_2), \tag{8.107}$$

$$i(E_{+1}S_{-1} - E_{-1}S_{+1})\, T(A_{2g}) \qquad (T_2\, M{=}0), \tag{8.108}$$

and

$$(E_{+1}S_{-1} + E_{-1}S_{+1} + 2E_0S_0)\, T(A_{2g}) \qquad (T_1\, M{=}0). \tag{8.109}$$

Then, the final form of the effective Hamiltonian is obtained as

$$\begin{aligned}\mathscr{H}_{\text{eff}} = {} & \Delta S_Z T(A_{2g}) + \alpha E_0 V(A_{1g})/2 + i\beta(E_{-1}V_+ - E_{+1}V_-)/\sqrt{2} \\ & + \beta'(E_{-1}V_+ + E_{+1}V_-)/\sqrt{2} - i\gamma(E_{+1}S_{-1} - E_{-1}S_{+1})\, T(A_{2g}) \\ & + \gamma' E_0 S_0 T(A_{2g}) + \gamma''(E_{+1}S_{-1} + E_{-1}S_{+1})\, T(A_{2g}).\end{aligned} \tag{8.110}$$

It should be remarked that factor i in the terms proportional to β and γ is important, as, due to the relations

$$\begin{aligned} KS_{\pm 1}K^{-1} &= S_{\mp 1} , \\ KE_{\pm 1}K^{-1} &= -E_{\mp 1} , \end{aligned} \tag{8.111}$$

these terms are time-reversal invariants only if factor i is included. The first relation in (8.111) differs from (8.11) because of the different definitions of $S_{\pm 1}$ and $S_{\pm}$. The base functions $\tilde{\Psi}(S\Gamma M_s M)$ of the effective Hamiltonian will be chosen so that the matrices of $V_{\pm}(E_g)$ are given by (8.77).

The secular matrices of (8.110) for $\mathbf{E} \parallel Z$ and $\mathbf{E} \parallel X$ with bases $\tilde{\Psi}(S\Gamma M_s M)$ are calculated as

$$\begin{array}{cccc} \tfrac{1}{2}u_+ & -\tfrac{1}{2}u_- & \tfrac{1}{2}u_- & -\tfrac{1}{2}u_+ \end{array}$$

$$\left[\begin{array}{cc:cc} \Delta+(\alpha+\gamma')E_Z & 0 & & \\ 0 & \Delta+(\alpha+\gamma')E_Z & \multicolumn{2}{c}{0} \\ \hdashline & & -\Delta+(\alpha-\gamma')E_Z & 0 \\ \multicolumn{2}{c:}{0} & 0 & -\Delta+(\alpha-\gamma')E_Z \end{array}\right] \times \tfrac{1}{2} \tag{8.112}$$

and

$$\left[\begin{array}{cc:cc} \Delta & 0 & (i\beta-\beta')E_X & -(i\gamma+\gamma'')E_X \\ 0 & \Delta & -(i\gamma-\gamma'')E_X & -(i\beta+\beta')E_X \\ \hdashline -(i\beta+\beta')E_X & (i\gamma+\gamma'')E_X & -\Delta & 0 \\ (i\gamma-\gamma'')E_X & (i\beta-\beta')E_X & 0 & -\Delta \end{array}\right] \times \tfrac{1}{2}. \tag{8.113}$$

These secular matrices show that a linear Stark shift is expected only when an electric field is parallel to the trigonal axis. If the symmetry of the system is D_3, the rotation invariants should transform only as the bases of A_1 and $T_2M = 0$, so that $\alpha = \beta' = \gamma' = \gamma'' = 0$ in (8.110). In this case no linear Stark shift can be expected even if the system lacks inversion symmetry.

Parameter α in (8.110) is given by the second-order perturbation process involving odd-parity potential $V_0(T_{1u})$ and $\mathscr{H}_S$, while γ' is given

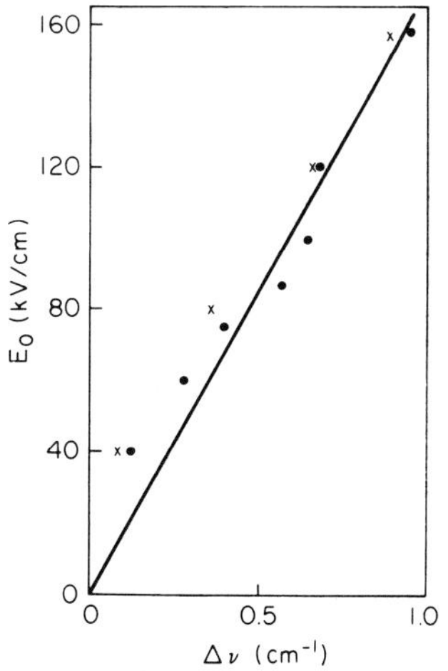

FIG. 8.4. Pseudo-Stark splitting $\Delta\nu$ of ruby versus applied electric field E_0 parallel to the trigonal axis. Data are obtained from the R_1(●) and R_2(×) lines.

by the third-order process involving an odd-parity optential, $\mathscr{H}_{so}$ and $\mathscr{H}_S$. Therefore α may be expected to be larger than γ'. In ruby there are two different sites of Cr^{3+} where the odd-parity potentials are different only in sign, so that we have to assume α with different sign for Cr^{3+} ions in two sites. Then, the linear Stark shifts in opposite directions are superposed and observed as if it were a linear Stark splitting. This is called ***pseudo-Stark splitting***, which has been observed[‡] as shown in Fig. 8.4. Since α comes from odd-parity potential $V_0(T_{1u})$, the observation of the pseudo-Stark splitting in ruby emphasizes the importance of $V_0(T_{1u})$. In the experimental data the presence of γ' has not been confirmed as seen in Fig. 8.4.

‡ W. Kaiser, S. Sugano, and D. L. Wood, *Phys. Rev. Letters* **6**, 605 (1961).

Chapter IX INTERACTION BETWEEN ELECTRON AND NUCLEAR VIBRATION

So far we have assumed that the nuclear framework of the systems of interest is at rest. However, in real problems the systems undergo the vibration of the nuclear framework even at 0°K (the zero-point vibration), and the interaction of electrons with the nuclear vibration causes the release of the parity-selection rule as briefly discussed in Section 5.2.1, the broadening of spectral lines, and so forth. In particular, as seen in the dynamical Jahn–Teller effect which we will discuss later on, the separation of the electron and nuclear motions in degenerate states is sometimes impossible. These problems will be discussed briefly in this chapter.

9.1 Nuclear Vibrations

9.1.1 Adiabatic Approximation

The nonrelativistic Hamiltonian of the system having N electrons and N_0 nuclei is given by

$$\mathscr{H} = \mathscr{H}_{\mathrm{e}} + \mathscr{H}_{\mathrm{n}} + \mathscr{H}_{\mathrm{en}}\,, \tag{9.1a}$$

where

$$\mathscr{H}_{\mathrm{e}} = -\frac{\hbar^2}{2m}\sum_{i=1}^{N} \Delta_{ei} + \sum_{i>j=1}^{N} \frac{e^2}{r_{ij}}\,, \tag{9.1b}$$

$$\mathscr{H}_{\mathrm{n}} = -\sum_{k=1}^{N_0} \frac{\hbar^2}{2M_k} \Delta_{nk} + \sum_{k>l=1}^{N_0} \frac{Z_k Z_l e^2}{R_{kl}}, \tag{9.1c}$$

$$\mathscr{H}_{\mathrm{en}} = -\sum_{k=1}^{N_0} \sum_{i=1}^{N} \frac{Z_k e^2}{r_{ik}}. \tag{9.1d}$$

In (9.1) Δ_{ei} is the Laplacian operator of the ith electron coordinates; Δ_{nk} is the Laplacian operator of the kth nuclear coordinates: R_{kl} is the nuclear distance between nuclei k and l; r_{ik} is the distance between electron i and nucleus k; $Z_k e$ is the charge of nucleus k; and M_k is the mass of nucleus k. We observe that $\mathscr{H}_{\mathrm{e}}$ involves the electron coordinates only, $\mathscr{H}_{\mathrm{n}}$ the nuclear coordinates only, and $\mathscr{H}_{\mathrm{en}}$ both the electron and nuclear coordinates. To solve the Schrödinger equation with the Hamiltonian (9.1), we first assume that the electronic wavefunction $\Psi(\mathbf{r}\,;\mathbf{R})$, obtained by keeping the nuclei fixed at R, has a physical meaning. This assumption seems reasonable if one considers a great difference between the velocities of the electron and nuclear motions due to the difference between the electron and nuclear masses; the electron velocity is of the order of 10^8 cm/sec while the velocity of the nuclear motion is of the order of 10^5 cm/sec. The wavefunction $\Psi(\mathbf{r}\,;\mathbf{R})$ satisfies the equation

$$\Big(\mathscr{H}_{\mathrm{e}} + \sum_{k>l} \frac{Z_k Z_l e^2}{R_{kl}} + \mathscr{H}_{\mathrm{en}}\Big) \Psi_\mu(\mathbf{r}:\mathbf{R}) = U_\mu(\mathbf{R})\, \Psi_\mu(\mathbf{r}:\mathbf{R}), \tag{9.2}$$

which is derived by dropping out the nuclear kinetic energy terms in (9.1). Here $\mathbf{R}$ represents the positions of nuclei, $\mathbf{R}_1, \mathbf{R}_2, \ldots, \mathbf{R}_{N_0}$, and $\mathbf{r}$ the electron coordinates, $\mathbf{r}_1, \mathbf{r}_2, \ldots, \mathbf{r}_N$. Subscript μ is a set of quantum numbers. We regard $\Psi(\mathbf{r}\,;\mathbf{R})$ as a function of $\mathbf{r}$ with parameter $\mathbf{R}$. By using this wavefunction we express the total wavefunction for the Hamiltonian (9.1) as

$$\Phi(\mathbf{rR}) = \Psi(\mathbf{r}:\mathbf{R})\, \chi(\mathbf{R}). \tag{9.3}$$

Then, $\chi(\mathbf{R})$ may be determined from

$$\Big[-\sum_k (\hbar^2/2M_k)\, \Delta_{nk} + U_\mu(\mathbf{R})\Big] \chi_{\mu v}(\mathbf{R}) = E_{\mu v} \chi_{\mu v}(\mathbf{R}), \tag{9.4}$$

only when the following replacement is permissible‡:

$$-\sum_k \frac{\hbar^2}{2M_k} \Delta_{nk} \Psi(\mathbf{r}:\mathbf{R})\, \chi(\mathbf{R}) \rightarrow \Psi(\mathbf{r}:\mathbf{R}) \Big[-\sum_k \frac{\hbar^2}{2M_k} \Delta_{nk} \chi(\mathbf{R})\Big]. \tag{9.5}$$

‡ This replacement is permissible if $\nabla_n \Psi \cdot \nabla_n \chi$ and $\chi\, \Delta_n \Psi$ are negligible compared with $\Psi\, \Delta_n \chi$. To compare the magnitudes of these quantities, we notice that Ψ spreads

Now, as shown in (9.2) and (9.4), the equations of motion for electrons and nuclei are separated in the present approximation. This approximation is called the *adiabatic approximation* or *Born–Oppenheimer approximation.* As shown in (9.2), the displacement of nuclear positions may induce a deformation of the electron orbital but not an electron jump from one orbital to another. This means that electrons follow the nuclear motion adiabatically. As shown in (9.4), $U_\mu(\mathbf{R})$ serves as a potential for the nuclear motion so that it is called an *adiabatic potential.* The adiabatic potential is calculated as the eigenvalue of (9.2) with parameter $\mathbf{R}$. Therefore, its form depends upon the electronic state μ. It is important to note the Hamiltonian (9.1) as well as the Hamiltonian in (9.2) is invariant to any symmetry operation on the electron and nuclear system which brings the system into itself. Consequently, the adiabatic potential is also invariant to such a symmetry operation.

9.1.2 Normal Modes of Vibration

Now let us consider the problem of solving equation (9.4) for nuclear motion, which may be reexpressed in terms of nuclear displacements multiplied by $(M_k)^{1/2}$, $\mathbf{Q}_k = (M_k)^{1/2}\,(\mathbf{R}_k - \mathbf{R}_{0k})$, as follows ($\mathbf{R}_{0k}$ are the equilibrium positions):

$$\Big[-\frac{\hbar^2}{2}\sum_k\Big(\frac{\partial^2}{\partial Q_{kx}^2}+\frac{\partial^2}{\partial Q_{ky}^2}+\frac{\partial^2}{\partial Q_{kz}^2}\Big)+U_\mu(\mathbf{Q})\Big]\,\chi_\mu(\mathbf{Q}) = E_\mu\chi_\mu(\mathbf{Q}), \qquad (9.6)$$

where $\mathbf{Q}$ represents $\mathbf{Q}_1, \mathbf{Q}_2, ..., \mathbf{Q}_{N_0}$. For the purpose of solving (9.6), we first consider a $3N_0$-dimensional vector space whose basic unit vectors, $\hat{e}_{kx}, \hat{e}_{ky}, \hat{e}_{kz}$ ($k = 1, 2, ..., N_0$), are the basic vectors of the orthogonal coordinate system fixed at the equilibrium position of the kth nucleus.

---

over the distance of the order of 1 atomic unit while χ extends over the amplitude of nuclear vibration x_0. Then, it follows that

$$\nabla_n\Psi\cdot\nabla_n\chi/\Psi\,\Delta_n\chi \sim x_0,$$

$$\chi\,\Delta_n\Psi/\Psi\,\Delta_n\chi \sim x_0^2.$$

On the other hand, denoting the electronic energy as E^{e}, the vibrational energy as E^{v}, the vibrational frequency as ν, and the force constant responsible for the vibration as k, one has the relation, $E^{\mathrm{v}} = h\nu = \hbar(k/M)^{1/2} = \hbar(k/m)^{1/2}\,(m/M)^{1/2}$, which shows that $E^{\mathrm{v}}/E^{\mathrm{e}} \sim (m/M)^{1/2}$: in atomic unit $E^{\mathrm{e}} \sim 1$ and $k \sim 1$. Considering the relation, $E^{\mathrm{v}} \sim kx_0^2$, one obtains $x_0 \sim (m/M)^{1/4}$. Then, one learns that the adiabatic approximation is valid if $(m/M)^{1/4} \ll 1$. It is interesting to note that $v_{\mathrm{n}}/v_{\mathrm{e}} \sim (m/M)^{3/4}$ follows from $E^{\mathrm{e}} \sim mv_{\mathrm{e}}^2$ and $E^{\mathrm{v}} \sim Mv_{\mathrm{n}}^2$, where v_{e} and v_{n} are the electron and nuclear velocities. Therefore, the condition $x_0 \ll 1$ for the adiabatic approximation to be valid is stronger than $v_{\mathrm{n}}/v_{\mathrm{e}} \ll 1$ and $E^{\mathrm{v}}/E^{\mathrm{e}} \ll 1$.

In this space vector $\mathbf{Q}$ is given by its $3N_0$ components with respect to basic vectors $\hat{e}_{ki}$ ($i = x, y, z$) as

$$\mathbf{Q} = (Q_{1x}, Q_{1y}, Q_{1z}, Q_{2x}, Q_{2y}, ..., Q_{N_0z}). \tag{9.7}$$

For an arbitrary scaler λ, $\lambda\mathbf{Q}$ is defined as

$$\lambda\mathbf{Q} = (\lambda Q_{1x}, \lambda Q_{1y}, ..., \lambda Q_{N_0z}), \tag{9.8}$$

and the sum of two vectors,

$$\mathbf{Q}^{(\mathrm{I})} = (Q_{1x}^{(\mathrm{I})}, Q_{1y}^{(\mathrm{I})}, ..., Q_{N_0z}^{(\mathrm{I})}),$$

$$\mathbf{Q}^{(\mathrm{II})} = (Q_{1x}^{(\mathrm{II})}, Q_{1y}^{(\mathrm{II})}, ..., Q_{N_0z}^{(\mathrm{II})}),$$

is defined as

$$\mathbf{Q}^{(\mathrm{I})} + \mathbf{Q}^{(\mathrm{II})} = (Q_{1x}^{(\mathrm{I})} + Q_{1x}^{(\mathrm{II})}, Q_{1y}^{(\mathrm{I})} + Q_{1y}^{(\mathrm{II})}, ..., Q_{N_0z}^{(\mathrm{I})} + Q_{N_0z}^{(\mathrm{II})}). \tag{9.9}$$

This vector space is called the *displacement vector space*. Now we consider the transformation of basic vectors $\hat{e}_{ki}$ by operation R which brings the system into itself:

$$R\hat{e}_{ki} = \sum_{lj} \hat{e}_{lj} A_{ji}^{(lk)}(R). \tag{9.10}$$

Matrix $\mathbf{A}$ whose elements are $A_{ji}^{(lk)}$ is considered to consist of three dimensional matrices $\mathbf{A}^{(lk)}$ as shown in Fig. 9.1. Matrix $\mathbf{A}^{(lk)}$ is nonzero

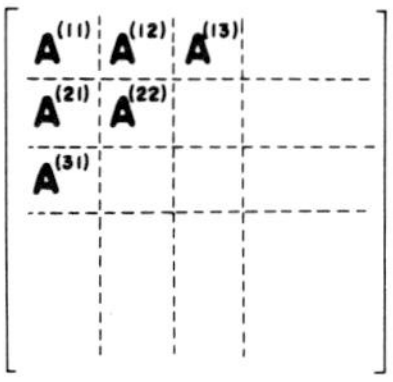

FIG. 9.1. Matrix $\mathbf{A}$ consisting of three-dimensional matrices $\mathbf{A}^{(lk)}$.

only when nucleus k is transformed to nucleus l by R, and all the nonzero matrices $\mathbf{A}^{(lk)}$ are the same and equal to the transformation matrix for the basic vectors of a three-dimensional space: for example, when R is the rotation around the z-axis by angle φ, the nonzero matrix $\mathbf{A}^{(lk)}$ is given as

$$\mathbf{A}^{(lk)} = \begin{bmatrix} \cos\varphi & -\sin\varphi & 0 \\ \sin\varphi & \cos\varphi & 0 \\ 0 & 0 & 1 \end{bmatrix}, \tag{9.11}$$

which is independent of l and k. We can show that the aggregate of the

$3N_0$-dimensional matrices $\mathbf{A}(R)$ for various R's forms a representation of the group to which the R's belong. However, this representation A, in general, is reducible. In order to learn what irreducible representations appear in reducing representation A, we use the property of characters similar to (1.78).

For example, let us consider an XY_6 molecule with the O_h-symmetry in which atom X is surrounded by atoms Y, octahedrally. The basic vectors in this system are shown in Fig. 9.2. The characters of representa-

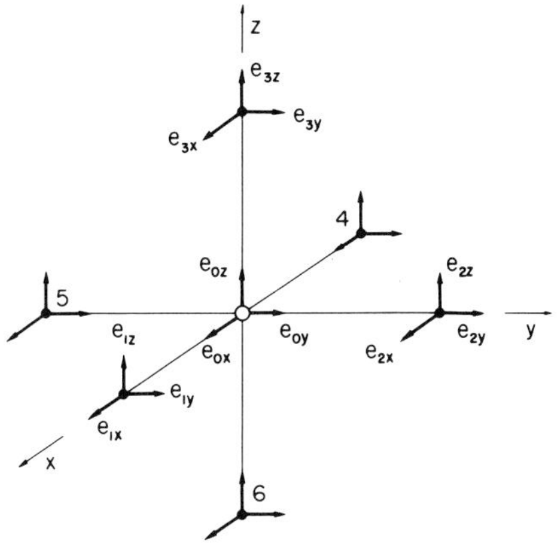

FIG. 9.2. The basic vectors in a XY_6 molecule. ○, X atom; ●, Y atom.

tion matrices $\mathbf{A}(R)$ for pure rotations in classes $\hat{E}$, $\hat{C}_4$, $\hat{C}_4^2$, $\hat{C}_3$, and $\hat{C}_2$ of O_h are obtained from

$$\chi(R) = N_R(1 + 2\cos\varphi), \tag{9.12}$$

where N_R is the number of the atoms which do not move on rotation R, and φ is the rotation angle around the symmetry axis.‡ The character of $\mathbf{A}$ for rotation-inversions in classes $\hat{I}$, $\hat{C}_4 I$, $\hat{C}_4^2 I$, $\hat{C}_3 I$, and $\hat{C}_2 I$ are obtained from

$$\chi(R) = -N_R(1 + 2\cos\varphi), \tag{9.13}$$

which may be derived if one considers the that, for inversion after rotation around the z-axis by angle φ, $\mathbf{A}^{(lk)}$ is given by

$$\mathbf{A}^{(lk)} = \begin{bmatrix} -\cos\varphi & \sin\varphi & 0 \\ -\sin\varphi & -\cos\varphi & 0 \\ 0 & 0 & -1 \end{bmatrix}. \tag{9.14}$$

‡ If the symmetry axis is the z-axis, (9.12) is evident from (9.11). When the symmetry axis is not the z-axis, the matrix corresponding to (9.11) is given by $\mathbf{U}\mathbf{A}^{(lk)}\mathbf{U}^{-1}$ where $\mathbf{U}$ is an orthogonal transformation matrix. Since the character of $\mathbf{U}\mathbf{A}^{(lk)}\mathbf{U}^{-1}$ is the same as that of $\mathbf{A}^{(lk)}$, (9.12) is valid for the rotation around any symmetry axis.

By using (9.12) and (9.13), one can calculate the characters of **A** for all the symmetry operations in the O_h-group as shown in Table 9.1, in which N_R and $\chi(R)$ for atom X and the group of atoms Y_6 are separately given; the sum of these $\chi(R)$ is the character for XY_6. Comparing Table 9.1 with Table 1.7, one sees that representation A is reduced to

$$A_{1g} + E_g + T_{1g} + 3T_{1u} + T_{2g} + T_{2u}\,. \tag{9.15}$$

TABLE 9.1

CHARACTERS OF REPRESENTATION A IN THE O_h-GROUP

		$\hat{E}$	$\hat{C}_4$	$\hat{C}_4^2$	$\hat{C}_3$	$\hat{C}_2$	$\hat{I}$	$\hat{C}_4 I$	$\hat{C}_4^2 I$	$\hat{C}_3 I$	$\hat{C}_2 I$
	φ	0	$\pi/2$	π	$2\pi/3$	π	0	$\pi/2$	π	$2\pi/3$	π
Y_6	N_R	6	2	2	0	0	0	0	4	0	2
	$\chi(R)$	18	2	−2	0	0	0	0	4	0	2
X	N_R	1	1	1	1	1	1	1	1	1	1
	$\chi(R)$	3	1	−1	0	−1	−3	−1	1	0	1
XY_6	$\chi(R)$	21	3	−3	0	−1	−3	−1	5	0	3

Note that $\chi(R)$ for atom X is the character of T_{1u}. The number of bases of the irreducible representation is $1 + 2 + (3 \times 6) = 21$, which agrees with $3N_0 = 21$.

The transformation of $\mathbf{A}(R)$ into an irreducible form is achieved by an orthogonal transformation $\mathbf{C}$ as follows:

$$\sum_{lj,ki} C^{(\alpha)}_{\Gamma\gamma,lj} A^{(lk)}_{ji}(R)\, C^{(\alpha')}_{ki,\Gamma'\gamma'} = D^{(\Gamma)}_{\gamma\gamma'}(R)\,\delta(\alpha\alpha')\,\delta(\Gamma\Gamma'), \tag{9.16}$$

where α is introduced to distinguish the same irreducible representation appearing more than once. When the same irreducible representation appears more than once, the orthogonal transformation is not uniquely determined from (9.16) only. Now, we denote the unit basic vectors of the irreducible representations as $\hat{e}^{(\alpha)}_{\Gamma\gamma}$, which are given by

$$\hat{e}^{(\alpha)}_{\Gamma\gamma} = \sum \hat{e}_{ki} C^{(\alpha)}_{ki,\Gamma\gamma}\,. \tag{9.17}$$

With respect to these basic vectors, a displacement vector in the $3N_0$-dimensional space is now given as

$$\mathbf{Q} = (\ldots, Q^{(\alpha)}_{\Gamma\gamma} \ldots), \tag{9.18}$$

where new components $Q_{\Gamma\gamma}^{(\alpha)}$ are given by

$$Q_{\Gamma\gamma}^{(\alpha)} = \sum_{ki} C_{\Gamma\gamma,ki}^{(\alpha)} Q_{ki} \,, \tag{9.19}$$

and they transform by operation R in the following way[‡]:

$$\begin{aligned} RQ_{\Gamma\gamma}^{(\alpha)} &= \sum_{ki} C_{\Gamma\gamma,ki}^{(\alpha)} RQ_{ki} = \sum_{ki,lj} C_{\Gamma\gamma,ki}^{(\alpha)} A_{ij}^{(kl)}(R)\, Q_{lj} \\ &= \sum_{\substack{ki,lj,\\ \alpha'\Gamma'\gamma'}} C_{\Gamma\gamma,ki}^{(\alpha)} A_{ij}^{(kl)}(R)\, C_{lj,\Gamma'\gamma'}^{(\alpha')} Q_{\Gamma'\gamma'}^{(\alpha')} = \sum_{\gamma'} D_{\gamma\gamma'}^{(\Gamma)} Q_{\Gamma\gamma'}^{(\alpha)} . \end{aligned} \tag{9.20}$$

In deriving (9.20), we used the inverse relation of (9.19),

$$Q_{ki} = \sum_{\alpha\Gamma\gamma} C_{ki,\Gamma\gamma}^{(\alpha)} Q_{\Gamma\gamma}^{(\alpha)} . \tag{9.21}$$

Returning to the problem of solving (9.6), we reexpress (9.6) in terms of $Q_{\Gamma\gamma}^{(\alpha)}$. Since $U_\mu(\mathbf{Q})$ is invariant to any operation R in the group to which the system belongs, $U_\mu(\mathbf{Q})$ may be expanded in powers of $Q_{\Gamma\gamma}^{(\alpha)}$ as follows:

$$U_\mu(\mathbf{Q}) = U_\mu(0) + \tfrac{1}{2} \sum_{\substack{\alpha\alpha'\\ \Gamma\gamma}} K_{\mu\Gamma}^{(\alpha\alpha')} Q_{\Gamma\gamma}^{(\alpha)} Q_{\Gamma\gamma}^{(\alpha')}, \tag{9.22}$$

in which $K_{\mu\Gamma}^{(\alpha\alpha')}$ are numerical constants, and only the terms up to the second power of $Q_{\Gamma\gamma}^{(\alpha)}$ are retained. In (9.22) no term linear to $Q_{\Gamma\gamma}^{(\alpha)}$ appears as the system is in equilibrium at $\mathbf{Q} = 0$. The fact that (9.22) is invariant to R may be seen in the following: $U_\mu(0)$ is obviously invariant to R and the second term in (9.22) is transformed under the operation R as

$$\begin{aligned} R\Big[\sum_{\alpha\alpha'\Gamma\gamma} K_{\mu\Gamma}^{(\alpha\alpha')} Q_{\Gamma\gamma}^{(\alpha)} Q_{\Gamma\gamma}^{(\alpha')}\Big] &= \sum_{\alpha\alpha'\Gamma\gamma} K_{\mu\Gamma}^{(\alpha\alpha')} RQ_{\Gamma\gamma}^{(\alpha)} RQ_{\Gamma\gamma}^{(\alpha')} \\ &= \sum_{\gamma'\gamma''} \sum_{\alpha\alpha'\Gamma\gamma} K_{\mu\Gamma}^{(\alpha\alpha')} D_{\gamma\gamma'}^{(\Gamma)}(R)\, D_{\gamma\gamma''}^{(\Gamma)}(R)\, Q_{\Gamma\gamma'}^{(\alpha)} Q_{\Gamma\gamma''}^{(\alpha')} \\ &= \sum_{\alpha\alpha'\Gamma\gamma} K_{\mu\Gamma}^{(\alpha\alpha')} Q_{\Gamma\gamma}^{(\alpha)} Q_{\Gamma\gamma}^{(\alpha')}, \end{aligned} \tag{9.23}$$

in which we used (9.19) and the orthogonality relation

$$\sum_{\gamma} D_{\gamma\gamma'}^{(\Gamma)}(R)\, D_{\gamma\gamma''}^{(\Gamma)}(R) = \delta(\gamma'\gamma''). \tag{9.24}$$

[‡] Strictly speaking, $RQ_{\Gamma\gamma}^{(\alpha)}$ should be written as $RQ_{\Gamma\gamma}^{(\alpha)}R^{-1}$, because $Q_{\Gamma\gamma}^{(\alpha)}$ is an operator.

The kinetic energy term in (9.6) is reexpressed in terms of $Q^{(\alpha)}_{\Gamma\gamma}$ as

$$-\frac{\hbar^2}{2}\sum_{ki}\frac{\partial^2}{\partial Q^2_{ki}} = -\frac{\hbar^2}{2}\sum_{\substack{\alpha\Gamma\gamma\\ \alpha'\Gamma'\gamma'}}\frac{\partial}{\partial Q^{(\alpha')}_{\Gamma'\gamma'}}\frac{\partial}{\partial Q^{(\alpha)}_{\Gamma\gamma}}\sum_{ki} C^{(\alpha')}_{\Gamma'\gamma',ki}C^{(\alpha)}_{\Gamma\gamma,ki}$$

$$= -\frac{\hbar^2}{2}\sum_{\alpha\Gamma\gamma}\frac{\partial^2}{\{\partial Q^{(\alpha)}_{\Gamma\gamma}\}^2}. \qquad (9.25)$$

By using (9.22) and (9.25), (9.6) is now given as

$$\Big[-\frac{\hbar^2}{2}\sum_{\alpha\Gamma\gamma}\frac{\partial^2}{\{\partial Q^{(\alpha)}_{\Gamma\gamma}\}^2} + \frac{1}{2}\sum_{\alpha\alpha'\Gamma\gamma} K^{(\alpha\alpha')}_{\mu\Gamma} Q^{(\alpha)}_{\Gamma\gamma} Q^{(\alpha')}_{\Gamma\gamma}\Big]\chi_\mu(\mathbf{Q})$$
$$= [E_\mu - U_\mu(0)]\,\chi_\mu(\mathbf{Q}), \qquad (9.26)$$

which is separable into the equations with different $\Gamma\gamma$ as

$$\Big[-\frac{\hbar^2}{2}\sum_{\alpha}\frac{\partial^2}{\{\partial Q^{(\alpha)}_{\Gamma\gamma}\}^2} + \frac{1}{2}\sum_{\alpha\alpha'} K^{(\alpha\alpha')}_{\mu\Gamma}\, Q^{(\alpha)}_{\Gamma\gamma}\, Q^{(\alpha')}_{\Gamma\gamma}\Big]$$
$$\times\, \chi_{\mu\Gamma\gamma}(Q^{(1)}_{\Gamma\gamma},\ldots, Q^{(g_\Gamma)}_{\Gamma\gamma})$$
$$= \epsilon_{\mu\Gamma}\chi_{\mu\Gamma\gamma}(Q^{(1)}_{\Gamma\gamma},\ldots, Q^{(g_\Gamma)}_{\Gamma\gamma}), \qquad (9.27)$$

if $\chi_\mu(\mathbf{Q})$ is assumed to have the form

$$\chi_\mu(\mathbf{Q}) = \prod_{\Gamma\gamma}\chi_{\mu\Gamma\gamma}(Q^{(1)}_{\Gamma\gamma},\ldots, Q^{(g_\Gamma)}_{\Gamma\gamma}). \qquad (9.28)$$

Functions $\chi_{\mu\Gamma\gamma}$ involve $Q^{(1)}_{\Gamma\gamma}$, $Q^{(2)}_{\Gamma\gamma},\ldots, Q^{(g_\Gamma)}_{\Gamma\gamma}$ when the Γ irreducible representation appears g_Γ times. In (9.27), $\epsilon_{\mu\Gamma}$ is independent of γ and E_μ is related to $\epsilon_{\mu\Gamma}$ as

$$\sum_{\Gamma}(\Gamma)\,\epsilon_{\mu\Gamma} = E_\mu - U_\mu(0), \qquad (9.29)$$

where (Γ) is the dimension of Γ.

When a certain Γ appears only once, (9.27) is the wave equation for a simple harmonic oscillator with angular frequency $\omega_{\mu\Gamma} = (K_{\mu\Gamma})^{1/2}$; in this case $K_{\mu\Gamma}$ is always positive as the equilibrium configuration is stable. Then, the eigenvalue of (9.27) is

$$\epsilon^{v}_{\mu\Gamma} = (v + \tfrac{1}{2})\,\hbar(K_{\mu\Gamma})^{1/2}, \qquad v = 0, 1, 2,\ldots, \qquad (9.30)$$

and the eigenfunction is known to be

$$\chi^{(v)}_{\mu\Gamma\gamma}(\xi) = N_v H_v(\xi)\exp(-\tfrac{1}{2}\xi^2), \qquad (9.31)$$

where $H_v(\xi)$ is the vth Hermite polynomial given by

$$H_v(\xi) = (-1)^v \exp(\xi^2) \frac{\partial^v}{\partial \xi^v} \exp(-\xi^2), \tag{9.32}$$

$$\xi = \alpha Q_{\Gamma\gamma}, \qquad \alpha^4 = K_{\mu\Gamma}/\hbar^2, \tag{9.33}$$

and normalization constant $N_v^{-1/2}$ is given as

$$N_v = \left(\frac{\alpha}{\sqrt{\pi}\, 2^v v!}\right)^{1/2}. \tag{9.34}$$

When a certain Γ appears g_Γ times, we apply an orthogonal transformation to $Q_{\Gamma\gamma}^{(\alpha)}$'s;

$$Q_{\Gamma\gamma}^{(\beta)} = \sum_{\alpha=1}^{g_\Gamma} B_{\beta\alpha} Q_{\Gamma\gamma}^{(\alpha)}, \tag{9.35}$$

so that $U_\mu(\mathbf{Q}) - U_\mu(0)$ has the following form:

$$U_\mu(\mathbf{Q}) - U_\mu(0) = \tfrac{1}{2} \sum_{\beta\Gamma\gamma} K_{\mu\Gamma}^{(\beta)} Q_{\Gamma\gamma}^{(\beta)} Q_{\Gamma\gamma}^{(\beta)}. \tag{9.36}$$

Now $K_\mu^{(\beta)}$ and the orthogonal transformation can be determined by solving

$$\sum_{\alpha\alpha'} K_{\mu\Gamma}^{(\alpha\alpha')} B_{\alpha\beta} B_{\alpha'\beta'} = K_{\mu\Gamma}^{(\beta)}\, \delta(\beta\beta') \tag{9.37}$$

or

$$\sum_{\alpha'} K_{\mu\Gamma}^{(\alpha\alpha')} B_{\alpha'\beta} = K_{\mu\Gamma}^{(\beta)} B_{\alpha\beta}, \tag{9.38}$$

which gives the g_Γ-dimensional secular equation. It should be emphasized that the additional orthogonal transformation **B** can be determined without violating (9.16) as (9.16) does not determine the transformation **C**, uniquely. Transformation **B** depends on the physical property of the adiabatic potential, and is not determined by symmetry considerations only. Even when the $Q_{\Gamma\gamma}^{(\beta)}$'s are used the kinetic energy term has the same form as that given in (9.25), so that with the potential-energy term of (9.36) the wave equation is now reduced to that for a single harmonic oscillator having angular frequency $\omega_{\mu\Gamma}^{(\beta)} = (K_{\mu\Gamma}^{(\beta)})^{1/2}$.

The oscillation specified by $\beta\Gamma$ is called the $\beta\Gamma$ *normal-mode*, and $Q_{\Gamma\gamma}^{(\beta)}$ is called the γ *normal-coordinate* belonging to the $\beta\Gamma$ normal-mode. In the XY_6 molecule described in Fig. 9.2, eight kinds of normal-modes as given in (9.15) are obtained. However, it is evident from a physical point of view that in general six degrees of freedom out of $3N_0$ are ascribed to the free translation and rotation of the system. For example, in the XY_6 molecule one T_{1u} mode belongs to the free translation and one T_{1g} to the free rotation. The normal coordinates of the free translation and rotations are easily obtained by insight and the $K_{\mu\Gamma}^{(\beta)}$'s for these modes are, of course, zero. Therefore, the procedure of determining transformation **B** is simplified if the normal-coordinates of the free translation and rotation are subtracted from the $Q_{\Gamma\gamma}^{(\beta)}$'s from the beginning.

9.1.3 Explicit Forms of Normal Coordinates

In principle, the explicit forms of the normal coordinates are obtained by determining both the transformation matrix **C** from (9.16) or from its modified form,

$$\sum_{ki} A_{ji}^{(lk)}(R)\, C_{ki,\Gamma\gamma}^{(\alpha)} = \sum_{\gamma'} C_{\Gamma\gamma',lj}^{(\alpha)} D_{\gamma'\gamma}^{(\Gamma)}(R), \tag{9.39}$$

and the transformation matrix **B** from (9.38). However, it is much simpler for the systems of high symmetry to use geometrical insight in determining the normal coordinates.

In what follows we consider the example of the XY_6 molecule whose normal modes have already been classified as; A_{1g}, E_g, T_{1g}, $3T_{1u}$, T_{2g}, and T_{2u}. Among these normal modes the modes of internal vibrations besides the free translation and rotation are A_{1g}, E_g, $2T_{1u}$, T_{2g}, and T_{2u}. To determine the normal coordinates of the internal vibrations, we first note that in the internal vibrations both momentum and angular momentum of the system are zero. From the fact that the velocity components are in the ratio of the displacement components, the conditions of vanishing momentum and angular momentum are expressed in terms of the displacement components $S_{ki} = Q_{ki}/(M_k)^{1/2}$ as

$$\sum_k M_k S_{ki} = 0, \tag{9.40}$$

$$\sum_k M_k(\mathbf{S}_k \times \mathbf{R}_{0k})_i = 0 \qquad (i = x, y, z) \tag{9.41}$$

in which

$$\mathbf{S}_k = \sum_i S_{ki} \hat{e}_{ki} \,.$$

The normal coordinate of the A_{1g} mode is easily obtained as shown in Fig. 9.3. It is given by the displacements which keep the original symmetry, O_h. These displacements clearly satisfy (9.40) and (9.41).

Therefore, the basic unit vector $\hat{e}_{A_{1g}}$ of the normal coordinate $Q_{A_{1g}}$ is given as

$$\hat{e}_{A_{1g}} = \frac{1}{\sqrt{6}}(\hat{e}_{1x} + \hat{e}_{2y} + \hat{e}_{3z} - \hat{e}_{4x} - \hat{e}_{5y} - \hat{e}_{6z}). \tag{9.42}$$

From the arguments given in Chapter VI, it is clear that the presence of the $E_g u$ normal-mode reduces the symmetry of the system to D_{4h} whose fourfold symmetry axis is the z-axis, as $Q_{E_g u}$ transforms as the u component of the E_g irreducible representation. From such a symmetry consideration, $Q_{E_g u}$ is immediately given as shown in Fig. 9.3 with

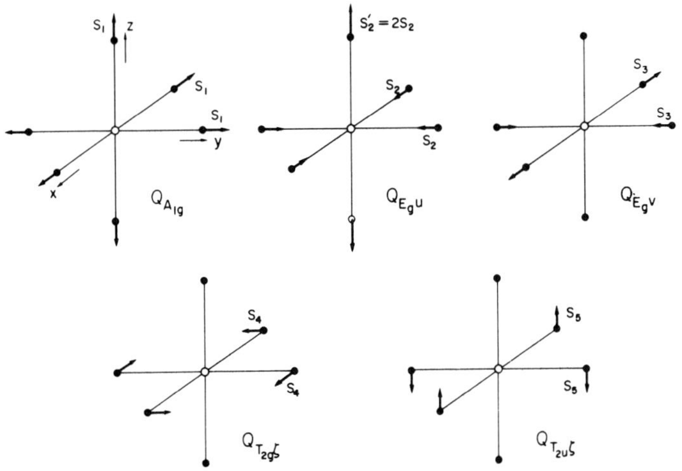

FIG. 9.3. Normal coordinates of a XY_6 molecule.

displacement vectors having different lengths S_2 and S_2'. Then, the orthogonality relation between $Q_{A_{1g}}$ and $Q_{E_g u}$ determines S_2' to be $2S_2$. It is needless to show that these displacements satisfy (9.40) and (9.41). After all, one obtains

$$\hat{e}_{E_g u} = \frac{1}{\sqrt{12}}[2\hat{e}_{3z} - 2\hat{e}_{6z} - (\hat{e}_{1x} + \hat{e}_{2y} - \hat{e}_{4x} - \hat{e}_{5y})]. \tag{9.43}$$

By a similar consideration, it is easy to obtain

$$\hat{e}_{E_g v} = \tfrac{1}{2}(\hat{e}_{1x} - \hat{e}_{2y} - \hat{e}_{4x} + \hat{e}_{5y}), \tag{9.43'}$$

$$\hat{e}_{T_{2g}\zeta} = \tfrac{1}{2}(\hat{e}_{1y} + \hat{e}_{2x} - \hat{e}_{4y} - \hat{e}_{5x}), \tag{9.44}$$

$$\hat{e}_{T_{2u}\zeta} = \tfrac{1}{2}(\hat{e}_{1z} - \hat{e}_{2z} + \hat{e}_{4z} - \hat{e}_{5z}). \tag{9.45}$$

The displacement vectors of these normal-modes are shown in Fig. 9.3. For obtaining the displacement vectors in the $T_{2u}\zeta$ normal-mode, it is convenient to use the fact that the ζ base of T_{2u} transforms like $z(x^2 - y^2)$ as shown in Table 6.4. In normal-modes A_{1g}, E_g, T_{2g}, and T_{2u}, the displacements of the central atom are zero. They contribute only to the T_{1u} mode as shown in Table 9.1.

Some complication arises in determining the normal coordinates of the two T_{1u} vibrational normal-modes. As seen from Table 9.1 two T_{1u} modes and one T_{1u} mode are obtained from the displacements of six Y nuclei and a single X nucleus, respectively. Therefore, if one ignores conditions (9.40) and (9.41), it is a simple matter to obtain three kinds of mutually orthogonal displacements of symmetry $T_{1u}\gamma$ as shown in Fig. 9.4. The free translation is obtained by superposing the displacements in (a), (b), and (c) of Fig. 9.4 with the equal weight and phase. The displacements in the two $T_{1u}\gamma$ vibrational normal-modes may be those given in Fig. 9.5. From the orthogonality between $Q^{(I)}_{T_{1u}\gamma}$ and $Q^{(II)}_{T_{1u}\gamma}$, one obtains the relation

$$4\alpha\alpha' M + 2\beta\beta' M - M_0 = 0, \tag{9.46}$$

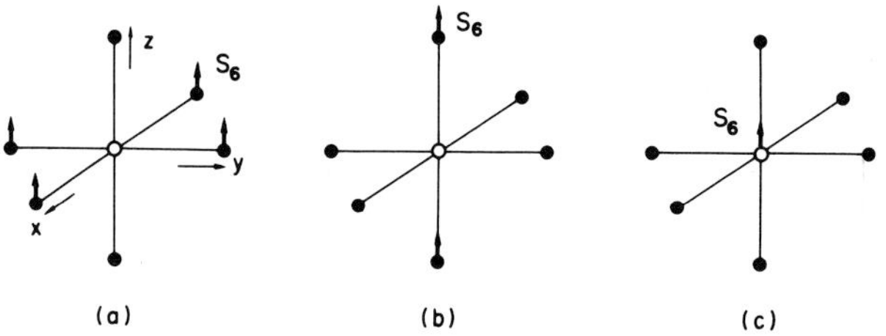

FIG. 9.4. Three independent modes of nuclear displacements of T_{1u}.

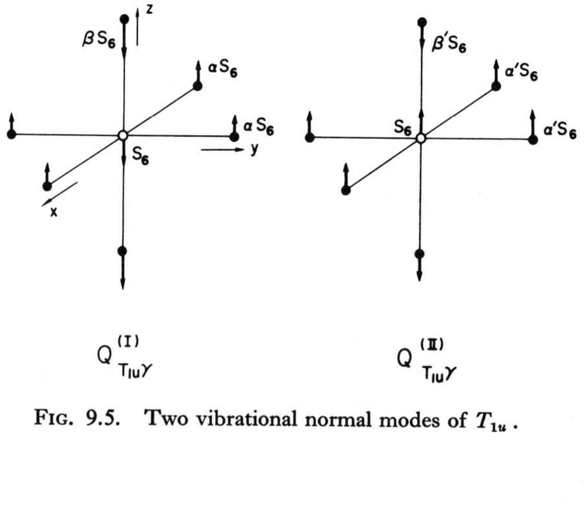

FIG. 9.5. Two vibrational normal modes of T_{1u}.

and from (9.40)

$$M_0 + 2\beta M - 4\alpha M = 0, \tag{9.47}$$

$$M_0 - 2\beta' M + 4\alpha' M = 0, \tag{9.48}$$

where M_0 and M are the nuclear masses of nuclei X and Y, respectively, and α, β, α', and β' the parameters determining the lengths and signs of the displacements as shown in Fig. 9.5. Relations (9.47) and (9.48) are equivalent to the orthogonality relations between the free translation mode and $Q^{(\mathrm{I})}_{T_{1u}\gamma}$ and those between the translation mode and $Q^{(\mathrm{II})}_{T_{1u}\gamma}$. Since there are four unknown parameters and three equations that they should satisfy, the values of the parameters cannot uniquely be determined. To determine the parameters completely, it is necessary to know the explicit form of the adiabatic potential. It has been known that a simple choice of $\beta = 1$ nearly diagonalizes a reasonably assumed form of the adiabatic potential in some cases.‡ If $\beta = 1$ is assumed, one may determine all the parameters from (9.46)–(9.48) as follows:

$$\begin{aligned} \alpha &= (M_0 + 2M)/4M, & \beta &= 1. \\ \alpha' &= 0, & \beta' &= M_0/2M. \end{aligned} \tag{9.49}$$

The displacements with the parameters given in (9.49) are visualized in Fig. 9.6. The basic vectors corresponding to the normal coordinates in this case are given as

$$\hat{e}^{(\bar{\mathrm{I}})}_{T_{1u}\gamma} = \left[\frac{4M}{(M_0 + 6M)(M_0 + 2M)}\right]^{1/2} \times \left\{\left[\frac{M_0 + 2M}{4M^{1/2}}\right](\hat{e}_{1z} + \hat{e}_{2z} + \hat{e}_{4z} + \hat{e}_{5z}) - M^{1/2}(\hat{e}_{3z} + \hat{e}_{6z}) - (M_0)^{1/2}\hat{e}_{0z}\right\} \tag{9.50}$$

and

$$\hat{e}^{(\overline{\mathrm{II}})}_{T_{1u}\gamma} = \left[\frac{2M}{M_0(2M + M_0)}\right]^{1/2} \times \left[(M_0)^{1/2}\hat{e}_{0z} - \frac{M_0}{2\,M^{1/2}}(\hat{e}_{3z} + \hat{e}_{6z})\right]. \tag{9.51}$$

Normal coordinates $Q_{\Gamma\gamma}$ are obtained simply from $\hat{e}_{\Gamma\gamma}$ by substituting Q_{ki} for $\hat{e}_{ki}$.

Problem 9.1. Derive the normal-coordinates of a X_3 molecule with the shape of a regular triangle (D_{3h}-symmetry) by using (9.39). ◇

Problem 9.2. Derive the normal-coordinates of a tetrahedral MX_4 molecule by symmetry considerations. ◇

‡ S. Koide and M. H. L. Pryce, *Phil. Mag.* 3, 607 (1958).

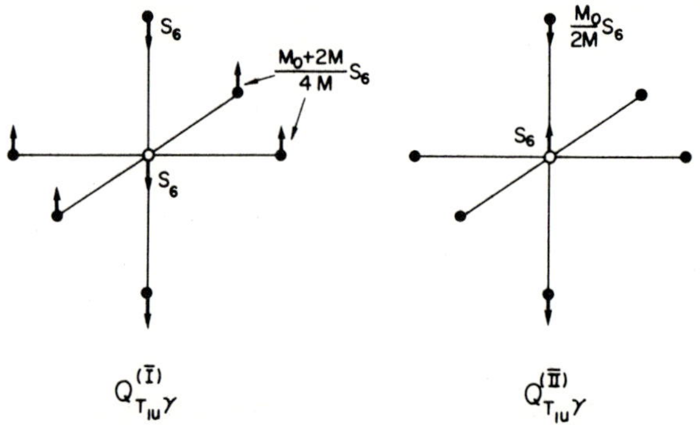

FIG. 9.6. Two T_{1u} normal-modes for a particular adiabatic potential.

9.2 Linear Interaction in Nondegenerate Electronic States

9.2.1 INTERACTION HAMILTONIAN

The interaction Hamiltonian between electrons and nuclei is given by $\mathscr{H}_{\text{en}}$ in (9.1) or (9.2). This Hamiltonian is invariant to any operation on the electron plus nuclear system which brings the system into itself. Therefore, it may be expanded in powers of the normal coordinates as

$$\mathscr{H}_{\text{en}}(\mathbf{r}\ \mathbf{Q}) = V_0(\mathbf{r}\ \mathbf{Q} = 0) + \sum_{\beta\Gamma\gamma} V^{(\beta)}_{\Gamma\gamma}(\mathbf{r})\, Q^{(\beta)}_{\Gamma\gamma} + \cdots, \tag{9.52}$$

in which $V_0(\mathbf{r}\ \mathbf{Q} = 0)$ is the ligand-field potential that we have discussed so far. In (9.52) the terms linear to $Q^{(\beta)}_{\Gamma\gamma}$, except the term with the identity representation, are vanishing within the μ electronic state when the state is nondegenerate as shown by

$$\int d\tau \Psi_\mu(\mathbf{r} : \mathbf{Q} = 0)^* V_{\Gamma\gamma}(\mathbf{r})\, \Psi_\mu(\mathbf{r} : \mathbf{Q} = 0) = 0 \tag{9.53}$$

($\Gamma \neq$ the identity representation). In (9.53) $\Psi_\mu(\mathbf{r} : \mathbf{Q} = 0)$ is the unperturbed nondegenerate electronic wavefunction obtained at the equilibrium position $\mathbf{Q} = 0$, and it is the base of the one-dimensional irreducible representation Γ_μ. Equation (9.53) holds as $\Gamma_\mu \times \Gamma_\mu$ is the identity representation. The nonvanishing linear term of the identity representation within the nondegenerate electronic state should be canceled by the linear term $U_A Q_A$ in the expansion of $\sum Z_k Z_l e^2/R_{kl}$:

$$\sum_{k>l} Z_k Z_l e^2/R_{kl} = U_0 + U_A Q_A + \sum U^{(\beta\beta')}_{\Gamma\gamma} Q^{(\beta)}_{\Gamma\gamma} Q^{(\beta')}_{\Gamma\gamma}, \tag{9.54}$$

in which U_0, U_A, and $U_{\Gamma\gamma}^{(\beta\beta')}$ are numerical coefficients and Q_A is the normal coordinate of the identity representation. Expansion (9.54) is mainly based on the symmetry property that $\sum Z_k Z_l e^2/R_{kl}$ is invariant to any operation of the group to which the system belongs. The cancellation

$$\int d\tau \Psi_\mu^*(\mathbf{r} : \mathbf{Q} = 0)\, V_A(\mathbf{r})\, \Psi_\mu(\mathbf{r} : \mathbf{Q} = 0) + U_A = 0 \tag{9.55}$$

is necessary as the adiabatic potential in the μ electronic state,

$$U_\mu(\mathbf{Q}) = \int d\tau \Psi_\mu(\mathbf{r} : \mathbf{Q} = 0) \left[\mathscr{H}_e + \sum \frac{Z_k Z_l e^2}{R_{kl}} + \mathscr{H}_{en}\right] \Psi_\mu(\mathbf{r} : \mathbf{Q} = 0), \tag{9.56}$$

is quadratic to $Q_{\Gamma\gamma}^{(\beta)}$. In degenerate-electronic states the situation is quite different, and in the next section we will discuss why nonvanishing linear terms are important in degenerate electronic states.

Although the matrix elements of $V_{\Gamma\gamma}^{(\beta)}$ in nondegenerate states are vanishing, the nondiagonal matrix elements of $V_{\Gamma\gamma}^{(\beta)}$ between a nondegenerate state and others are, in general, nonvanishing and play important roles in giving the intensities of parity-forbidden transitions.

9.2.2 Intensities of Parity-Forbidden Transitions

In Section 5.2.1 we discussed the parity-selection rule that may be released slightly by the instantaneous distortions of the system induced by nuclear vibrations. In what follows we will explain this mechanism on a more rigorous theoretical basis.

Let us consider the parity-forbidden transition, $\mu \to \nu$. The electronic wavefunctions $\Psi_\mu(\mathbf{r} : \mathbf{R}_0)$ and $\Psi_\nu(\mathbf{r} : \mathbf{R}_0)$ are assumed to be obtained by solving (9.2) with a fixed value of $\mathbf{R}$, $\mathbf{R} = \mathbf{R}_0$. These wavefunctions are nothing but those employed in the ligand-field theory‡; in calculating $\Psi(\mathbf{r} : \mathbf{R}_0)$ the first term $V_0(\mathbf{r} : \mathbf{R}_0)$ in the expansion of $\mathscr{H}_{en}$ as given in (9.52) has already been taken into account as a ligand-field potential. The second term in (9.52) which is linear to nuclear displacements from $\mathbf{R}_0$ may be regarded as a perturbation to the solution of the ligand-field theory.

For simplicity we deal with a rather particular case in which the equilibrium nuclear positions as well as the angular frequencies of the corresponding normal-modes in both the μ and ν electronic states are the same; $\mathbf{R}_{\mu 0} = \mathbf{R}_{\nu 0} = \mathbf{R}_0$, and $\omega_{\mu\Gamma\gamma}^{(\beta)} = \omega_{\nu\Gamma\gamma}^{(\beta)} = \omega_{\Gamma\gamma}^{(\beta)}$. Then, it is evident

‡ In many cases we use as $\mathbf{R}_0$ an equilibrium-nuclear configuration determined by x-ray- or neutron-diffraction experiments in the ground state.

that $V_{\Gamma\gamma}^{(\beta)}(\mathbf{r})$ with odd-parity Γ in the second term of (9.52) gives the instantaneous odd-parity field induced by the nuclear vibration of the $\beta\Gamma$-mode. In this case the effective electric-dipole transition moment of the parity-forbidden $\mu v \rightarrow \nu v'$ transition is given as

$$\begin{aligned}\mathbf{P}_{\text{eff}}(\mu v \rightarrow \nu v') = & \sum_{\lambda_{\text{odd}}} \langle \mu v \mid \sum_{\text{odd}\beta\Gamma\gamma} V_{\Gamma\gamma}^{(\beta)}(\mathbf{r})\, Q_{\Gamma\gamma}^{(\beta)} \mid \lambda_{\text{odd}} v' \rangle \\ & \times \langle \lambda_{\text{odd}} v' \mid \mathbf{P} \mid \nu v' \rangle [E(\mu v) - E(\lambda_{\text{odd}} v')]^{-1} \\ & + \sum_{\lambda_{\text{odd}}} \langle \mu v \mid \mathbf{P} \mid \lambda_{\text{odd}} v \rangle \langle \lambda_{\text{odd}} v \mid \sum_{\text{odd}\beta\Gamma\gamma} V_{\Gamma\gamma}^{(\beta)}(\mathbf{r})\, Q_{\Gamma\gamma}^{(\beta)} \mid \nu v' \rangle \\ & \times [E(\nu v') - E(\lambda_{\text{odd}} v)]^{-1}, \end{aligned} \tag{9.57}$$

in which the λ_{odd}'s are the electronic states of odd-parity and $E(\mu v)$ is the energy of the electron-nuclear system given by

$$E(\mu v) = U_{\mu}(0) + \sum_i \epsilon_i^{v_i}, \tag{9.58}$$

where number v_i is a quantum number of the ith vibrational normal mode ($i = \beta\Gamma\gamma$). In (9.58) $\epsilon_v^{v_i}$ is the energy of the harmonic oscillator i as given in (9.30). As we did in (6.60), (9.57) may be simplified in the closure approximation as

$$\mathbf{P}_{\text{eff}}(\mu v \rightarrow \nu v') = (2/\Delta E) \langle \mu v \mid \mathbf{P} \sum_{\text{odd}\beta\Gamma\gamma} V_{\Gamma\gamma}^{(\beta)}(\mathbf{r})\, Q_{\Gamma\gamma}^{(\beta)} \mid \nu v' \rangle, \tag{9.59}$$

in which ΔE is a suitable average of the denominators in (9.57). By using the relation

$$\langle v \mid Q \mid v' \rangle = \left(\frac{\hbar}{2\omega}\right)^{1/2} [v^{1/2}\, \delta_{v',v-1} + (v+1)^{1/2}\, \delta_{v',v+1}], \tag{9.60}$$

Eq. (9.59) may be expressed as

$$\begin{aligned}\mathbf{P}_{\text{eff}}(\mu v \rightarrow \nu v') = & \frac{2}{\Delta E} \sum_i \left(\frac{\hbar}{2\omega_i}\right)^{1/2} \\ & \times \langle \mu \mid \mathbf{P} V_i(\mathbf{r}) \mid \nu \rangle [(v_i)^{1/2}\, \delta_{v_i',v_i-1} + (v_i+1)^{1/2}\, \delta_{v_i',v_i+1}]. \end{aligned} \tag{9.61}$$

Now, if μv and $\nu v'$ are the good quantum numbers of the system, we may expect nonvanishing electric-dipole transition lines at the photon energies $E_{\mu\nu} \pm \hbar\omega_i$ (i ; odd–parity normal modes) where $E_{\mu\nu}$ is the electronic excitation energy taking no account of nuclear vibrations. The dipole strengths, $S_k = |P_{\mathrm{eff},k}|^2$, of these lines for the k-polarization are given from (9.61) as

$$S_k(E_{\mu\nu} \pm \hbar\omega_i) = \left(\frac{2}{\Delta E}\right)^2 \frac{\hbar}{2\omega_i} |\langle\mu \mid P_k V_i(\mathbf{r})| \nu\rangle|^2$$
$$\times \begin{cases} \langle v_i\rangle + 1 \\ \langle v_i\rangle \end{cases} \quad \text{for absorption,}$$
$$\times \begin{cases} \langle v_i\rangle \\ \langle v_i\rangle + 1 \end{cases} \quad \text{for emission,} \tag{9.62}$$

where $\langle v\rangle$ is the thermal average of v given by

$$\langle v\rangle = \frac{\sum_{v=0}^{\infty} v \exp[-(\hbar\omega/kT)v]}{\sum_{v=0}^{\infty} \exp[-(\hbar\omega/kT)v]} = \frac{1}{\exp(\hbar\omega/kT) - 1}. \tag{9.63}$$

In deriving (9.62) the initial state is assumed to be in the thermal equilibrium. In the absorption the intensity of the $E_{\mu\nu} - \hbar\omega_i$ line is zero at $T = 0$, while in the emission the intensity of the $E_{\mu\nu} + \hbar\omega_i$ line is zero at $T = 0$ as $\langle v\rangle_{T=0} = 0$. At high temperatures the intensities of the both $E_{\mu\nu} \pm \hbar\omega_i$ lines are proportional to T, as

$$\frac{1}{\exp(\hbar\omega/kT) - 1} \approx \frac{kT}{\hbar\omega} \quad \text{for} \quad \frac{\hbar\omega}{kT} \ll 1. \tag{9.64}$$

So far we have considered a particular case in which the equilibrium nuclear positions as well as the angular frequencies of the corresponding normal-modes are the same in both the initial and final states. However, in many real problems this is not the case; The equilibrium position and the angular frequency of the harmonic oscillator of a certain mode associated with the ν electronic state are, in general, different from those of the corresponding mode associated with the μ electronic state. We denote the $\bar{v}$th vibrational function of the ν state for a certain vibrational mode as $\chi_\nu^{\bar{v}}(\mathbf{Q})$, and expand it in terms of the vibrational wavefunctions of the μ state as follows:

$$\chi_\nu^{\bar{v}}(\mathbf{Q}) = \sum_v a_{\bar{v}v} \chi_\mu^{v}(\mathbf{Q}). \tag{9.65a}$$

This is always possible as $\chi_\mu^{v}(\mathbf{Q})$ with $v = 0, 1,\ldots$, form a complete orthonormal set. In (9.65a) coefficients $a_{\bar{v}v}$ satisfy the relation

$$\sum_v |a_{\bar{v}v}|^2 = \sum_{\bar{v}} |a_{\bar{v}v}|^2 = 1. \tag{9.65b}$$

Then, $\mathbf{P}_{\text{eff}}(\mu v \to \nu \bar{v})$ involves the matrix element

$$\langle v \mid Q \mid \bar{v} \rangle = \sum_{v'} a_{\bar{v}v'} \langle v \mid Q \mid v' \rangle, \tag{9.66}$$

where v' indicates the v' vibrational state associated with the μ electronic state. As seen from (9.66), Q may connect any $\bar{v}$ with v as far as $a_{\bar{v},v-1}$ and $a_{\bar{v},v+1}$ are nonzero. Therefore, we expect many nonvanishing electric-dipole transition lines for a certain mode at the photon-energies, $E_{\mu\nu} + \hbar(\omega v - \bar{\omega}\bar{v})$ where $\bar{\omega}$ is the angular frequency of the harmonic oscillator of the corresponding mode in the ν state. In many cases these lines are difficult to resolve and observed as a broad band. The integrated dipole strength of this band induced by the ith odd-parity mode is obtained from

$$S_k(\mu \to \nu) = \left(\frac{2}{\Delta E}\right)^2 \frac{\hbar}{2\omega_i} |\langle \mu \mid P_k V_i(\mathbf{r}) \mid \nu \rangle|^2$$
$$\times \sum_{\bar{v}} \sum_{vv'} \exp\left[-\frac{\hbar\omega_i}{kT} v\right] | a_{\bar{v}v'} |^2 |\langle v \mid Q \mid v' \rangle|^2 \Big/ \sum_v \exp\left[-\frac{\hbar\omega_i}{kT} v\right], \tag{9.67}$$ ‡

which is identical to the sum of $S_k(E_{\mu\nu} + \hbar\omega_i)$ and $S_k(E_{\mu\nu} - \hbar\omega_i)$ given in (9.62) because of (9.65). Therefore, the temperature dependence of $S_k(\mu \to \nu)$ is given by the factor

$$2\langle v_i \rangle + 1 = \coth \frac{\hbar\omega_i}{2kT}, \tag{9.68}$$

which is again proportional to T at high temperatures.

9.2.3 Spectral Line Shapes

As seen from (5.5) and (5.16), the absorption coefficient $k(\nu)$ at the photon energy $h\nu$ is given as

$$k(\nu) = \frac{8\pi^3 N}{3c} \nu \frac{| P(\nu)|^2}{h\, \Delta\nu}, \tag{9.69}$$ §

‡ In deriving (9.67), the relation

$$\sum_{\bar{v}} \left| \sum_{v'} a_{\bar{v}v'} \langle v \mid Q \mid v' \rangle \right|^2 = \sum_{\bar{v}v'} | a_{\bar{v}v'} |^2 |\langle v \mid Q \mid v' \rangle|^2$$

was used. This relation may be proved by using the relation

$$\sum_{\bar{v}} a^*_{\bar{v}v'} a_{\bar{v}v''} = \delta_{v'v''}.$$

§ In dielectric media of refractive index n, the right-hand side of (9.69) should be divided by n.

where N is the number of absorption centers per cubic centimeter and $|P(\nu)|^2$ is the dipole strength in the photon energy range $h\,\Delta\nu$ at $h\nu$. The dipole strength $|P(\nu)|^2$ divided by $h\,\Delta\nu$ is called a spectral shape function and is denoted by $F(\nu)$. This function gives the spectral line shape. The spectral shape function for the $\mu \rightarrow \bar{\mu}$ electronic transition at temperature T is given by

$$F_{\mu\bar{\mu}}(\nu) = (1/\hbar\,\Delta\nu) \sum_{v} \sum_{\bar{v}} |\langle \mu v \mid \tilde{P} \mid \bar{\mu}\bar{v} \rangle|^2 \rho_{\mu v}\,, \tag{9.70}$$

where the summation over v and $\bar{v}$ should be performed with the restriction, $h\nu < E_{\bar{\mu}\bar{v}} - E_{\mu v} < h(\nu + \Delta\nu)$,

$$\rho_{\mu v} = \exp(-E_{\mu v}/kT) \Big/ \sum_{v'} \exp(-E_{\mu v'}/kT),$$

and $\tilde{P}$ is the effective electric-dipole moment operator for a certain polarization. The restriction on the summation in (9.70) may be removed if one uses the delta function in the following way:

$$F_{\mu\bar{\mu}}(\nu) = \sum_{v} \sum_{\bar{v}} \rho_{\mu v}\, |\langle \mu v \mid \tilde{P} \mid \mu v \rangle|^2\, \delta(E_{\bar{\mu}\bar{v}} - E_{\mu v} - h\nu). \tag{9.71}$$

Elaborate theoretical works have been done on the calculation of the shape function in the systems with a single vibrational mode[‡] and also with many vibrational modes.[§] In order to simplify the problem without losing the physical insight, we adopt a semiclassical approximation for the calculation of $F(\nu)$, which is valid at high temperatures.

Let us suppose that the adiabatic potentials $U_\mu(Q)$ and $U_{\bar{\mu}}(Q)$ are given as shown in Fig. 9.7. Note that the following argument is valid only when Q_0 is much larger than the vibrational amplitude of the ground state. For simplicity, we assume that the system has a single nondegenerate vibrational mode whose normal coordinate is Q. The absorption takes place around $Q = 0$, and the important final states for the transition will be the excited vibrational states with large vibrational quantum numbers if the dipole strength does not depend upon Q appreciably. Since these vibrational states can be treated classically, the final vibrational state $\chi_{\bar{\mu}}^{\bar{v}}(Q)$ will oscillate rapidly except near the classical turning points, i.e., the points Q for which $U_{\bar{\mu}}(Q) = E_{\bar{\mu}\bar{v}}$. Therefore, a good approximation to the sum over $\bar{v}$ can be obtained by replacing $E_{\bar{\mu}\bar{v}}$ in the delta function

‡ K. Huang and A. Rhys, *Proc. Roy. Soc.* **A204**, 406 (1950).

§ M. Lax, *J. Chem. Phys.* **86**, 929 (1952). R. Kubo and Y. Toyozawa, *Progr. Theor. Phys. (Kyoto)* **13**, 160 (1955).

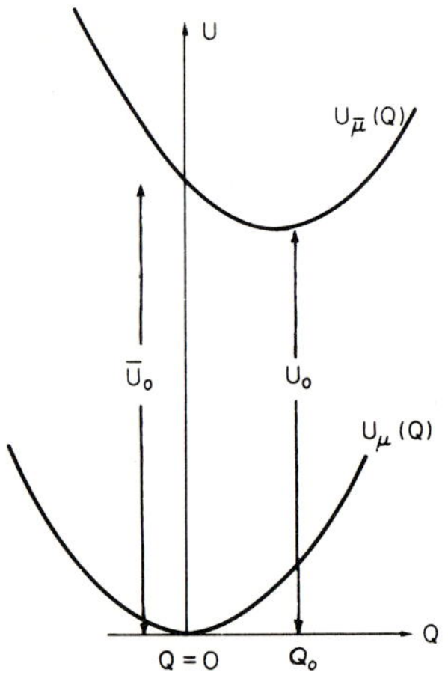

FIG. 9.7. Adiabatic potentials $U_\mu(Q)$ and $U_{\bar\mu}(Q)$.

in (9.71) by $U_{\bar\mu}(Q)$. The summation over $\bar v$ can now be performed with the closure relationship

$$\sum_{\bar v} \chi_{\bar\mu}{}^{\bar v}(Q')\,\chi_{\bar\mu}{}^{\bar v}(Q) = \delta(Q - Q'),$$

and (9.71) simplifies to the form,

$$F_{\mu\bar\mu}(\nu) = \sum_v \rho_{\mu v} \int dQ\, |\tilde P_{\mu\bar\mu}(Q)|^2\, |\chi_\mu{}^v(Q)|^2\, \delta[U_{\bar\mu}(Q) - E_{\mu v} - h\nu], \quad (9.72)$$

where

$$\tilde P_{\mu\bar\mu}(Q) = \int d\boldsymbol{\tau} \Psi_\mu(\mathbf{r} : Q{=}0)^* \tilde P \Psi_{\bar\mu}(\mathbf{r} : Q{=}0).$$

Furthermore, if the temperature is high enough so that many initial vibrational levels v are populated, $E_{\mu v}$ in (9.72) may be replaced by $U_\mu(Q)$ with the result

$$F_{\mu\bar\mu}(\nu) = \int dQ\, |\tilde P_{\mu\bar\mu}(Q)|^2 \rho_\mu(Q)\, \delta[\Delta U(Q) - h\nu], \quad (9.73)$$

where

$$\Delta U(Q) = U_{\bar\mu}(Q) - U_\mu(Q), \quad (9.74)$$

$$\rho_\mu(Q) = \exp[-U_\mu(Q)/kT] \Big/ \int dQ \exp[-U_\mu(Q)/kT]. \quad (9.75)$$

The semiclassical approximation, under which (9.37) was derived, is called the *Franck–Condon approximation*. The physical implication of this approximation is intuitively clear. It shows that the electronic transition takes place so rapidly that the nuclear positions do not change during the transition.

In the treatment of the parity-forbidden transitions in the ligand-field theory, $\tilde{P}_{\mu\bar{\mu}}(Q)$ is calculated as

$$\tilde{P}_{\mu\bar{\mu}}(Q) = \frac{2Q}{\Delta E} \int d\boldsymbol{\tau} \Psi_{\mu}(\mathbf{r} : Q{=}0)^* V_{\mathrm{odd}}(\mathbf{r})\, \Psi_{\bar{\mu}}(\mathbf{r} : Q{=}0)$$
$$= Q\tilde{P}^{0}_{\mu\bar{\mu}}, \tag{9.76}$$

where $V_{\mathrm{odd}}(\mathbf{r})$ is the coefficient of odd-parity normal coordinate Q in the expansion of $\mathscr{H}_{\mathrm{en}}$. Assuming the adiabatic potentials‡

$$U_{\mu}(Q) = \tfrac{1}{2}KQ^2, \tag{9.77}$$

$$U_{\bar{\mu}}(Q) = U_0 + \tfrac{1}{2}K(Q - Q_0)^2, \tag{9.78}$$

one can calculate the semiclassical shape function (9.73) which is now given in the form

$$F_{\mu\bar{\mu}}(Q) = \frac{|\tilde{P}^{0}_{\mu\bar{\mu}}|^2 \int dQ\, Q^2 \exp(-KQ^2/2kT)\, \delta(U_0 - KQ_0Q + \tfrac{1}{2}KQ_0^{\,2} - h\nu)}{\int dQ \exp(-KQ^2/2kT)}. \tag{9.79}$$

By using the formulas for the delta function,

$$\delta(ax) = (1/|\,a\,|)\,\delta(x),$$
$$\int_{-\infty}^{\infty} f(x)\,\delta(x - b)\,dx = f(b), \tag{9.80}$$

and performing the integration in (9.79), one obtains

$$F_{\mu\bar{\mu}}(\nu) = \frac{|\tilde{P}^{0}_{\mu\bar{\mu}}|^2(\bar{U}_0 - h\nu)^2}{8(2\pi)^{1/2}K^{5/2}Q_0^{\,3}(kT)^{1/2}} \exp\left[-\frac{(\bar{U}_0 - h\nu)^2}{8KQ_0^{\,2}kT}\right], \tag{9.81}$$

in which $\bar{U}_0$ is the excitation energy at $Q = 0$ as shown in Fig. 9.7. The shape function obtained in (9.81) is schematically illustrated in Fig. 9.8.

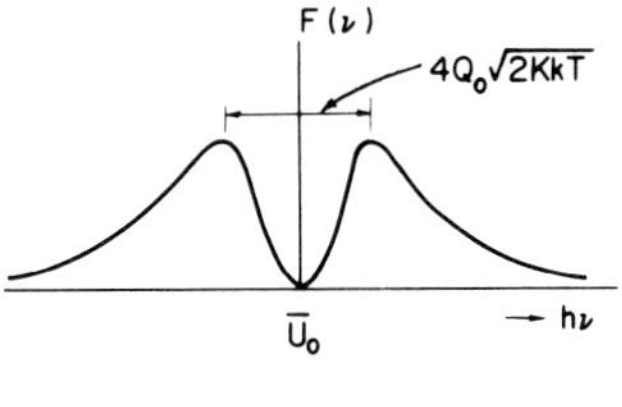

FIG. 9.8. The calculated line shape of a parity-forbidden transition.

‡ Here, to make the calculation simple, the same force constant K is assumed for both the μ and $\bar{\mu}$ states. This simplification does not change any essential result of the calculation.

The spectral shape thus calculated has a deep dip at $h\nu = \bar{U}_0$. This reflects the fact that the parity-forbidden transitions are allowed only at the moment when an instantaneous odd-parity distortion occurs: At $Q = 0$ where no distortion occurs the intensity should be zero. The separation of the two peaks is given by $4Q_0(2KkT)^{1/2}$ which is proportional to $\sqrt{T}$. If the shape function is integrated over ν, it is proportional to T in agreement with (9.68) at high temperatures. Although the calculated spectral shape exhibits a very interesting feature, it is premature to apply the result to real systems‡ which are subjected to the nuclear vibrations of even-parity modes as well as odd-parity modes. The inclusion of the even-parity modes will change the spectral shape, although it will not change the integrated intensity. Detailed theoretical studies of the spectral shapes of parity-forbidden transitions have not been worked out so far, and it is highly desirable to calculate the spectral line shapes in the systems having both even- and odd-parity vibrational modes. Moreover, in the systems with electronic degeneracy, the dynamical Jahn–Teller effect plays important roles in giving the spectral shapes. A brief account of the dynamical Jahn–Teller effect on the spectral shapes will be found in a later section.

In concluding this subsection it is interesting to compare the shape function in (9.81) with that for the parity-allowed transition, $\mu \to \bar{\mu}$. In the parity-allowed transition, the single vibrational mode to be considered is the even-parity mode and the main part of $\tilde{P}_{\mu\bar{\mu}}(Q)$ is independent of Q;

$$\tilde{P}_{\mu\bar{\mu}}(Q) \approx P^0_{\mu\bar{\mu}} = \int d\tau \Psi_\mu(\mathbf{r} : \mathbf{Q}{=}0)^* P \Psi_{\bar{\mu}}(\mathbf{r} : \mathbf{Q}{=}0). \tag{9.82}$$

We assume the same adiabatic potentials as those given in (9.77) and (9.78), although Q in the present case is the normal coordinate of the even-parity mode. Then, the shape function to be calculated is

$$F^0_{\mu\bar{\mu}}(\nu) = \frac{|P^0_{\mu\bar{\mu}}|^2 \int dQ \exp(-KQ^2/2kT)\, \delta(U_0 - KQ_0 Q + \tfrac{1}{2}KQ_0^2 - h\nu)}{\int dQ \exp(-KQ^2/2kT)}. \tag{9.83}$$

This integral function can be integrated in the same way as in the case of the forbidden transition. The result is given as

$$F^0_{\mu\bar{\mu}}(\nu) = \frac{|P^0_{\mu\bar{\mu}}|^2}{Q_0(2\pi KkT)^{1/2}} \exp\left[-\frac{(\bar{U}_0 - h\nu)^2}{2KQ_0^2kT}\right], \tag{9.84}$$

‡ If the system has inversion-symmetry, Q_0 in (9.78) is zero. Therefore, the argument mentioned here may be applied only to the systems without inversion-symmetry.

which is schematically illustrated in Fig. 9.9. If this shape function is integrated over ν, it is $|P^0_{\mu\bar{\mu}}|^2$ and independent of temperature in contrast to the case of the forbidden transition.

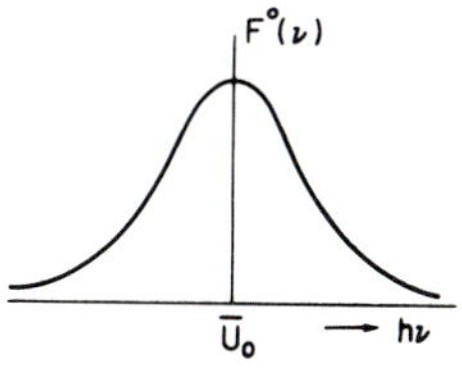

FIG. 9.9. The line shape of allowed transitions.

9.3 Static Jahn–Teller Effect

9.3.1 Jahn–Teller Theorem

In Section 9.2.1 we mentioned that the linear coupling between electrons and nuclear motion is absent in nondegenerate electronic states. However, this is not the case for degenerate states. We consider the degenerate orbital wavefunction involving $\Psi_{\Gamma\gamma}(\mathbf{r} : \mathbf{Q}_0)$ which is labeled with degenerate irreducible representation Γ. By using (6.89), the strength of the linear coupling in the Γ degenerate state is given as

$$\int d\tau \Psi_{\Gamma\gamma}(\mathbf{r} : \mathbf{Q}_0)^* V^{(\beta)}_{\bar{\Gamma}\bar{\gamma}} \Psi_{\Gamma\gamma'}(\mathbf{r} : \mathbf{Q}_0)$$
$$= (\Gamma)^{-1/2} \langle \Gamma \| V^{(\beta)}_{\bar{\Gamma}} \| \Gamma \rangle \langle \Gamma\gamma \mid \Gamma\gamma' \bar{\Gamma}\bar{\gamma} \rangle \quad (9.85)$$

($(\bar{\Gamma}) > 1$, and $\bar{\Gamma} \neq$ the identity representation) in which $V^{(\beta)}_{\bar{\Gamma}\bar{\gamma}}(\mathbf{r})$ is a real function in the interaction terms in (9.52) except the term with the identity representation. In the case of O-symmetry, integral (9.85) is nonvanishing if $\bar{\Gamma}$ indicating a normal mode is neither A_2 nor T_1, and if it appears in reducing $\Gamma \times \Gamma$. The first condition is derived from (6.93). More explicitly, (9.85) is nonvanishing in the following cases:

$$\begin{aligned} \bar{\Gamma} &= E, & \Gamma &= E, T_1, T_2, \\ \bar{\Gamma} &= T_2, & \Gamma &= T_1, T_2. \end{aligned} \quad (9.86)$$

The same result can be obtained in a more general way. Note that (9.85) may be reexpressed as

$$(9.85) = \frac{1}{2} \int d\tau [\Psi_{\Gamma\gamma}(\mathbf{r} : \mathbf{Q}_0)\, \Psi_{\Gamma\gamma'}(\mathbf{r} : \mathbf{Q}_0) + \Psi_{\Gamma\gamma'}(\mathbf{r} : \mathbf{Q}_0)\, \Psi_{\Gamma\gamma}(\mathbf{r} : \mathbf{Q}_0)]\, V^{(\beta)}_{\bar{\Gamma}\bar{\gamma}}(\mathbf{r}), \quad (9.87)$$

because the orbital wavefunction may be chosen to be real. We will show in Appendix IX that the function $[\Psi_{\Gamma\gamma}\Psi_{\Gamma\gamma'} + \Psi_{\Gamma\gamma'}\Psi_{\Gamma\gamma}]$ in (9.87) transforms like the base of the *symmetric product representation* $[\Gamma \times \Gamma]$, which is defined as

$$D^{[\Gamma\times\Gamma]}_{\gamma_1\gamma_2,\gamma\gamma'}(R) = \tfrac{1}{2}[D^{(\Gamma)}_{\gamma_1\gamma}(R)\, D^{(\Gamma)}_{\gamma_2\gamma'}(R) + D^{(\Gamma)}_{\gamma_1\gamma'}(R)\, D^{(\Gamma)}_{\gamma_2\gamma}(R)]. \tag{9.88}$$

In (9.88) R is a symmetry operation in the group of interest. Then, we can conclude that (9.85) is nonvanishing if $\bar{\Gamma}$ indicating a normal mode appears in reducing symmetric product representation $[\Gamma \times \Gamma]$. In the case of O-symmetry, $[\Gamma \times \Gamma]$ with degenerate Γ can be reduced as follows:

$$\begin{aligned} [E \times E] &= A_1 + E, \\ [T_1 \times T_1] &= A_1 + E + T_2\,, \\ [T_2 \times T_2] &= A_1 + E + T_2\,. \end{aligned} \tag{9.89}$$

Therefore, we obtain the same result as given in (9.86).

Jahn and Teller[‡] have shown by examining the systems belonging to all the point-groups that in molecular systems, except in linear molecules, we can always find the normal vibrational mode with symmetry $\bar{\Gamma}$ which appears in reducing the symmetric product of any degenerate irreducible representation Γ. If (9.85) is nonvanishing for a certain normal mode $\bar{\Gamma}$, at least a linear term with $\bar{\Gamma}$ appears in the adiabatic potential, which in turn tells us that the initially assumed nuclear configuration $\mathbf{Q}_0$ is unstable because the adiabatic potential is not minimum at $\mathbf{Q}_0$. Therefore, Jahn and Teller's finding may be stated as follows; Except linear molecules, degenerate orbital states in molecules are unstable. This statement is called the *Jahn–Teller theorem.*

Later Jahn[§] extended Jahn–Teller's work to the case in which spins are involved in the wavefunctions. He found that the degenerate states

[‡] H. A. Jahn and E. Teller, *Proc. Roy. Soc.* (*London*) **A161**, 220 (1937).

[§] H. A. Jahn, *Proc. Roy. Soc.* (*London*) **A164**, 117 (1938). If the system has an even number of electrons, the electronic state may be labeled with a single-valued irreducible representation of the group. Jahn has shown that in this case (9.85) involving spin is nonvanishing if $\bar{\Gamma}$ appears in $[\Gamma \times \Gamma]$. If the system has an odd number of electrons, the electronic state may be labeled with a double-valued irreducible representation $\tilde{\Gamma}$ of a double-group. Jahn has shown that in this case (9.85) involving spin is nonvanishing if $\bar{\Gamma}$ appears in *antisymmetric product representation* $\{\tilde{\Gamma} \times \tilde{\Gamma}\}$ which we will explain in Appendix IX. Thus, the problem is reduced to showing whether one can always find normal mode $\bar{\Gamma}$ which appears in $[\Gamma \times \Gamma]$ for any Γ $((\Gamma) > 1)$ in the system with an even number of electrons and in $\{\tilde{\Gamma} \times \tilde{\Gamma}\}$ for any $\tilde{\Gamma}$ $((\tilde{\Gamma}) > 2)$ in the system of an odd number of electrons. When $\tilde{\Gamma}$ is two-dimensional $\{\tilde{\Gamma} \times \tilde{\Gamma}\}$ involves only an identity representation. Therefore, in Kramers doublets (9.85) is always vanishing as $\bar{\Gamma}$ is not the identity representation.

including spin, which are not Kramers doublets, are unstable in molecules except linear ones.

9.3.2 E_g-State in a Cubic System

As a simple example showing Jahn–Teller instability, we shall consider here the E_g-state in the system with O_h-symmetry. We ignore spin because the instability due to spin always occurs through the spin-orbit interaction and thus is a small effect. Since $[E_g \times E_g] = A_{1g} + E_g$, the interaction term linear to the E_g normal coordinates are nonvansishing. In this case the adiabatic potential $U(Q_1Q_2)$ for normal coordinates, $Q_1 = Q_{E_gu}$ and $Q_2 = Q_{E_gv}$ is calculated from (9.56) by the use of formula (6.89) as

$$U(Q_1Q_2) = \tfrac{1}{2}\omega^2(Q_1^2 + Q_2^2)\,\mathbf{1} + A\begin{bmatrix} -Q_1 & Q_2 \\ Q_2 & Q_1 \end{bmatrix}, \tag{9.90}$$

in which ω is the angular frequency of the E_g vibrational mode and $\mathbf{1}$ is a two-dimensional unit matrix whose bases are u and v in this order. In (9.90) a constant term arising from the matrix element of $\mathscr{H}_e$ is discarded. The constant A of the linear coupling is given as

$$A = \tfrac{1}{2}\langle E_g \| V_{E_g}(\mathbf{r}) \| E_g \rangle. \tag{9.91}$$

It should be noted that, since the electronic state is doubly degenerate, the adiabatic potential is given in the form of a two-dimensional matrix. This adiabatic potential matrix can be diagonalized if one uses bases Ψ_1 and Ψ_2 given as follows:

$$\begin{aligned} \Psi_1 &= \Psi_u \cos\alpha - \Psi_v \sin\alpha, \\ \Psi_2 &= \Psi_u \sin\alpha + \Psi_v \cos\alpha, \end{aligned} \tag{9.92}$$

where

$$\tan 2\alpha = Q_2/Q_1\,. \tag{9.93}$$

If Q_1 and Q_2 are expressed in terms of the polar coordinates in the two-dimensional space as

$$Q_1 = \rho\cos\theta, \qquad Q_2 = \rho\sin\theta, \qquad (\rho \geqslant 0,\ \ 0 \leqslant \theta < 2\pi) \tag{9.94}$$

Eq. (9.93) shows that

$$\alpha = \theta/2. \tag{9.95}$$

With the bases given in (9.92) and (9.95), two surfaces of the adiabatic potential are given by

$$U_{1,2}(\rho\theta) = \tfrac{1}{2}\omega^2\rho^2 \pm A\rho, \tag{9.96}$$

which is independent of θ. The minus and plus signs in (9.96) are associated with U_1 and U_2, respectively, and bases Ψ_1 and Ψ_2 are associated with U_1 and U_2, respectively. These two surfaces are illustrated in Fig. 9.10.

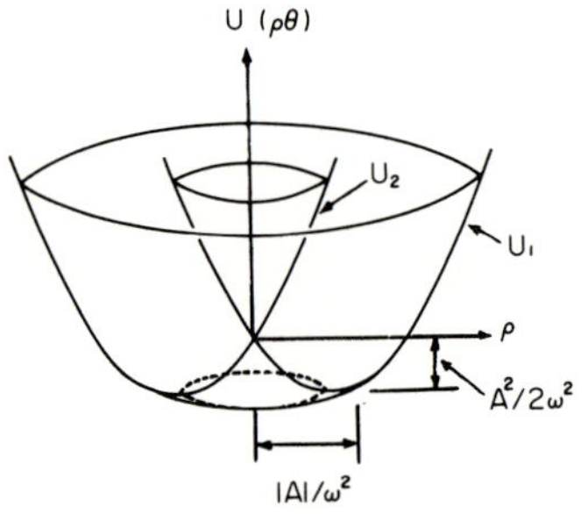

FIG. 9.10. Two energy surfaces of the E_g state.

As shown in Fig. 9.10, the minimum of the lower surface occurs at $\rho_{\min} = |A|/\omega^2$ with an arbitrary value of θ. Therefore, the stable nuclear configuration of the system cannot be determined from (9.90). To determine the stable nuclear configuration, we have to go a step further and take into account the anharmonicity of the vibration. The anharmonicity may be expressed by the cubic term $\sum B_{nm}Q_1{}^nQ_2{}^m$ with $n + m = 3$ and should be invariant to any operation in the O_h-group. By using the same method as that used for obtaining low-symmetry ligand fields (Section 6.1.1), the anharmonic term is readily obtained as

$$B(Q_1{}^3 - 3Q_1Q_2{}^2) = B\rho^3 \cos 3\theta. \tag{9.97}$$

Adding this anharmonic term to (9.96), one obtains three minima (maxima for $B > 0$) at

$$\left.\begin{array}{ll} \theta_{\min} = \dfrac{2\pi}{3} n, & n = 0, 1, 2 \\ \rho_{\min} \doteqdot |A|/\omega^2 & (12\,|AB|/\omega^4 \ll 1) \end{array}\right\} \quad \text{for} \quad B < 0 \tag{9.98}$$

and the maxima (minima for $B > 0$) at

$$\left.\begin{array}{ll} \theta_{\max} = \dfrac{2\pi}{3} n, & n = \dfrac{1}{2}, \dfrac{3}{2}, \dfrac{5}{2} \\ \rho_{\max} \doteqdot \rho_{\min} & (12\,|AB|/\omega^4 \ll 1) \end{array}\right\} \quad \text{for} \quad B < 0. \tag{9.99}$$

The stable nuclear configuration corresponding to these minima of the energy surface for $B < 0$ are given from (9.98) as

$$Q_1\,(n{=}0) = \rho_{\min}\,, \qquad Q_2\,(n{=}0) = 0; \tag{9.100a}$$

$$Q_1\,(n{=}1) = -\frac{1}{2}\rho_{\min}\,, \qquad Q_2\,(n{=}1) = \frac{\sqrt{3}}{2}\rho_{\min}; \tag{9.100b}$$

$$Q_1\,(n{=}2) = -\frac{1}{2}\rho_{\min}\,, \qquad Q_2\,(n{=}2) = -\frac{\sqrt{3}}{2}\rho_{\min}; \tag{9.100c}$$

and for $B > 0$ by

$$Q_1\left(n{=}\frac{1}{2}\right) = \rho_{\min}\,, \qquad Q_2\left(n{=}\frac{1}{2}\right) = \frac{\sqrt{3}}{2}\rho_{\min}; \tag{9.101a}$$

$$Q_1\left(n{=}\frac{3}{2}\right) = -\rho_{\min}\,, \qquad Q_2\left(n{=}\frac{3}{2}\right) = 0; \tag{9.101b}$$

$$Q_1\left(n{=}\frac{5}{2}\right) = \frac{1}{2}\rho_{\min}\,, \qquad Q_2\left(n{=}\frac{5}{2}\right) = -\frac{\sqrt{3}}{2}\rho_{\min}\,. \tag{9.101c}$$

The stable configurations for $B < 0$ are illustrated in Fig. 9.11. The stable configurations for $B > 0$ with $n = \frac{1}{2}$, $\frac{3}{2}$, and $\frac{5}{2}$ are obtained, respectively, from those with $n = 2$, 0, and 1 for $B < 0$ by changing the sign of the displacement vector, **S**. As seen in Fig. 9.11, three stable configurations correspond to the three equivalent tetragonal distortions along the x-, y-, and z-axes.

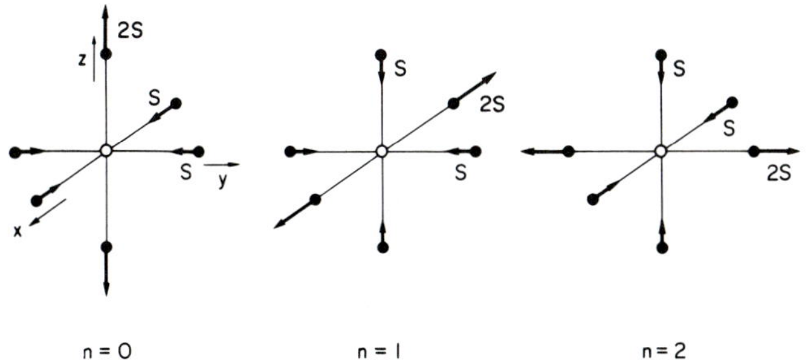

FIG. 9.11. Three equivalent tetragonal distortions.

Problem 9.3. Calculate the stable nuclear configuration of the T_1 and T_2 states in the cubic system. ◇

9.3.3 Experimental Evidence

One of the experimental evidences for the static Jahn-Teller effect has been provided by the paramagnetic resonance measurements‡ in a $CuSiF_6 \cdot 6H_2O$ crystal. The unit cell of this crystal determined by x-ray experiments contains one $[Cu(H_2O)_6]^{2+}$ complex ion. However, the observed paramagnetic resonance spectrum of the Cu^{2+} ion at low temperatures indicates the presence of three kinds of Cu^{+2} ions with different tetragonal symmetry axes; $g_\perp = 2.11$ and $g_\parallel = 2.46$. At high temperatures the spectrum shows an isotropic g-value, $g = 2.24$.

Since the ground state of the Cu^{2+} ion is $t_{2g}^6 e_g^3\ {}^2E_g$, one immediately notices that the low temperature spectrum may be explained by introducing the static Jahn–Teller distortions of the Cu^{2+} sites, which have to be tetragonal as discussed in the previous subsection. The presence of the three kinds of Cu^{2+} ions corresponds to the presence of three equivalent tetragonal distortions. The anisotropic g-value in the low temperature spectrum may be calculated by using the basic function Ψ_1 in (9.92) associated with the lower surface of the adiabatic potential, U_1. Taking into account the nondiagonal matrix elements of the spin-orbit interaction and the Zeeman term as done in the example of Section 8.4.2, one may evaluate the g-values in three directions as

$$g_x = 2 + \frac{2\zeta}{10Dq}\left(\sin\frac{\theta}{2} - \sqrt{3}\cos\frac{\theta}{2}\right)^2, \tag{9.102a}$$

$$g_y = 2 + \frac{2\zeta}{10Dq}\left(\sin\frac{\theta}{2} + \sqrt{3}\cos\frac{\theta}{2}\right)^2, \tag{9.102b}$$

$$g_z = 2 + \frac{8\zeta}{10Dq}\left(\sin\frac{\theta}{2}\right)^2, \tag{9.102c}$$

where ζ is the spin-orbit coupling constant for a single electron. If the system is stabilized at the potential minima at low temperatures, the insertion of $\theta_{\min}$ given in (9.98) and (9.99) into (9.102) leads to the g-values, for example,

$$\left.\begin{aligned} g_x = g_y = g_\perp &= 2 + \frac{6\zeta}{10Dq} \quad &(9.103a)\\ g_z = g_\parallel &= 2 \quad &(9.103b)\end{aligned}\right\} \quad \text{for} \quad B < 0,$$

$$\left.\begin{aligned} g_x = g_y = g_\perp &= 2 + \frac{2\zeta}{10Dq} \quad &(9.104a)\\ g_z = g_\parallel &= 2 + \frac{8\zeta}{10Dq} \quad &(9.104b)\end{aligned}\right\} \quad \text{for} \quad B > 0,$$

‡ B. Bleany and D. J. E. Ingram, *Proc. Phys. Soc.* (*London*) **A63**, 408 (1950). B. Bleany and K. D. Bowers, *Proc. Phys. Soc.* (*London*) **A65**, 667 (1952).

corresponding to the distortion along the z-axis. At high temperatures the system begins to rotate taking all the possible values of θ. Thus, the high temperature g-value is calculated by averaging (9.102) over θ. The result is

$$g_x = g_y = g_z = 2 + (4\zeta/10Dq). \tag{9.105}$$

The experimentally determined g-shifts, $\Delta g = g - 2$, in the low-temperature spectrum are $\Delta g_{\parallel}{}^L = 0.46$ and $\Delta g_{\perp}{}^L = 0.11$ and the g-shift at the high-temperatures spectrum is $\Delta g^H = 0.24$, which is almost equal to $\frac{1}{2}\Delta g_{\parallel}{}^L$ and $2\,\Delta g_{\perp}{}^L$ in agreement with the results given in (9.104) and (9.105). The result in (9.103) is not supported by the experimental data. From this one may conclude that parameter B is positive in this crystal.

9.4 Dynamical Jahn–Teller Effect

9.4.1 Quantum-Mechanical Treatment

In the previous section we showed that in the E_g state the introduction of anharmonicity may bring the system into stable nuclear configurations. However, in the absence of such anharmonicity or in the case in which the zero-point energy of nuclear vibration exceeds the stabilization energy due to the anharmonicity, the system undergoes a motion from one equilibrium configuration to another. In this case the motions of electrons and nuclei are strongly coupled. This effect is called the *dynamical Jahn–Teller effect*.

To illustrate the dynamical Jahn–Teller effect, we first consider a simple example of the E state in which the Jahn–Teller motion is confined to the lower surface of the adiabatic potential, U_1. This is the case in which the energy quantum of the oscillation, $\hbar\omega$, is much smaller than the Jahn–Teller stabilization energy, $A^2/2\omega^2$, and only the lower lying vibrational levels are dealt with. For simplicity the anharmonic term will be neglected.

It is convenient to express the basic functions in (9.92) in terms of the electronic wavefunctions u_+ and u_- defined in (6.25):

$$\begin{aligned} \Psi_1 &= -\frac{1}{\sqrt{2}}(u_+e^{i\theta/2} - u_-e^{-i\theta/2}), \\ \Psi_2 &= \frac{1}{\sqrt{2}}(u_+e^{i\theta/2} + u_-e^{-i\theta/2}). \end{aligned} \tag{9.106}$$

These functions are considered to be the eigenfunctions of an equation similar to (9.2). Following the adiabatic approximation, we assume that the total eigenfunction of the electron plus nuclei system has the form

$$\Phi(\mathbf{rQ}) = \Psi_1(\mathbf{r} : \mathbf{Q})\chi(\mathbf{Q}), \tag{9.107}$$

because the motion is mostly confined to the lower surface. Then, using the relation

$$\begin{aligned}
&-\frac{\hbar^2}{2}\left(\frac{\partial^2}{\partial Q_1{}^2} + \frac{\partial^2}{\partial Q_2{}^2}\right)\Psi_1(\mathbf{r} : \mathbf{Q})\,\chi(\mathbf{Q}) \\
&= -\frac{\hbar^2}{2}\left(\frac{\partial^2}{\partial \rho^2} + \frac{1}{\rho}\frac{\partial}{\partial \rho} + \frac{1}{\rho^2}\frac{\partial^2}{\partial \theta^2}\right)\Psi_1(\mathbf{r} : \mathbf{Q})\,\chi(\mathbf{Q}) \\
&= -\frac{\hbar^2}{2}\left[\Psi_1(\mathbf{r} : \mathbf{Q})\frac{\partial^2}{\partial \rho^2} + \frac{1}{\rho}\Psi_1(\mathbf{r} : \mathbf{Q})\frac{\partial}{\partial \rho} - \frac{1}{4\rho^2}\Psi_1(\mathbf{r} : \mathbf{Q})\right. \\
&\qquad \left. - \frac{1}{\rho^2}\Psi_2(\mathbf{r} : \mathbf{Q})\frac{\partial}{\partial \theta} + \frac{1}{\rho^2}\Psi_1(\mathbf{r} : \mathbf{Q})\frac{\partial^2}{\partial \theta^2}\right]\chi(\mathbf{Q}),
\end{aligned} \tag{9.108}$$

one obtains the equation for $\chi(\mathbf{Q})$ similar to (9.4) as follows:

$$\begin{aligned}
&\int d\tau \Psi_1{}^*(\mathbf{r} : \mathbf{Q})\left[-\frac{\hbar^2}{2}\left(\frac{\partial^2}{\partial Q_1{}^2} + \frac{\partial^2}{\partial Q_2{}^2}\right) + U_1(\mathbf{Q})\right]\Psi_1(\mathbf{r} : \mathbf{Q})\,\chi(\mathbf{Q}) \\
&= \left[-\frac{\hbar^2}{2}\left(\frac{\partial^2}{\partial \rho^2} + \frac{1}{\rho}\frac{\partial}{\partial \rho} + \frac{1}{\rho^2}\frac{\partial^2}{\partial \theta^2} - \frac{1}{4\rho^2}\right) + \frac{\omega^2}{2}\rho^2 - A\rho\right]\chi(\mathbf{Q}) \\
&= E\chi(\mathbf{Q}).
\end{aligned} \tag{9.109}$$

In deriving (9.109) the orthogonality relation between Ψ_1 and Ψ_2 was used. By setting

$$\chi(\mathbf{Q}) = \rho^{-1/2}F(\rho)\, e^{il\theta}, \tag{9.110}$$

Eq. (9.109) gives the equation which $F(\rho)$ satisfies:

$$\left[\frac{d^2}{d\rho^2} - \frac{l^2}{\rho^2} - \omega^2\rho^2 + 2A\rho + 2E\right]F(\rho) = 0. \tag{9.111}$$

Here, for simplicity, $\hbar$ is taken to be unity. Furthermore, by changing the variable ρ into η which is related to ρ by

$$\eta = \rho - (A/\omega^2) \tag{9.112}$$

and setting

$$F(\rho) = G(\eta), \tag{9.113}$$

Eq. (9.111) can be reexpressed in the form

$$\left[\frac{d^2}{d\eta^2} - \frac{l^2}{(\eta + A/\omega^2)^2} - \omega^2\eta^2 + 2E'\right] G(\eta) = 0, \tag{9.114}$$

in which

$$E' = E + (A^2/2\omega^2). \tag{9.115}$$

Since we are dealing with the lower lying levels associated with the U_1 potential surface, it is a good approximation to neglect η in the denominator of the second term of (9.114). Then, the approximate equation for $G(\eta)$ is

$$\left(\frac{d^2}{d\eta^2} - \omega^2\eta^2 + 2E''\right) G(\eta) = 0, \tag{9.116}$$

where

$$E'' = E' - \frac{l^2}{2(A/\omega^2)^2}. \tag{9.117}$$

Equation (9.116) is the well-known equation for a simple harmonic oscillator with eigenvalues $(n + \frac{1}{2})\hbar\omega$ with $n = 0, 1, 2, \ldots$. Therefore, the eigenvalues of (9.109) are now found to be

$$E = \left(n + \frac{1}{2}\right)\hbar\omega + \frac{l^2}{2(A/\omega^2)^2} - \frac{A^2}{2\omega^2} \qquad (n = 0, 1, 2, \ldots). \tag{9.118}$$

In (9.118) possible values of l are determined from the condition that the total wavefunction $\Phi(\mathbf{rQ})$ is periodic with respect to θ with the period 2π. It follows from (9.106) and (9.110) that if

$$\left.\begin{aligned} l \pm \tfrac{1}{2} &= \text{integers}, \\ \text{or} \qquad l &= \pm\tfrac{1}{2}, \pm\tfrac{3}{2}, \pm\tfrac{5}{2}, \ldots \end{aligned}\right\}, \tag{9.119}$$

$\Phi(\mathbf{rQ})$ is periodic with the period 2π.

The energy eigenvalues given by (9.118) correspond to those of a one-dimensional harmonic oscillator whose center undergoes a rotatory motion in the circular orbit of radius $|A|/\omega^2$ and with angular momentum l at the bottom of the U_1 surface. The last term is the energy at the bottom of the U_1 surface. The degeneracy of each level is twofold corresponding to the plus and minus signs of l.

Next we consider the case in which the zero-point energy of the oscillation, $\frac{1}{2}\hbar\omega$, is larger than the stabilization energy $A^2/2\omega^2$. In this case the Jahn–Teller motion is no longer confined only to the lower surface of the adiabatic potential. For treating this case, we start with the zeroth order wavefunctions obtained by assuming $A = 0$. We use the electronic wavefunctions, $u_+(\mathbf{r} : \mathbf{Q}=0)$ and $u_-(\mathbf{r} : \mathbf{Q}=0)$, and the vibrational wavefunctions $\chi(\mathbf{Q})$ which satisfy the equation

$$\left[-\frac{\hbar^2}{2}\left(\frac{\partial^2}{\partial Q_1{}^2}+\frac{\partial^2}{\partial Q_2{}^2}\right)+\frac{\omega^2}{2}(Q_1{}^2+Q_2{}^2)-E\right]\chi(\mathbf{Q})=0. \tag{9.120}$$

As the solution of (9.120) we choose the form[‡]

$$\chi_{nm}(\rho\theta) = F_{n|m|}(\rho)\, e^{im\theta}, \tag{9.121}$$

$$E_{nm} = \hbar\omega n, \tag{9.122}$$

where $n = 1, 2, \ldots$, and $m = n-1, n-3, \ldots, -n+1$. For the function $F_{n|m|}(\rho)$, see Pauling and Wilson's book.[‡] Thus, the zeroth order wavefunctions are given by

$$\Phi^{\pm}_{nm} = u_\pm(\mathbf{r})\,\chi_{nm}(\rho\theta), \tag{9.123}$$

and the corresponding energy level has $2n$-fold degeneracy.

The Jahn–Teller interaction, which is linear in $\mathbf{Q}$ and treated as a perturbation, is expressed as

$$\begin{aligned}\mathscr{H}_{J-T} &= V_u(\mathbf{r})Q_1 + V_v(\mathbf{r})Q_2\\ &= \frac{\rho}{\sqrt{2}}\,[V_{u_-}(\mathbf{r})\,e^{i\theta} - V_{u_+}(\mathbf{r})\,e^{-i\theta}].\end{aligned} \tag{9.124}$$

By using the matrix elements

$$\begin{aligned}\langle u_\pm \mid V_{u_+}(\mathbf{r}) \mid u_\pm\rangle &= \langle u_\pm \mid V_{u_-}(\mathbf{r}) \mid u_\pm\rangle = 0\\ \langle u_\mp \mid V_{u_\pm}(\mathbf{r}) \mid u_\pm\rangle &= \mp\sqrt{2}A\end{aligned} \tag{9.125}$$

and the fact[§] that the only nonvanishing matrix elements of $\rho e^{\pm i\theta}$ are

$$\begin{aligned}\langle\chi_{n,m} \mid \rho e^{-i\theta} \mid \chi_{n+1,m+1}\rangle &= \langle\chi_{n+1,m+1} \mid \rho e^{i\theta} \mid \chi_{n,m}\rangle = \left[\frac{\hbar}{2\omega}(n+m+1)\right]^{1/2},\\ \langle\chi_{n,m} \mid \rho e^{-i\theta} \mid \chi_{n-1,m+1}\rangle &= \langle\chi_{n-1,m+1} \mid \rho e^{i\theta} \mid \chi_{n,m}\rangle = \left[\frac{\hbar}{2\omega}(n-m-1)\right]^{1/2},\end{aligned} \tag{9.126}$$

‡ L. Pauling and E. B. Wilson, "Introduction to Quantum Mechanics." McGraw-Hill, New York, 1935.

§ H. C. Longuett-Higgins, U. Öpik, M. H. L. Pryce, and R. A. Sack, *Proc. Roy. Soc. (London)* **A244**, 1 (1958).

one obtains the matrix elements of $\mathscr{H}_{J-T}$ as follows:

$$\langle \Phi^{+}_{n,m} | \mathscr{H}_{J-T} | \Phi^{-}_{n',m'} \rangle = A \langle \chi_{n,m} | \rho e^{i\theta} | \chi_{n',m'} \rangle = A \left\{ \frac{\hbar}{2\omega} [n \pm (m-1)] \right\}^{1/2} \delta_{n',n\mp 1} \, \delta_{m',m-1} , \tag{9.127a}$$

$$\langle \Phi^{-}_{n,m} | \mathscr{H}_{J-T} | \Phi^{+}_{n',m'} \rangle = A \langle \chi_{n,m} | \rho e^{-i\theta} | \chi_{n',m'} \rangle = A \left\{ \frac{\hbar}{2\omega} [n \pm (m+1)] \right\}^{1/2} \delta_{n',n\pm 1} \, \delta_{m',m+1} . \tag{9.127b}$$

The matrix elements in (9.127a and b) show that, if we assign the quantum numbers‡ $j = \pm 1$ to $\Phi^{\pm}_{n,m}$, $\mathscr{H}_{J-T}$ connects the states with the same quantum number,§ $l = m - (1/2) j$ ($j = \pm 1$). The interaction scheme between the $\Phi^{\pm}_{n,m}$ levels due to $\mathscr{H}_{J-T}$ is indicated in Fig. 9.12, where the levels connected by broken lines interact with each other.

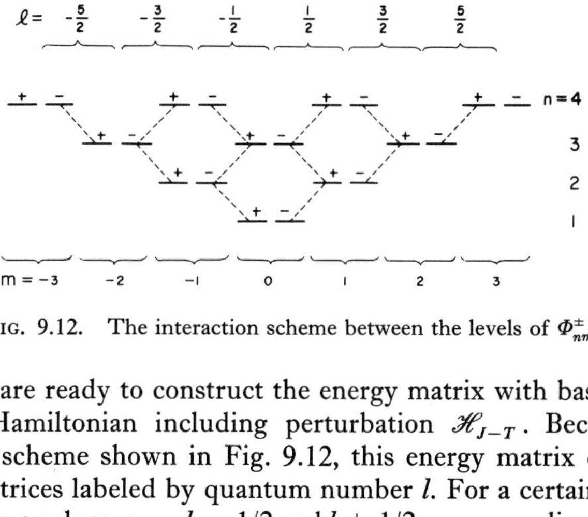

FIG. 9.12. The interaction scheme between the levels of $\Phi^{\pm}_{nm}$.

Now we are ready to construct the energy matrix with bases $\Phi^{\pm}_{n,m}$ for the total Hamiltonian including perturbation $\mathscr{H}_{J-T}$. Because of the interaction scheme shown in Fig. 9.12, this energy matrix decomposes into the matrices labeled by quantum number l. For a certain value of l, m can take two values, $m = l - 1/2$ and $l + 1/2$ corresponding to $j = -1$ and $+1$, respectively. For $m = l - 1/2$, the possible values of n are $|l - 1/2| + 1$, $|l - 1/2| + 3$, $|l - 1/2| + 5, \ldots$, and for $m = l + 1/2$ they are $|l + 1/2| + 1$, $|l + 1/2| + 3$, $|l + 1/2| + 5, \ldots$, as seen from (9.121) and (9.122). We arrange the bases in the order, $n = |l| + 1/2$, $|l| + 3/2$, $|l| + 5/2, \ldots$, irrespective of the sign of l. Then, from (9.122)

‡ Note that this j is equivalent to $-j$ in the paper by H. C. Longuett-Higgins *et al.*, *Proc. Roy. Soc. (London)* **A244**, 1 (1958).

§ For a more general argument, see W. Moffitt and W. Thorson, *Phys. Rev.* **108**, 1251 (1957).

and (9.127), one sees that the energy matrices with quantum member $\pm l$ are identical (therefore the energy levels are always doubly degenerate), and are given by

$$
\begin{array}{cccccc}
n = |l| + \frac{1}{2}, & |l| + \frac{3}{2}, & |l| + \frac{5}{2}, & |l| + \frac{7}{2}, & |l| + \frac{9}{2}, & \cdots
\end{array}
$$

$$
\begin{bmatrix}
m_0+1 & k\sqrt{m_0+1} & & & & \\
k\sqrt{m_0+1} & m_0+2 & k\sqrt{1} & & & \\
& k\sqrt{1} & m_0+3 & k\sqrt{m_0+2} & & \\
& & k\sqrt{m_0+2} & m_0+4 & k\sqrt{2} & \\
& & & k\sqrt{2} & m_0+5 & \cdots \\
& & & & \cdots & \cdots
\end{bmatrix}, \tag{9.128}
$$

in which

$$
\begin{aligned}
m_0 &= |l| - \tfrac{1}{2}, \\
k &= A/\omega(\hbar\omega)^{1/2},
\end{aligned} \tag{9.129}
$$

and all the matrix elements are given in the unit of $\hbar\omega$.

In our case of $(1/2)\,\hbar\omega \gg (A^2/2\omega^2)$ which is equivalent to $k^2 \ll 1$, the eigenvalues of (9.128) are obtained by using the second-order perturbation theory as

$$
\begin{aligned}
E_{pl} &= p + |l| - \tfrac{1}{2} - (|l| + \tfrac{1}{2})\,k^2 \qquad (p = 1, 3, 5,\ldots), \\
E_{pl} &= p + |l| - \tfrac{1}{2} + (|l| - \tfrac{1}{2})\,k^2 \qquad (p = 2, 4, 6,\ldots).
\end{aligned} \tag{9.130}
$$

This result shows that the $2n$-fold degeneracy of the nth level in the absence of the Jahn–Teller coupling is removed by $\mathscr{H}_{J-T}$ and only twofold degeneracy remains for all the energy levels. This twofold degeneracy corresponds to the same degeneracy found in the opposite case of $k^2 \gg 1$ [see (9.118) and (9.119)]. The energy of the lowest level, $l = \pm\frac{1}{2}$, $p = 1$, is given by

$$
E_{1,\pm(1/2)} = 1 - k^2, \tag{9.131a}
$$

or multiplying by $\hbar\omega$,

$$
E_{1,\pm(1/2)} = \hbar\omega - (A^2/\omega^2), \tag{9.131b}
$$

which means that the zero-point energy of a one-dimensional oscillator, $(\frac{1}{2})\,\hbar\omega$, is reduced by the amount equal to the Jahn–Teller stabilization energy, $A^2/2\omega^2$.

For various values of k^2, the eigenvalues of the secular matrices given in (9.128) with several values of l have been numerically calculated by Longuett-Higgins *et al.*

9.4.2 Spectral Line Shapes of Allowed Transitions

As one of the observable phenomena in which the dynamical Jahn–Teller effect plays an important role, we briefly mention here the spectral line shape of a parity-allowed transition in which the final state has orbital degeneracy while the initial state is an orbital singlet.

The simplest problem of this sort is the parity-allowed $A \rightarrow E$ transition in which only the E_g vibrational mode couples with the electronic states. The quantum mechanical calculation of the spectral line shape in this problem is possible if the eigenvectors of the secular equation obtained from (9.128) are known. For simplicity we assume that the same vibrational functions $\chi_{nm}(\mathbf{Q})$ in (9.121) are associated with both the initial and the final states, although in real problems there is no reason for the frequencies of the both states to be the same. Then, by using the orthogonality relations among the $\chi_{nm}(\mathbf{Q})$'s, it is straightforward to calculate the relative intensities of the $A_{nm} \rightarrow E_{pl}$ transitions if the eigenvectors of the E_{pl} levels are available. The calculation has been made by Longuett-Higgins, *et al.*, and the results for the $A_{10} \rightarrow E_{p(1/2)}$ transitions with $k^2 = 5$, 10, and 20 are shown in Fig. 9.13. The intensities of the $A_{10} \rightarrow E_{pl}$ transitions with $|l| \neq 1/2$ are zero, as $\chi_{10}(\mathbf{Q})$ is not mixed in the eigenvectors of the pl levels with $|l| \neq 1/2$ as seen from Fig. 9.12. The appearance of two intensity maxima is a characteristic feature of the results, and reflects the existence of two energy surfaces in the final state.

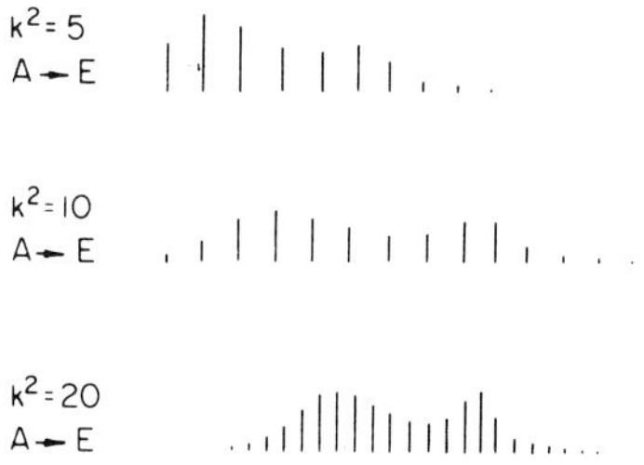

FIG. 9.13. The calculated intensities of the $A_{10} \rightarrow E_p$ allowed transitions [H. C. Longuett-Higgins, U. Öpik, M. H. L. Pryce, and R. A. Sack, *Proc. Roy. Soc.* (*London*) **A244**, 1 (1959)].

Such a characteristic feature may also be reproduced in a much simpler calculation‡ by using the semiclassical Franck–Condon approximation. In the semiclassical calculation we assume that

$$U_{E(1,2)}(Q) = U_0 + \tfrac{1}{2}K\rho^2 \mp A\rho, \qquad U_A(Q) = \tfrac{1}{2}K\rho^2. \tag{9.132}$$

‡ Y. Toyozawa and M. Inoue, *J. Phys. Soc. Japan* **21**, 1663 (1966).

Then the spectral line shape function as given in (9.73) is calculated as

$$F_{AE}(\nu) = \frac{1}{2}\sum_{1,2} F_{A,E(1,2)}(\nu)$$

$$= \frac{|P^0_{AE}|^2 \sum_{1,2} \int_0^{2\pi} d\theta \int_0^\infty d\rho\, \rho \exp[-K\rho^2/2kT]\, \delta(U_0 \mp A\rho - h\nu)}{2\int_0^{2\pi} d\theta \int_0^\infty d\rho\, \rho \exp[-K\rho^2/2kT]}$$

$$= \frac{|P^0_{AE}|^2 K\, |h\nu - U_0|}{2A^2kT} \exp\left[-\frac{K(h\nu - U_0)^2}{2A^2kT}\right], \qquad (9.133)$$

in which P^0_{AE} is the electric-dipole transition moment as given in (9.82). The shape function in (9.133) is plotted in Fig. 9.14. This shape function has two maxima corresponding to those found in the quantum mechanical calculation (Fig. 9.13). The central dip of the spectral shape comes from the mathematical nature of the adiabatic potential surfaces at $\mathbf{Q} = 0$.

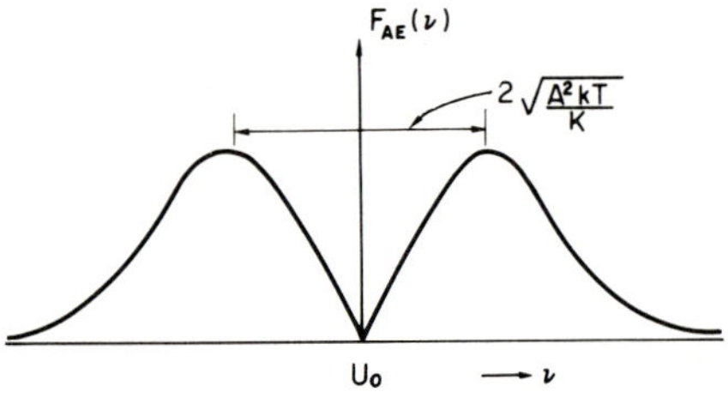

FIG. 9.14. The calculated line shape of the $A \rightarrow E$ allowed transition [Y. Toyozawa and M. Inoue, *J. Phys. Soc. Japan* **21**, 1663 (1966)].

Toyozawa and Inoue‡ have shown that such splitting of the absorption band appears when the point $\mathbf{Q} = 0$ is the *branch point* of adiabatic potential surfaces. When the point $\mathbf{Q} = 0$ is the *intersection point*, no splitting is expected. It should be remarked that, if the degeneracies of the electronic state and the coupled vibration are the same, the point $\mathbf{Q} = 0$ is always a branch point and, if the degeneracy of the coupled vibration is smaller than that of the electronic state, the point $\mathbf{Q} = 0$ is the intersection point. The splittings of the absorption bands due to the $A_{1g} \rightarrow T_{1u}$ transition have been observed in alkali halide phosphors with heavy metal impurities and they have been nicely explained on the bases of the dynamical Jahn–Teller effect described here.

‡ Y. Toyozawa and M. Inoue, *J. Phys. Soc. Japan* **21**, 1663 (1966).

Chapter X MOLECULAR ORBITAL AND HEITLER–LONDON THEORIES

10.1 Strong- and Weak-Field Schemes

In the ligand-field theory developed in this book, we started with accommodating a certain number of electrons in the one-electron orbitals which are the bases of the irreducible representation of the symmetry group of the system. However, as long as the physical quantities such as reduced matrices are left as parameters to be determined experimentally, it was unnecessary to know the detailed functional forms of these one-electron orbitals. To simplify the theory we sometimes used the d-function approximation in which the t_{2g} and e_g orbitals have only the d-character, and we estimated the values of the physical parameters by using the radial functions of free ions. In doing this the estimated values for some parameters are found to be in fair agreement with those experimentally determined, but, as we will show later, at the same time we meet serious difficulties: the sign of the calculated crystal-field splitting parameter is wrong and the calculated unpaired spin densities on the atomic parts of ligands differs by orders of magnitudes from that found by experiments. These difficulties show that the purely ionic model employed in Bethe's crystalline-field theory, in which the electron transfer between the metal ion and the ligand is completely ignored, cannot be the physical model of the ligand-field theory. In this circumstance it has become quite necessary to discuss further the detailed functional forms of the one-electron orbitals appearing in the ligand-field theory to provide a sound physical basis of the ligand-field theory.

The ligand-field theory called a *strong-field scheme* is formulated in this book so that it gives a good description of the system with a very large cubic-field splitting, and its nature is similar to that of the molecular orbital (MO) theory: the t_{2g} and e_g orbitals may be regarded as one-electron molecular orbitals with given symmetry. Therefore, the MO treatment of these orbitals makes it possible to go a step beyond the purely ionic model.

On the other hand, a theory may be formulated so as to give a good description of the system with a very small cubic-field splitting, which, though not discussed in this book, is called a *weak-field scheme*. In the weak-field scheme the basic functions are those for free ions, $d^N SLM_s M_L$, while they are $t_{2g}^n e_g{}^m S\Gamma M_s \gamma$ in the strong-field approximation. Here L is the resultant orbital angular momentum and M_L is its component. Naturally the weak-field scheme is faithful to the purely ionic model employed in Bethe's crystalline-field theory. In this scheme the matrices of the Coulomb interaction are already diagonal, and both the diagonal and nondiagonal matrix elements of the cubic-field potential appear in and between the $SL\Gamma$ and $SL'\Gamma$ states with the same Γ. However, both the strong- and weak-field schemes are identical when, in addition to the d-function approximation, all the configuration interactions between the $t_{2g}^n e_g{}^m S\Gamma$ terms and $t_{2g}^{n'} e^{m'} S\Gamma$ terms are completely taken into account in the strong-field scheme, and all the nondiagonal matrix elements between the $SL\Gamma$ and $SL'\Gamma$ states are taken into account in the weak-field scheme. This is because both methods use the same subspace spanned by different sets of basic functions and take into account the same interactions.

We mentioned that in the strong-field scheme the MO treatment of the t_{2g} and e_g basic functions provides a natural way of getting out of the purely ionic model. Then, a question arises as to what kind of treatment will play the same role in the weak-field scheme. The answer is that the Heitler–London method provides a natural way of accounting for the deviation from the purely ionic model, when a small mixture of the excited configuration in which an electron is transferred between the metal and ligand ions is taken into account.

Since the ligand-field theory discussed so far is formulated in the strong-field scheme, in this chapter we will mainly stand by the MO model and see how the physical quantities appearing in the ligand-field theory may be interpreted. However, we will show later that the MO model is identical to the HL model at a certain stage, so that the interrelationship between these two models will always be kept in mind and discussed in some detail. The next two sections will be devoted to a simple description of the MO method for the systems with closed-shell and open-shell electron configurations.

10.2 Simple Description of MO Theory

10.2.1 Molecular Orbitals

Molecular orbitals (MO) are one-electron orbitals in molecules. In the MO theory a single electron is considered to move in an average field of the nuclei and the electrons, and its motion is described by an MO. The theoretical foundation of this picture is provided by the self-consistent field (SCF) theory of Hartree or Hartree–Fock, which we will discuss in the next subsection.

As seen from the argument given in Chapter I, the MO's should be the bases of the irreducible representations of the symmetry group to which the system belongs. Apart from the symmetry property, the intuitively obvious characteristic of MO is derived from the consideration that when an electron is near one nucleus, the forces exerted on it are those chiefly from the nucleus and the other electrons near it. In other words, the most important terms of the Hamiltonian for the electron near nucleus A are those which comprise the Hamiltonian for an electron in an isolated atom A. Such a consideration leads us to the approximate method of expressing MO by a suitable linear combination of atomic orbitals (AO), called the LCAO (Linear Combination of Atomic Orbitals) method.

For example, we consider a molecule consisting of two atoms A and B with atomic orbitals φ_A and φ_B, respectively. The LCAO MO in this case is

$$\psi = C_A\varphi_A + C_B\varphi_B, \tag{10.1}$$

in which C_A and C_B are numerical coefficients. For simplicity we assume that ψ has no orbital degeneracy, and that the molecule has two electrons accommodated in ψ, i.e., it has a closed-shell configuration. In this case we will show in the next subsection that the ψ with the lowest orbital energy satisfies the equation

$$h\psi = \epsilon\psi, \tag{10.2}$$

where h is an appropriate Hamiltonian for a single electron and ϵ is the orbital energy. In general, the occupied molecular orbitals of the system satisfy Eq. (10.2) in the case of a closed-shell electron configuration, but they do not when the system has an open-shell configuration. The open-shell case will be discussed in the next section.

Now, ϵ, C_A, and C_B are determined by the following secular equations derived from (10.2):

$$\begin{aligned} C_A(\epsilon_A - \epsilon) + C_B(\beta - S\epsilon) &= 0, \\ C_A(\beta - S\epsilon) + C_B(\epsilon_B - \epsilon) &= 0, \end{aligned} \tag{10.3}$$

where

$$\epsilon_A = \langle \varphi_A \mid h \mid \varphi_A \rangle, \qquad \epsilon_B = \langle \varphi_B \mid h \mid \varphi_B \rangle,$$

$$\beta = \langle \varphi_A \mid h \mid \varphi_B \rangle = \langle \varphi_B \mid h \mid \varphi_A \rangle, \qquad S = \langle \varphi_A \mid \varphi_B \rangle.$$

From (10.3) one may obtain the equation to determine ϵ,

$$(\epsilon_A - \epsilon)(\epsilon_B - \epsilon) - (\beta - S\epsilon)^2 = 0. \tag{10.4}$$

Since the second term of (10.4) is positive, the first term should also be positive. This means that the two roots of (10.4), ϵ^a and ϵ^b, are located as

$$\begin{aligned} \epsilon^a &> \epsilon_A \quad \text{and} \quad \epsilon^a > \epsilon_B, \\ \epsilon^b &< \epsilon_A \quad \text{and} \quad \epsilon^b < \epsilon_B. \end{aligned} \tag{10.5}$$

Relation (10.5) is visualized in Fig. 10.1, in which $\epsilon_A < \epsilon_B$ is assumed.

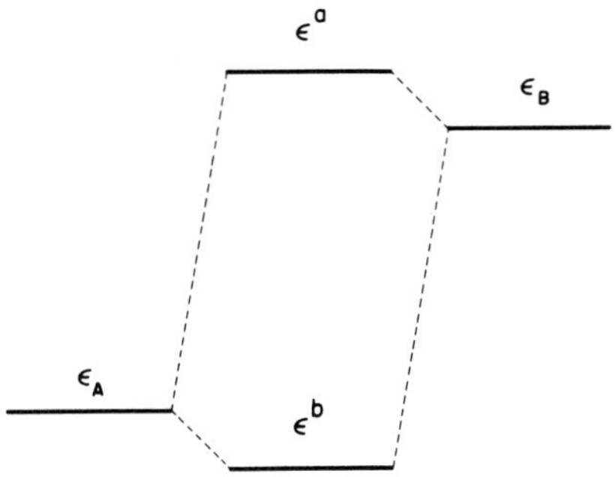

FIG. 10.1. Formation of bonding and antibonding energy levels.

The MO's, ψ^b and ψ^a having orbital energies ϵ^b and ϵ^a, are called *bonding and antibonding molecular orbitals*, respectively. These names come from the fact that the system is stabilized by accommodating electrons in the bonding MO while its energy is increased by placing electrons in the antibonding MO. The bonding and antibonding orbitals are orthogonal to each other.‡ It should be noted that ϵ_A and ϵ_B are not the orbital energies of atomic orbitals φ_A and φ_B, but do correspond quite closely. Here β is called the *resonance integral* between φ_A and φ_B and, together

‡ Strictly speaking, since ψ^a is not the occupied orbital, it is not clear that ψ^a satisfies Eq. (10.2). As shown later, if ψ^a is assumed to satisfy Eq. (10.2) as done here, one can prove that ψ^a is orthogonal to ψ^b. A theoretically rigorous way of introducing ψ^a is to assume the orthogonality between ψ^a and ψ^b at the beginning and to derive ψ^a from ψ^b by assuming that ψ^a is given by a linear combination of the same set of AO's as in ψ^b.

with *overlap integral* S, is responsible for the formation of the MO's; if $\beta = S = 0$, one has the relations, $\epsilon^b = \epsilon_A$, $\epsilon^a = \epsilon_B$, $\psi^b = \varphi_A$, and $\psi^a = \varphi_B$, as seen from (10.4).

If $\beta/(\epsilon_B - \epsilon_A)$ and S are small compared with unity, the amount of the energy gained by the molecular orbital formation, $\epsilon_A - \epsilon^b$, is small compared to $\epsilon_B - \epsilon_A$. The same is true for $\epsilon^a - \epsilon_B$. Then, the bonding and antibonding orbital energies are simply calculated from (10.4) as

$$\begin{aligned} \epsilon^b &\approx \epsilon_A - \frac{(\beta - \epsilon_A S)^2}{\epsilon_B - \epsilon_A}, \\ \epsilon^a &\approx \epsilon_B + \frac{(\beta - \epsilon_B S)^2}{\epsilon_B - \epsilon_A}. \end{aligned} \tag{10.6}$$

The MO's in this case are given by

$$\begin{aligned} \psi^b &= N_b^{-1/2}(\varphi_A + \gamma\varphi_B), \\ \psi^a &= N_a^{-1/2}(\varphi_B - \lambda\varphi_A), \end{aligned} \tag{10.7}$$

in which coefficients γ and λ are small compared with unity. From the orthogonality between ψ^b and ψ^a, λ is related to γ as

$$\lambda = \frac{\gamma + S}{1 + \gamma S}. \tag{10.8}$$

In (10.7) $N_b^{-1/2}$ and $N_a^{-1/2}$ are normalization constants. By inserting (10.6) into (10.3) and neglecting small quantities of higher order, the relation

$$\gamma \approx -\frac{\beta - \epsilon_A S}{\epsilon_B - \epsilon_A} \tag{10.9}$$

may be derived. To the same approximation λ is given as

$$\lambda \approx -\frac{\beta - \epsilon_B S}{\epsilon_B - \epsilon_A} \approx \gamma + S. \tag{10.10}$$

Now, in the ground state of our two-electron system both electrons are placed in the bonding orbital. Therefore, the ground state is a singlet and the wavefunction is given by the Slater determinant

$$\Psi = |\,\psi^b\bar{\psi}^b\,|, \tag{10.11}$$

in which a normalization factor is included as defined in (2.15) and the meaning of the upper bar was explained at the beginning of Section 2.1.3.

By substituting (10.7), (10.11) may be expressed in terms of the AO's as

$$\Psi = N_b^{-1}[| \varphi_A\bar{\varphi}_A | + \gamma(| \varphi_A\bar{\varphi}_B | + | \varphi_B\bar{\varphi}_A |) + \gamma^2 | \varphi_B\bar{\varphi}_B |]. \qquad (10.12)$$

In (10.12) the first term $| \varphi_A\bar{\varphi}_A |$ represents the state in which two electrons are located on atom A. This state corresponds to a purely ionic configuration. The second term, $| \varphi_A\bar{\varphi}_B | + | \varphi_B\bar{\varphi}_A |$, represents the state in which one of the electrons on atom A in the ionic configuration is transferred to atom B. This term may be interpreted as corresponding to a covalent configuration in which the transferred electron makes a covalent bond with the other electron having the opposite spin. Parameter γ is determined so as to minimize the total energy. The third term, $| \varphi_B\bar{\varphi}_B |$, represents the state in which two electrons are transferred to atom B and may be neglected if γ is much smaller than unity. Neglecting the third term, we may express (10.12) as

$$\Psi = \Psi_{\text{ion}} + \gamma\Psi_{\text{cov}} \, . \qquad (10.13)$$

This is equivalent to the wavefunction used in the HL method in which the covalent configuration Ψ_{cov} is mixed with the dominant ionic configuration.

To compare the present problem with the opposite case of a small ionic contribution to a predominantly covalent bond, we cite the example of a hydrogen molecule, in which atoms A and B are identical. In this case $\epsilon_A = \epsilon_B$ and (10.4) gives two eigenvalues,

$$\epsilon^{\pm} = \frac{\epsilon_A \pm \beta}{1 \pm S} = \epsilon_A \pm \frac{\beta - \epsilon_A S}{1 \pm S} \, . \qquad (10.14)$$

By inserting these energy eigenvalues into (10.3), the MO's ψ^+ and ψ^- associated with ϵ^+ and ϵ^-, respectively, are found to be

$$\psi^{\pm} = [2(1 \pm S)]^{-1/2}(\varphi_A \pm \varphi_B). \qquad (10.15)$$

In this case the coefficients of the atomic orbitals are independent of β, S, and ϵ_A. They are determined by the symmetry of the molecule. Since $\beta - \epsilon_A S$ is negative according to a numerical estimate, ψ^+ has an orbital energy lower than that of ψ^-. Therefore, the wavefunction of the ground state is given by

$$\begin{aligned}\Psi(H_2) &= | \psi^+\bar{\psi}^+ | \\ &= \frac{1}{2(1 + S)} [| \varphi_A\bar{\varphi}_B | + | \varphi_B\bar{\varphi}_A | + | \varphi_A\bar{\varphi}_A | + | \varphi_B\bar{\varphi}_B |]. \qquad (10.16)\end{aligned}$$

The first two terms in (10.16) represent a covalent configuration and the last two a purely ionic configuration. Thus, the wavefunction in the MO theory contains the covalent and ionic configurations equally:

$$\Psi^{\mathrm{MO}}(H_2) = \Psi_{\mathrm{cov}} + \Psi_{\mathrm{ion}} . \tag{10.17}$$

On the other hand, in the HL theory without ionic configurations, the wavefunction of the ground singlet is given by

$$\begin{aligned}\Psi^{\mathrm{HL}}(H_2) &= [2(1 + S^2)]^{-1/2}[\varphi_A(1)\,\varphi_B(2) + \varphi_B(1)\,\varphi_A(2)] \\ &\quad \times \frac{1}{\sqrt{2}}\,[\alpha(1)\,\beta(2) - \beta(1)\,\alpha(2)] \\ &= [2(1 + S^2]^{-1/2}[|\,\varphi_A\bar{\varphi}_B\,| + |\,\varphi_B\bar{\varphi}_A\,|],\end{aligned} \tag{10.18}$$

which corresponds to Ψ_{cov}. The better wavefunction lies between (10.17) and (10.18) and is known to be rather close to (10.18). Therefore, in the hydrogen problem, the simple MO theory gives too much weight to the ionic configuration as a result of the coefficients of the linear combination being fixed by the symmetry.

10.2.2 Hartree–Fock Equations

The theoretical foundation for the concept of one-electron orbitals in a many-electron system is provided by the SCF (Self-Consistent Field) theory of Hartree–Fock, which leads us to the Hartree–Fock (HF) equations for one electron-orbitals. In order to describe this theory in a simple fashion, we consider the molecule having two electrons as discussed in the previous subsection, and assume that ψ is the MO of the lowest orbital energy with no orbital degeneracy. As given in (10.11), the total wavefunction is given by

$$\Psi = |\,\psi\bar{\psi}\,|. \tag{10.19}$$

The HF equation is derived by minimizing the total energy by setting

$$\delta\langle\Psi\,|\,\mathscr{H}\,|\,\Psi\rangle = 0, \tag{10.20}$$

with a subsidiary condition,

$$\langle\psi\,|\,\psi\rangle = 1, \tag{10.21}$$

which assures the normalization of Ψ. The total Hamiltonian in (10.20) is

$$\mathscr{H} = \sum_{i=1,2} f_i + g_{12} , \tag{10.22}$$

where f_i is a one-electron operator acting on electron i including the kinetic energy and the nuclear attraction potential, and g_{12} is the Coulomb interaction operator between electrons 1 and 2. One obtains from (10.19) and (10.22)

$$\langle \Psi \mid \mathscr{H} \mid \Psi \rangle = 2\langle \psi \mid f \mid \psi \rangle + \langle \psi\psi \parallel \psi\psi \rangle, \tag{10.23}$$

in which the abbreviated notation for the Coulomb integral, $\langle \psi\psi \parallel \psi\psi \rangle$, was defined in (2.76).

Now by using (10.23), our variation problem reduces to finding Ψ which satisfies the equation

$$\delta[\langle \Psi \mid \mathscr{H} \mid \Psi \rangle - 2\epsilon\langle \psi \mid \psi \rangle] = 4\langle \delta\psi \mid f \mid \psi \rangle + 4\langle \delta\psi\psi \parallel \psi\psi \rangle - 4\epsilon\langle \delta\psi \mid \psi \rangle = 0 \tag{10.24}$$

for an arbitrary $\delta\psi$ and the normalization condition (10.21) at the same time. In (10.24) 2ϵ is Lagrange's undetermined multiplier for the condition (10.21). It is clear that (10.24) is satisfied for an arbitrary $\delta\psi$ if the following equation is satisfied:

$$h\psi = \epsilon\psi, \tag{10.25}$$

where

$$h = f + \langle \psi \parallel \psi \rangle. \tag{10.26}$$

Here, we have used the abbreviation

$$\langle \psi_i \parallel \psi_j \rangle = \int d\boldsymbol{\tau}_2 \psi_i^*(2)\, g_{12} \psi_j(2).$$

Equation (10.25) is called the HF equation and has already been used in the previous subsection. Equations (10.25) and (10.26) show that an electron moves in the averaged potential $\langle \psi \parallel \psi \rangle$ due to another electron, in addition to the nuclear potential included in f.

The present example for a two-electron system can easily be extended to the cases of closed shells containing more electrons. Also in such a many-electron problem, the HF equation like (10.25) for each occupied orbital may be derived by applying the variation principle. The HF Hamiltonian h in this case involves, in addition to f, the energy due to the averaged Coulomb and exchange potentials due to other electrons. Since the self-Coulomb and -exchange interaction energies cancel, one may formally include them in the HF Hamiltonian. Then, the HF equations for different orbitals are expressed by the use of the same HF

Hamiltonian. This assures the orthogonality between different orbitals.‡ For example, for the system of four electrons in nondegenerate orbitals ψ_1 and ψ_2, the HF equations are given by

$$h\psi_1 = \epsilon_1\psi_1, \qquad h\psi_2 = \epsilon_2\psi_2, \tag{10.27}$$

where

$$h = f + 2[\langle\psi_1 \| \psi_1\rangle + \langle\psi_2 \| \psi_2\rangle] - [\langle\psi_1 \| P_{12}\psi_1\rangle + \langle\psi_2 \| P_{12}\psi_2\rangle]. \tag{10.28}$$

In (10.28) P_{12} is the permutation operator between electrons 1 and 2, so that

$$\langle\psi_i \| P_{12}\psi_j\rangle = \int d\boldsymbol{\tau}_2 \psi_i^*(2)\, g_{12} P_{12}\psi_j(2).$$

By using the relation, $\langle\psi_1 | h | \psi_2\rangle = \langle\psi_2 | h | \psi_1\rangle^*$, it follows from (10.27) that

$$\langle\psi_1 | \psi_2\rangle(\epsilon_1 - \epsilon_2) = 0, \tag{10.29}$$

which shows that

$$\langle\psi_1 | \psi_2\rangle = 0 \qquad \text{if} \qquad \epsilon_1 \neq \epsilon_2 . \tag{10.30}$$

Problem 10.1. Derive (10.27) and (10.28). ◇

10.3 MO Theory for Open Shells

10.3.1 Hartree–Fock Equations

The MO theory or the HF theory for open-shell systems is slightly different from that for closed-shell systems described in the previous section. In the ligand-field theory magnetic electrons or chromophoric electrons are placed in open shells, so that it is necessary to describe here the theory for open shells.

For this purpose we consider a molecule having three electrons. Suppose in the ground state two electrons are spin-paired in a nondegenerate ψ_2 and a remaining electron with an up-spin is in another nondegenerate orbital ψ_1. The total wavefunction of this ground state is given by a Slater determinant as

$$\Psi = |\, \psi_1 \psi_2 \bar{\psi}_2 \,|. \tag{10.31}$$

‡ The orthogonality between different orbitals, as well as the normalization, is often imposed as subsidiary conditions in the variation problem. This variation procedure leads to the same result as ours if the system has a closed-shell configuration. In the present variation problem, the orthogonality conditions are not included in the subsidiary conditions, although the trial one-electron functions are assumed to be orthogonal to each other to retain the simple form of $\langle\Psi | \mathscr{H} | \Psi\rangle$.

The variation procedure to be taken is similar to that for closed shells, but an important difference is that one has to impose an additional subsidiary condition

$$\langle \psi_1 \mid \psi_2 \rangle = 0 \tag{10.32}$$

besides

$$\langle \psi_1 \mid \psi_1 \rangle = \langle \psi_2 \mid \psi_2 \rangle = 1. \tag{10.33}$$

It should be remembered that the orthogonality conditions between different orbitals are automatically fulfilled in the case of closed shells. The importance of the subsidiary orthogonality condition (10.32) in the case of open shells will be realized later.

By using the total Hamiltonian for the present system,

$$\mathscr{H} = \sum_{i=1}^{3} f_i + \sum_{j>i=1}^{3} g_{ij}, \tag{10.34}$$

it is straightforward to derive the following HF equations:

$$\begin{aligned} h_1\psi_1 &= \epsilon_1\psi_1 + \lambda_0\psi_2, \\ h_2\psi_2 &= \epsilon_2\psi_2 + \tfrac{1}{2}\lambda_0\psi_1, \end{aligned} \tag{10.35}$$

where

$$\begin{aligned} h_1 &= f + 2\langle \psi_2 \| \psi_2 \rangle - \langle \psi_2 \| P_{12}\psi_2 \rangle, \\ h_2 &= f + \langle \psi_1 \| \psi_1 \rangle + \langle \psi_2 \| \psi_2 \rangle - \tfrac{1}{2}\langle \psi_1 \| P_{12}\psi_1 \rangle. \end{aligned} \tag{10.36}$$

Here $2\epsilon_1$ and $2\epsilon_2$ are Lagrange's undetermined multiplier for conditions (10.33) and $4\lambda_0$ for condition (10.32). Assuming that ψ_1 and ψ_2 are real one can determine λ_0 from (10.35) and (10.36) as

$$\lambda_0 = -\langle \psi_1\psi_1 \| \psi_1\psi_2 \rangle. \tag{10.37}$$

Equations (10.35) are different from the HF equations for closed shells in two important points: (1) They are not Schrödinger type equations for one-electron orbitals because of the presence of the terms, $\frac{1}{2}\lambda_0\psi_1$ and $\lambda_0\psi_2$; (2) The one-electron Hamiltonians, h_1 and h_2, are different from each other while the same Hamiltonian can be employed in the case of closed shells. The first point seems to invalidate the orbital picture in which an electron moves in the average potential field coming from the other electrons and nuclei. Therefore, in this case the physical meaning of ϵ_1 and ϵ_2 is not clear. The second point together with the first does not assure the orthogonality between ψ_1 and ψ_2. For this reason the subsidiary orthogonality condition (10.32) had to be imposed.

10.3.2 Orbital Energies

In the previous subsection a limitation upon the orbital picture in the open-shell case was pointed out. However, here we will show that a physical meaning may be given to ϵ_1 and ϵ_2 in (10.37), which to some extent restores the validity of the concept of one-electron orbitals.

In order to define orbital energies in the open-shell problem, we use *Koopmans' theorem*, which states that, if the orbitals are assumed not to change their forms by the ionization of the molecule, the ionization potential of an electron in a given orbital is equal to the negative of the orbital energy of that orbital. In our problem the total wavefunction of the state in which an electron in the ψ_1 orbital is ionized is given by

$$\Psi_1 = | \psi_2 \bar{\psi}_2 |, \tag{10.38}$$

which represents a singlet ($S = 0$) state. The energy difference between states Ψ and Ψ_1 is related to the ionization potential, I_1, as follows:

$$-I_1 = \langle \Psi | \mathscr{H} | \Psi \rangle - \langle \Psi_1 | \mathscr{H}' | \Psi_1 \rangle, \tag{10.39}$$

where Ψ is the wavefunction given in (10.31), $\mathscr{H}$ the Hamiltonian in (10.34), and $\mathscr{H}'$ the two-electron Hamiltonian in (10.22). From the relations

$$\begin{aligned} \langle \Psi | \mathscr{H} | \Psi \rangle = 2\langle \psi_2 | f | \psi_2 \rangle + \langle \psi_1 | f | \psi_1 \rangle + 2\langle \psi_1\psi_2 \| \psi_1\psi_2 \rangle \\ - \langle \psi_1\psi_2 \| \psi_2\psi_1 \rangle + \langle \psi_2\psi_2 \| \psi_2\psi_2 \rangle, \end{aligned} \tag{10.40}$$

$$\langle \Psi_1 | \mathscr{H}' | \Psi_1 \rangle = 2\langle \psi_2 | f | \psi_2 \rangle + \langle \psi_2\psi_2 \| \psi_2\psi_2 \rangle, \tag{10.41}$$

and

$$\epsilon_1 = \langle \psi_1 | h_1 | \psi_1 \rangle = \langle \psi_1 | f | \psi_1 \rangle + 2\langle \psi_1\psi_2 \| \psi_1\psi_2 \rangle - \langle \psi_1\psi_2 \| \psi_2\psi_1 \rangle, \tag{10.42}$$

one obtains

$$-I_1 = \langle \psi_1 | f | \psi_1 \rangle + 2\langle \psi_1\psi_2 \| \psi_1\psi_2 \rangle - \langle \psi_1\psi_2 \| \psi_2\psi_1 \rangle = \epsilon_1 . \tag{10.43}$$

Thus, ϵ_1 turns out to be the orbital energy of ψ_1.

When an electron in ψ_2 is ionized, there appear a singlet and a triplet ($S = 1$) state. The singlet state is given by the wavefunction

$$\Psi_2^{\text{sing}} = \frac{1}{\sqrt{2}} [| \psi_1 \bar{\psi}_2 | - | \bar{\psi}_1 \psi_2 |], \tag{10.44}$$

and the triplet state with $M_S = 1$ is given by

$$\Psi_2^{\text{trip}} = | \psi_1 \psi_2 |. \tag{10.45}$$

The negatives of the ionization potential for the singlet and triplet states are given as

$$-I_2^{\text{sing}} = \langle \Psi \mid \mathscr{H} \mid \Psi \rangle - \langle \Psi_2^{\text{sing}} \mid \mathscr{H}' \mid \Psi_2^{\text{sing}} \rangle, \tag{10.46}$$

$$-I_2^{\text{trip}} = \langle \Psi \mid \mathscr{H} \mid \Psi \rangle - \langle \Psi_2^{\text{trip}} \mid \mathscr{H}' \mid \Psi_2^{\text{trip}} \rangle. \tag{10.47}$$

From the relations

$$\langle \Psi_2^{\text{sing}} \mid \mathscr{H}' \mid \Psi_2^{\text{sing}} \rangle = \langle \psi_1 \mid f \mid \psi_1 \rangle + \langle \psi_2 \mid f \mid \psi_2 \rangle + \langle \psi_1\psi_2 \parallel \psi_1\psi_2 \rangle + \langle \psi_1\psi_2 \parallel \psi_2\psi_1 \rangle, \tag{10.48}$$

$$\langle \Psi_2^{\text{trip}} \mid \mathscr{H}' \mid \Psi_2^{\text{trip}} \rangle = \langle \psi_1 \mid f \mid \psi_1 \rangle + \langle \psi_2 \mid f \mid \psi_2 \rangle + \langle \psi_1\psi_2 \parallel \psi_1\psi_2 \rangle - \langle \psi_1\psi_2 \parallel \psi_2\psi_1 \rangle, \tag{10.49}$$

and

$$\begin{aligned} \epsilon_2 &= \langle \psi_2 \mid h_2 \mid \psi_2 \rangle \\ &= \langle \psi_2 \mid f \mid \psi_2 \rangle + \langle \psi_2\psi_2 \parallel \psi_2\psi_2 \rangle + \langle \psi_1\psi_2 \parallel \psi_1\psi_2 \rangle - \tfrac{1}{2}\langle \psi_1\psi_2 \parallel \psi_2\psi_1 \rangle, \end{aligned} \tag{10.50}$$

one may show the relation

$$\tfrac{1}{4}(-3I_2^{\text{trip}} - I_2^{\text{sing}}) = \epsilon_2\,, \tag{10.51}$$

in which the left-hand side is the average of $-I_2^{\text{trip}}$ and $-I_2^{\text{sing}}$ taking into account the degeneracies associated with the states. In this way the physical meaning of ϵ_2 is clarified.

10.3.3 Covalency Parameter

So far we have not mentioned the details of ψ_1 and ψ_2. In the problem of transition metal compounds which we will mainly be concerned with, ψ_2 is predominantly of the character of the ligand orbitals with a small mixture of the d-orbital of the central metal ion, and ψ_1 is predominantly of the character of the d-orbital. The reason for this will be discussed in the next section. In this case ψ_1 and ψ_2 have the following forms:

$$\psi_1 = N_1^{-1/2}(\varphi - \lambda\chi), \tag{10.52}$$

$$\psi_2 = N_2^{-1/2}(\chi + \gamma\varphi), \tag{10.53}$$

where φ is the atomic d-orbital of the metal ion, χ a suitable linear combination of the atomic orbitals of the ligands, and $N_1^{-1/2}$ and $N_2^{-1/2}$ normalization constants. We assume that ψ_1 and ψ_2 given by (10.52) and (10.53) satisfy the symmetry requirement of the system. Because of the orthogonality relation (10.32), λ and γ are related to each other as

$$\lambda = (\gamma + S)/(1 + \gamma S), \tag{10.54}$$

where

$$S = \langle \varphi \mid \chi \rangle \tag{10.55}$$

is the overlap integral between φ and χ. Comparing (10.52) and (10.53) with (10.7), one notices that ψ_2 and ψ_1 correspond to the bonding and antibonding orbitals, respectively.

The physical meaning of γ becomes clear when one inserts (10.52) and (10.53) into the total wavefunction (10.31) with the result

$$\Psi = [N_2(1 - S^2)]^{-1/2}[\mid \varphi\chi\bar{\chi} \mid + \gamma \mid \varphi\chi\bar{\varphi} \mid]. \tag{10.56}$$

In deriving (10.56), use was made of (10.54) and

$$\begin{aligned} N_1 &= 1 - 2\lambda S + \lambda^2, \\ N_2 &= 1 + 2\gamma S + \gamma^2. \end{aligned} \tag{10.57}$$

The first term $\mid \varphi\chi\bar{\chi} \mid$ in (10.56) represents the state in which two electrons are placed in the ligand orbital. This state may be considered to correspond to the purely ionic configuration. The second term $\mid \varphi\chi\bar{\varphi} \mid$ represents the state in which a down-spin electron of the ligand orbital in the ionic configuration is transferred into the metal *d*-orbital. As mentioned in the previous section this corresponds to the covalent configuration. Therefore, the parameter γ measures the small degree of covalency in the ionic configuration. We should mention that (10.56) is equivalent to the HL wavefunction including a small admixture of the covalent configuration. The parameter γ is called *covalency parameter*.

The covalency parameter γ is determined by the variation principle as follows; Inserting the explicit forms of ψ_1 and ψ_2 given in (10.52) and (10.53) into the HF equations (10.35) derived from the variation principle, one obtains the equation to determine λ and γ as follows:

$$(N_1N_2)^{-1/2}[(B_1 - SA_1) + \lambda(A_1 - C_1) - \lambda^2(B_1 - SC_1)] = \lambda_0(1 - \lambda S), \tag{10.58}$$

$$2(N_1N_2)^{-1/2}[(B_2 - SC_2) + \gamma(A_2 - C_2) - \gamma^2(B_2 - SA_2)] = \lambda_0(1 + \gamma S), \tag{10.59}$$

where

$$A_\nu = \langle \varphi \mid h_\nu \mid \varphi \rangle, \qquad B_\nu = \langle \varphi \mid h_\nu \mid \chi \rangle, \qquad C_\nu = \langle \chi \mid h_\nu \mid \chi \rangle \qquad (\nu = 1, 2).$$

Then, replacing λ_0 in (10.58) and (10.59) by (10.37) and using the expression N_1 and N_2 given in (10.57) and the relation in (10.54), one is led, after some lengthy algebraic manipulation, to the equations

$$(\bar{B} - S\bar{A}) + \lambda(\bar{A} - \bar{C}) - \lambda^2(\bar{B} - S\bar{C}) = 0, \tag{10.60}$$

$$(\bar{B} - S\bar{C}) + \gamma(\bar{A} - \bar{C}) - \gamma^2(\bar{B} - S\bar{A}) = 0, \tag{10.61}$$

where

$$\bar{A} = \langle \varphi \mid \bar{h} \mid \varphi \rangle, \qquad \bar{B} = \langle \varphi \mid \bar{h} \mid \chi \rangle, \qquad \bar{C} = \langle \chi \mid \bar{h} \mid \chi \rangle,$$

and

$$\bar{h} = f + \langle \psi_1 \| \psi_1 \rangle + \langle \psi_2 \| \psi_2 \rangle. \tag{10.62}$$

It should be noted that Eqs. (10.60) and (10.61) are not independent of each other because of the relation (10.54). Expressing λ in (10.60) in terms of γ by using (10.54), one can show that (10.60) is automatically satisfied if (10.61) holds.

Neglecting the higher order small terms proportional to λ^2 and γ^2 in (10.60) and (10.61), one obtains

$$\lambda \approx \frac{-\bar{B} + S\bar{A}}{\bar{A} - \bar{C}}, \qquad \gamma \approx \frac{-\bar{B} + S\bar{C}}{\bar{A} - \bar{C}}; \tag{10.63}$$

hence

$$\lambda \approx \gamma + S.$$

It is important to note that h in (10.62) is the HF Hamiltonian for the down-spin electron in the bonding orbital ψ_2.

Problem 10.2. Prove (10.60) and (10.61) and show that these equations are not independent of each other. ◇

10.4 Covalency in Ligand-Field Theory

10.4.1 The t_{2g} and e_g Molecular Orbitals

At the beginning of this chapter we mentioned that the t_{2g} and e_g orbitals may be regarded as molecular orbitals with given symmetries. The purpose of this subsection is to derive the explicit forms of the t_{2g} and e_g MO's in LCAO.

First assume that the distance between the metal ion and the neigboring ligand is large enough so that there is essentially no overlap of the electron clouds of the metal ion and the ligand. In this case no electron is transferred between the metal ion and the ligand, and the system is purely ionic. Then, the t_{2g} and e_g orbitals are given in terms of the d-functions of the free metal ion as given in (1.31) and (1.32).

Now we bring the ligands closer to the metal ion keeping the octahedral symmetry of the system. Then, the electron clouds of the metal ion and of the ligands overlap and some of the electrons are transferred or exchanged between them. Suppose that the ligand ion has a closed-shell configuration $(1s)^2$ $(2s)^2$ $(2p)^6$ like an F^- ion when the metal and the ligand ions are sufficiently far apart. At such a large atomic distance, the energies of electrons in the atomic $1s$, $2s$, and $2p$ orbitals of the ligand are lower than those in the d-orbital of the metal ion. When the electron transfer begins to occur at a smaller distance, the metal and ligand orbitals are admixed and the energies of the electrons mainly in the ligand orbitals are depressed and those mainly in the d-orbitals are raised, resulting in the decrease of the total energy of the system. The stabilized orbital ψ^b is bonding and the destabilized orbital ψ^a antibonding. The bonding orbital has mainly the ligand character with a small admixture of the d-orbital, and the antibonding orbital mainly the d-orbital character with a small admixture of the ligand orbital.

In order to obtain the explicit forms of the bonding and antibonding MO's, we first have to construct from the ligand atomic orbitals the functions which are the bases of irreducible representations T_{2g} and E_g of the O_h-group. In doing this we can use the same method used for obtaining the normal coordinates of nuclear vibrations. When the ligand atomic orbitals are restricted to the p_x, p_y, and p_z type orbitals, we can consider a space similar to the displacement vector space but spanned by basic functions p_{ki} $(k = 1, 2,..., 6\,;\, i = x, y, z)$ and calculate transformation matrix $\mathbf{A}$ corresponding to that in (9.10). Since the basic functions of the p_{ki}'s transform exactly in the same way as the $\hat{e}_{ki}$'s do, it immediately follows that the A_{1g}, E_g, T_{1g}, $2T_{1u}$, T_{2g}, and T_{2u} irreducible representations are obtained by reducing representation A and the basic functions of these irreducible representations are simply given by replacing unit vectors $\hat{e}_{ki}$ in the normal coordinates in Section 9.1.2 by the ligand atomic orbitals p_{ki}. For convenience we denote the ligand atomic orbitals directed toward the central metal ion as $\varphi_{k\sigma}$ $(k = 1, 2,..., 6)$ whose portions at the metal ion side are always positive, and the other p_{ki} atomic orbitals as $\varphi_{ki}(k = 1, 2,..., 6\,;\, i = x, y, z)$ whose positive directions coincide with those of the x, y, and z-axes. Then, from (9.42), (9.43) and (9.44) the basic functions χ of irreducible representations T_{2g} and E_g are constructed as follows:

$$\begin{aligned} \chi_{u\sigma} &= (12)^{-1/2}(2\varphi_{3\sigma} + 2\varphi_{6\sigma} - \varphi_{1\sigma} - \varphi_{4\sigma} - \varphi_{2\sigma} - \varphi_{5\sigma}), \\ \chi_{v\sigma} &= \tfrac{1}{2}(\varphi_{1\sigma} + \varphi_{4\sigma} - \varphi_{2\sigma} - \varphi_{5\sigma}), \end{aligned} \tag{10.64}$$

$$\chi_{\xi\pi} = \tfrac{1}{2}(\varphi_{3y} - \varphi_{6y} + \varphi_{2z} - \varphi_{5z}), \tag{10.65a}$$

$$\chi_{\eta\pi} = \tfrac{1}{2}(\varphi_{1z} - \varphi_{4z} + \varphi_{3x} - \varphi_{6x}), \tag{10.65b}$$

$$\chi_{\zeta\pi} = \tfrac{1}{2}(\varphi_{1y} - \varphi_{4y} + \varphi_{2x} - \varphi_{5x}). \tag{10.65c}$$

As seen in (10.64) and (10.65) the basic functions $\chi_{u\sigma}$ and $\chi_{v\sigma}$ of irreducible representation E_g are constructed from the $\varphi_{k\sigma}$ orbitals and the basic functions $\chi_{\xi\pi}$, $\chi_{\eta\pi}$, and $\chi_{\zeta\pi}$ of T_{2g} from the π orbitals directing perpendicular to the $\varphi_{k\sigma}$'s.

Similarly, when the ligand atomic orbitals are the s orbitals, we consider a space spanned by basic functions φ_{ks} ($k = 1, 2, \ldots, 6$) and calculate transformation matrix $\mathbf{A}$ with these bases. In this case the small matrix corresponding to $\mathbf{A}^{(lk)}$ is a one-dimensional unit matrix, so that the character of representation A is simply given by N_R. Comparing N_R given in Table 9.1 with the characters in Table 1.7, one sees that A is reduced to irreducible representations, A_{1g}, E_g, and T_{1u}. Furthermore, considering that the transformation properties of $\varphi_{k\sigma}$ for symmetry operations in the O_h-group are the same as those of the φ_{ks}'s, one can obtain the basic functions χ_{us} and χ_{vs} of E_g from $\chi_{u\sigma}$ and $\chi_{v\sigma}$ by replacing $\varphi_{k\sigma}$ by φ_{ks}.

Now the MO's of the system are obtained by combining φ_γ and $\chi_{\gamma\kappa}$ ($\kappa = s, \sigma, \pi$) with the same γ as follows;

$$\begin{aligned} \psi_\gamma{}^a &= (N_\gamma{}^a)^{-1/2}(\varphi_\gamma - \sum_\kappa \lambda_\kappa \chi_{\gamma\kappa}), \\ \psi_\gamma{}^b &= (N_\gamma{}^b)^{-1/2}(\chi_{\gamma\kappa} + \gamma_\kappa \varphi_\gamma + \sum_{\kappa' \neq \kappa} \gamma_{\kappa\kappa'} \chi_{\gamma\gamma'}), \end{aligned} \tag{10.66}$$

where $\psi_\gamma{}^a$ and $\psi_\gamma{}^b$ are the antibonding and bonding MO's. For simplicity we denote $N_\gamma^{a,b}$, $\psi_\gamma^{a,b}$, φ_γ, and $\chi_{\gamma\kappa}$ for $\gamma = u$ and v as $N_e^{a,b}$, $\psi_e^{a,b}$, φ_e, and χ_κ ($\kappa = s, \sigma$), respectively, and those for $\gamma = \xi$, η, and ζ as $N_t^{a,b}$, $\psi_t^{a,b}$, φ_t and χ_π, respectively. Then, to the approximation in which only one kind of the s orbitals (for example, $2s$ in F^-) and only one kind of the p orbitals (for example, $2p$ in F^-) are taken into account, the antibonding orbitals are given as

$$\begin{aligned} \psi_e{}^a &= (N_e{}^a)^{-1/2}(\varphi_e - \lambda_s \chi_s - \lambda_\sigma \chi_\sigma), \\ \psi_t{}^a &= (N_t{}^a)^{-1/2}(\varphi_t - \lambda_\pi \chi_\pi), \end{aligned} \tag{10.67}$$

where

$$\begin{aligned} N_e{}^a &= 1 - 2\lambda_s S_s - 2\lambda_\sigma S_\sigma + \lambda_s{}^2 + \lambda_\sigma{}^2, \\ N_t{}^a &= 1 - 2\lambda_\pi S_\pi + \lambda_\pi{}^2, \end{aligned} \tag{10.68}$$

and the bonding orbitals are given as

$$\psi_{es}^b = (N_{es}^b)^{-1/2}(\chi_s + \gamma_s \varphi_e + \gamma_{s\sigma} \chi_\sigma), \tag{10.69a}$$

$$\psi_{e\sigma}^b = (N_{e\sigma}^b)^{-1/2}(\chi_\sigma + \gamma\, \varphi_e + \gamma_{\sigma s} \chi_s), \tag{10.69b}$$

$$\psi_t{}^b = (N_t{}^b)^{-1/2}(\chi_\pi + \gamma_\pi \varphi_t), \tag{10.69c}$$

where

$$N_{es}^b = 1 + 2\gamma_s S_s + \gamma_s^2 + 2\gamma_s \gamma_{s\sigma} S_\sigma , \tag{10.70a}$$

$$N_{e\sigma}^b = 1 + 2\gamma_\sigma S_\sigma + \gamma_\sigma^2 + 2\gamma_\sigma \gamma_{\sigma s} S_s , \tag{10.70b}$$

$$N_t^b = 1 + 2\gamma_\pi S_\pi + \gamma_\pi^2. \tag{10.70c}$$

In these expressions, the following overlap integrals are used:

$$S_s = \langle \varphi_e \mid \chi_s \rangle, \qquad S_\sigma = \langle \varphi_e \mid \chi_\sigma \rangle, \qquad S_\pi = \langle \varphi_t \mid \chi_\pi \rangle. \tag{10.71}$$

Assuming that λ, γ, and S's are small quantities of the order of ϵ ($\epsilon \ll 1$) and neglecting small quantities of higher order, one obtains the relations

$$\lambda_s \approx \gamma_s + S_s , \qquad \lambda_\sigma \approx \gamma_\sigma + S_\sigma , \qquad \lambda_\pi \approx \gamma_\pi + S_\pi \tag{10.72}$$

from the orthogonality relations

$$\langle \psi_e^a \mid \psi_{es}^b \rangle = \langle \psi_e^a \mid \psi_{e\sigma}^b \rangle = \langle \psi_t^a \mid \psi_t^b \rangle = 0. \tag{10.73}$$

From the remaining orthogonality relation

$$\langle \psi_{e\sigma}^b \mid \psi_{es}^b \rangle = 0, \tag{10.74}$$

one obtains

$$\gamma_{s\sigma} + \gamma_{\sigma s} = -(\lambda_s \lambda_\sigma - S_s S_\sigma) \approx 0, \tag{10.75a}$$

which shows

$$\gamma_{s\sigma} \approx -\gamma_{\sigma s} . \tag{10.75b}$$

In the purely ionic bonding where

$$\gamma_s = \gamma_\sigma = \gamma_\pi = 0, \tag{10.76}$$

one has the relations

$$\lambda_s = S_s , \qquad \lambda_\sigma = S_\sigma , \qquad \lambda_\pi = S_\pi \tag{10.77}$$

and

$$\gamma_{s\sigma} = \gamma_{\sigma s} = 0. \tag{10.78}$$

In this case the antibonding orbitals are simply the *d*-functions orthogonalized to the ligand orbitals.

10.4.2 Cubic-Field Splitting Parameter 10*Dq*

In the ligand-field theory the cubic-field splitting parameter, $10Dq$, is introduced as the difference of the e_g and t_{2g} orbital energies. Let us first consider a one-electron system. Since the energy of the antibonding orbital where an electron is placed is defined in Section 10.3.2 as the negative of the ionization energy of this electron, the energy required to excite one t_{2g} electron in the t_{2g} electron configuration to the e_g orbital is given by the difference between the ionization energy of an e_g electron in the e_g electron configuration and that of a t_{2g} electron in the t_{2g} configuration. This, of course, is the difference of the e_g and t_{2g} orbital energies in the one-electron system. In this case, $10Dq$ is given by

$$10Dq = \langle \psi_e{}^a \mid \bar{\bar{h}} \mid \psi_e{}^a \mid\rangle - \langle \psi_t{}^a \mid \bar{\bar{h}} \mid \psi_t{}^a \rangle, \tag{10.79}$$

in which

$$\bar{\bar{h}} = f + 2 \sum_{\substack{\text{core}\\\text{orbitals}}} \langle \psi_i \parallel \psi_i \rangle - \sum_{\substack{\text{core}\\\text{orbitals}}} \langle \psi_i \parallel P_{12} \psi_i \rangle. \tag{10.80}$$

We next consider the three-electron system in which paired electrons are placed in the t_{2g} orbital and an up-spin electron in the e_g orbital. The energy required to excite the down-spin electron to the e_g orbital is given by the difference of the ionization potential of the down-spin e_g electron in the $t_{2g}e_g{}^2$ configuration and that of the down-spin t_{2g} electron in the $t_{2g}^2 e_g$ configuration. Therefore, $10Dq$ in this case is given by the same expression as (10.79) with the one-electron Hamiltonian

$$\bar{\bar{h}} = f + \langle \psi_t \parallel \psi_t \rangle + \langle \psi_e \parallel \psi_e \rangle + 2 \sum_{\substack{\text{core}\\\text{orbitals}}} \langle \psi_i \parallel \psi_i \rangle - \sum_{\substack{\text{core}\\\text{orbitals}}} \langle \psi_i \parallel P_{12} \psi_i \rangle, \tag{10.81}$$

which corresponds to neither h_1 nor h_2 in (10.36) but to the HF Hamiltonian acting on the down-spin t_{2g} electron which is involved in the excitation.

After all, one can show that in any case the cubic-field splitting parameter is given by the expression (10.79) with the HF Hamiltonian acting on the electron involved in the excitation. This conclusion is intuitively understandable from the viewpoint of the MO theory. It is also possible to show that our expression of $10Dq$ in terms of one-electron energies is identical to the expression

$$10Dq = \langle \Psi^e \mid \mathscr{H} \mid \Psi^e \rangle - \langle \Psi^g \mid \mathscr{H} \mid \Psi^g \rangle, \tag{10.82}$$

in terms of the energies of a many-electron system. In (10.82) Ψ^g is the ground state wavefunction of the N-electron system with the $t_{2g}^n e_g{}^m$ $(n + m = N)$ electron configuration, and Ψ^e is the excited state wavefunction with the $t_g^{n-1} e_g^{m+1}$ electron configuration which is obtained by exciting one t_{2g} electron to the empty e_g orbital without rearranging the other electrons.

Now, if molecular orbitals ψ_e^a and ψ_t^a are assumed to have the forms given in (10.67), the cubic-field splitting parameter can be calculated numerically. In what follows we will cite an example of the numerical calculation for $(NiF_6)^{4-}$ cluster in a $KNiF_3$ crystal to point out the importance of the convalency. For this purpose we first consider the case in which the overlap integrals and the covalency are zero. In this case the cubic-field splitting parameter which will be denoted as Δ_{diag} is given by

$$\Delta_{\text{diag}} = \langle \varphi_e \mid \bar{\bar{h}} \mid \varphi_e \rangle - \langle \varphi_t \mid \bar{\bar{h}} \mid \varphi_t \rangle, \tag{10.83}$$

with the appropriate HF Hamiltonian $\bar{\bar{h}}$. It is instructive to divide Δ_{diag} into three parts:

$$\Delta_{\text{diag}} = \Delta_{\text{point}} + \Delta_K + \Delta_{\text{exch}} \,. \tag{10.84}$$

Here, Δ_{point} is the contribution obtained by shrinking the ligand-electron clouds into their nuclei, and treating the ligands as point charges as done in Section 1.1. Here Δ_K is the correction applied to Δ_{point} when the finite spread of the ligand electron clouds is taken into account. Therefore, this term represents the effect of the imperfect screening of the ligand nuclear charges by the ligand electrons. This term which was first discussed by Kleiner[‡] is called *Kleiner's correction.* Here, Δ_{exch}[§] represents the contribution from the exchange interaction between the *d*-electron and the ligand electron. Therefore, it has a quantum-mechanical origin. The calculated values of these contributions for $KNiF_3$ are listed in Table 10.1. In the calculation use was made of the HF 3*d*-wavefunction

TABLE 10.1[a]

CONTRIBUTIONS TO 10*Dq* FROM VARIOUS SOURCES

Δ_{point}	1390 cm^{-1}
Δ_K	—2080
Δ_{exch}	—2880
Δ_{diag}	—3570 cm^{-1}
10*Dq* (obs)	7250

[a] S. Sugano and R. G. Shulman, *Phys. Rev.* **130**, 517 (1963).

of a free Ni^{2+} ion and the HF 2*s*- and 2*p*-wavefunctions of a free F^- ion. The 1*s* orbital of the F^- ligand is shrunk into the ligand nucleus and the

‡ W. H. Kleiner, *J. Chem. Phys.* **20**, 1784 (1952).

§ Y. Tanabe and S. Sugano, *J. Phys. Soc. Japan* **11**, 864 (1956).

$1s$ electrons are treated as a part of the ligand nuclear charges. As seen in Table 10.1, the point-charge contribution gives a small value with a right sign, but this is canceled by the negative contributions of Δ_K and Δ_{exch}. As a result Δ_{diag} is negative in qualitative disagreement with the observed $10Dq$. Although the values in Table 10.1 are given for a specific system $KNiF_3$, it has generally been admitted that such a qualitative disagreement between the calculated Δ_{diag} and the observed $10Dq$ is always found in any cubic system.

The next step in the calculation of $10Dq$ is to take into account the effect of the nonorthogonality between the d-orbital and the ligand orbitals but still neglecting the covalency. The calculation at this stage corresponds to the exact calculation based on the purely ionic model. Denoting the contribution from the nonorthogonality to $10Dq$ as Δ_{nonorth}, one may divide the cubic-field-splitting parameter in the purely ionic model Δ_{ion} into two parts:

$$\Delta_{\text{ion}} = \Delta_{\text{diag}} + \Delta_{\text{nonorth}} . \tag{10.85}$$

The detailed calculation[‡,§] for $KNiF_3$ shows that the calculated value of Δ_{nonorth} is in the range, 5000–6000 cm^{-1}. Therefore, Δ_{ion} turns out to be $\sim$2000 cm^{-1} which is much smaller than the observed $10Dq \sim 7000$ cm^{-1}. It is interesting to note that the negative contributions from Δ_K and Δ_{exch} are almost canceled by Δ_{nonorth}, and Δ_{ion} is close to Δ_{point}, as was predicted on theoretical grounds by Phillips.[¶]

In our covalent model, the unexplained part of $10Dq$, which amount to $\sim$5000 cm^{-1} for $KNiF_3$, should be ascribed to the covalency. Denoting this part of $10Dq$ as Δ_{cov}, one expects that

$$10Dq = \Delta_{\text{ion}} + \Delta_{\text{cov}} . \tag{10.86}$$

Actually, if one uses the convalency parameters determined by the experiments on the transferred hyperfine interaction (see next subsection), one may hope to explain the observed $10Dq$.[♮]

10.4.3 Transferred Hyperfine Interaction

Fortunately, there is an almost direct way of measuring the covalency in predominantly ionic complexes of transition metal ions. It is to observe the transferred hyperfine interaction (THFI) at the ligands. The THFI

‡ R. E. Watson and A. J. Freeman, *Phys. Rev.* **134**, A1526 (1964).

§ S. Sugano and Y. Tanabe, *J. Phys. Soc. Japan* **20**, 1155 (1965).

¶ J. C. Phillips, *J. Phys. Chem. Solids* **11**, 226 (1959).

♮ J. Hubbard, D. E. Rimmer, and F. R. A. Hopgood, *Proc. Phys. Soc. (London)* **88**, 13 (1966).

arises from the unpaired spin density in the ligand atomic orbitals which interacts with the magnetic moment of the ligand nucleus. Neglecting small contributions to the unpaired spin density from the other sources for simplicity, the unpaired spin density in the ligand molecular orbital is given as

$$\langle \Psi \mid \Delta^s \mid \Psi \rangle, \tag{10.87}$$

where Ψ is the total wavefunction of the system. In (10.87) Δ^s is given as a sum of operators δ_i^s acting on electron i:

$$\Delta^s = \sum_i \delta_i^s, \tag{10.88}$$

in which, by using orbital operator p^s, δ_i^s is defined as

$$\begin{aligned} \delta_i^s &= 2p_i^s s_{zi}\,, \\ \langle \chi \mid p^s \mid \chi \rangle &= 1, \\ \langle \varphi \mid p^s \mid \chi \rangle = \langle \chi \mid p^s \mid \varphi \rangle &= \langle \varphi \mid p^s \mid \varphi \rangle = 0. \end{aligned} \tag{10.89}$$

For example, in the three-electron system described by Ψ in (10.56), (10.87) can be evaluated as follows: By introducing orbital φ_0 which is orthogonal to χ,

$$\varphi_0 = (1 - S^2)^{-1/2}(\varphi - S\chi), \tag{10.90}$$

Eq. (10.56) is reexpressed as

$$\Psi = (1 + 2\gamma S + \gamma^2)^{-1/2}[(1 + \gamma S)\mid \varphi_0\chi\bar{\chi} \mid + \gamma(1 - S^2)^{1/2} \mid \varphi_0\chi\bar{\varphi}_0 \mid]. \tag{10.91}$$

Then, by using the formulas given in Section 3.2.1., we can show that

$$\begin{aligned} \sum_\sigma \int d\tau \mid \varphi_0\chi\bar{\chi} \mid^* \Delta^s \mid \varphi_0\chi\bar{\chi} \mid &= \sum_{\sigma_1} \int d\tau_1 \varphi_0^*(1)\, \alpha^*(1)\, \delta_1^s \varphi_0(1)\, \alpha(1) \\ &\quad + \sum_{\sigma_1} \int d\tau_1 \chi^*(1)\, \alpha^*(1)\, \delta_1^s \chi(1)\, \alpha(1) \\ &\quad + \sum_{\sigma_1} \int d\tau_1 \chi^*(1)\, \beta^*(1)\, \delta_1^s \chi(1)\, \beta(1) \\ &= S^2/(1 - S^2), \end{aligned} \tag{10.92}$$

$$\sum_\sigma \int d\tau \mid \varphi_0\chi\bar{\varphi}_0 \mid^* \Delta^s \mid \varphi_0\chi\bar{\varphi}_0 \mid = 1, \tag{10.93}$$

$$\sum_\sigma \int d\tau \mid \varphi_0\chi\bar{\chi} \mid^* \Delta^s \mid \varphi_0\chi\bar{\varphi}_0 \mid = \frac{S}{(1 - S^2)^{1/2}}\,. \tag{10.94}$$

From these results one obtains

$$\langle \Psi \mid \Delta^s \mid \Psi \rangle = \frac{(\gamma + S)^2}{(1 - S^2)(1 + 2\gamma S + \gamma^2)} = \frac{\lambda^2}{1 - 2\lambda S + \lambda^2} = \frac{\lambda^2}{N_1}. \tag{10.95}$$

The same result is obtained simply by using the one-electron orbital picture. Since orbital ψ_2 is occupied by both up- and down-spins, no unpaired spin density is found in ψ_2. Therefore, the unpaired spin density comes from the electron only in ψ_1 and it is given by

$$\sum_{\sigma_1} \int d\boldsymbol{\tau}_1 \psi_1^*(1)\, \alpha(1)\, \delta_1^s \psi_1(1)\, \alpha(1) = \lambda^2/N_1, \tag{10.96}$$

in agreement with (10.95).

As seen from (10.64) and (10.65), the unpaired spin densities f_s, f_σ, and f_π in the ligand atomic orbitals s, σ, and π are related to the unpaired spin densities in the ligand molecular orbitals as

$$f_s = \lambda_s^2/3N_e^a, \qquad f_\sigma = \lambda_\sigma^2/3N_e^a, \qquad \text{and} \qquad f_\pi = \lambda_\pi^2/4N_t^a. \tag{10.97}$$

Therefore, the observation of the THFI is able to determine λ or γ, directly.

So far two different kinds of magnetic resonance experiments have been performed to measure the THFI: One is the nuclear magnetic resonance‡ (NMR) of the ligand nuclear spin and another is the electron spin resonance§ (ESR) of unpaired electrons. These experiments determine f_s and $f_\sigma - f_\pi$ separately. According to the NMR experiment by Shulman and Knox,

$$f_s \sim 0.5\% \qquad \text{for} \quad KMnF_3 \quad \text{and} \quad KNiF_3, \tag{10.98a}$$

$$f_\sigma - f_\pi \sim 4\% \qquad \text{for} \quad KNiF_3, \tag{10.98b}$$

$$f_\sigma - f_\pi \sim 5\% \qquad \text{for} \quad K_2NaCrF_6. \tag{10.98c}$$

The observed f_s is almost explained by assuming $\lambda_s \sim S_s$ which shows that the covalent mixing of the ligand $2s$ orbital is small. In the ground state of Ni^{2+} ions the t_{2g} shell is completely filled, so that no unpaired spin density f_π is expected. Therefore, we may conclude that $f_\sigma \sim 4\%$ for $KNiF_3$. In the ground state of Cr^{3+} ions in a strong-cubic field, the

‡ R. G. Shulman and K. Knox, *Phys. Rev. Letters* **4**, 603 (1960). R. G. Shulman, *Phys. Rev.* **121**, 125 (1961). R. G. Shulman and S. Sugano, *Phys. Rev.* **130**, 506 (1963).

§ J. H. E. Griffiths, J. Owen, and I. M. Ward, *Proc. Roy. Soc.* (*London*) **A219**, 526 (1953). T. P. P. Hall, W. Hayes, R. W. H. Stevenson, and J. Wilkins, *J. Chem. Phys.* **38**, 1977 (1963); **39**, 35 (1963).

e_g orbital is empty, so that we may conclude that $f_\pi \sim 5\%$ for K_2NaCrF_6. These observations clearly show that the covalent mixing of the ligand $2p_\sigma$ and $2p\pi$ orbitals are relatively large. For example, in $KNiF_3$, the observed values of the spin densities give

$$\begin{aligned} \gamma_s &= 0.04, \quad \lambda_s = 0.12 \quad (S_s = 0.08), \\ \gamma_\sigma &= 0.23, \quad \lambda_\sigma = 0.34 \quad (S_\sigma = 0.11). \end{aligned} \tag{10.99}$$

The fact that γ_σ is larger than S_σ emphasizes the importance of the covalency. It is important to point out again that we may hope to explain the observed cubic-field splitting parameter only when we use relatively large covelency parameters such as those determined by the observed THFI.

Theoretically the covalency parameter (or λ) may be calculated from (10.60), (10.61), or (10.63). However, at the present stage such a theoretical calculation cannot account for the observed large covalency and exposed to a fundamental difficulty related to the effect of the electron correlation. We will discuss this problem in the last section of this chapter.

10.4.4 Orbital Angular Momentum Reduction Factors

As shown in Section 7.1.1 the diagonal matrix element of the orbital angular momentum in the t_{2g} orbital and its nondiagonal matrix element between the e_g and t_{2g} orbitals are nonvanishing, and these matrix elements have been expressed by using the two orbital angular momentum reduction factors k and k' defined as

$$\langle t_2 \| \mathbf{l} \| t_2 \rangle = \sqrt{6}\, \langle t_2\xi \mid l_z \mid t_2\eta \rangle = \sqrt{6}\, ik, \tag{10.100}$$

$$\langle t_2 \| \mathbf{l} \| e \rangle = -\sqrt{3}\, \langle t_2\zeta \mid l_z \mid ev \rangle = -2\sqrt{3}\, ik'; \tag{10.101}$$

to the d-function approximation we have $k = k' = 1$. In this subsection we examine the effect of the covalency on k and k'.

If the t_{2g} and e_g molecular orbitals in (10.67) are used, the matrix elements of l_z in (10.100) and (10.101) are given as

$$\begin{aligned} \langle t_2\xi \mid l_z \mid t_2\eta \rangle = (N_t{}^a)^{-1}[&\langle \varphi_\xi \mid l_z \mid \varphi_\eta \rangle - \lambda_\pi \langle \varphi_\xi \mid l_z \mid \chi_\eta \rangle \\ &- \lambda_\pi \langle \chi_\xi \mid l_z \mid \varphi_\eta \rangle + \lambda_\pi{}^2 \langle \chi_\xi \mid l_z \mid \chi_\eta \rangle], \end{aligned} \tag{10.102}$$

$$\begin{aligned} \langle t_2\zeta \mid l_z \mid ev \rangle = (N_t{}^a N_e{}^a)^{-1/2}[&\langle \varphi_\zeta \mid l_z \mid \varphi_v \rangle \\ &- \lambda_s \langle \varphi_\zeta \mid l_z \mid \chi_{sv} \rangle - \lambda_\sigma \langle \varphi_\zeta \mid l_z \mid \chi_{\sigma v} \rangle - \lambda_\pi \langle \chi_\zeta \mid l_z \mid \varphi_v \rangle \\ &+ \lambda_s \lambda_\pi \langle \chi_\zeta \mid l_z \mid \chi_{sv} \rangle + \lambda_\sigma \lambda_\pi \langle \chi_\zeta \mid l_z \mid \chi_{\sigma v} \rangle]. \end{aligned} \tag{10.103}$$

In these expressions, it has already been shown that

$$\langle \varphi_\xi \mid l_z \mid \varphi_\eta \rangle = i, \qquad \langle \varphi_\zeta \mid l_z \mid \varphi_v \rangle = 2i. \tag{10.104}$$

By using the fact that the matrix of l_z with real bases is Hermitian and purely imaginary and also using the relations

$$l_z\varphi_\xi = -i\varphi_\eta\,, \qquad l_z\varphi_\eta = i\varphi_\xi\,, \qquad l_z\varphi_\zeta = -2i\varphi_v\,, \qquad l_z\varphi_v = 2i\varphi_\zeta\,, \tag{10.105}$$

one may show that

$$\langle \varphi_\xi \mid l_z \mid \chi_\eta \rangle = \langle \chi_\xi \mid l_z \mid \varphi_\eta \rangle = iS_\pi\,, \tag{10.106a}$$

$$\langle \varphi_\zeta \mid l_z \mid \chi_{sv} \rangle = 2iS_s\,, \qquad \langle \varphi_\zeta \mid l_z \mid \chi_{\sigma v} \rangle = 2iS_\sigma\,, \tag{10.106b}$$

$$\langle \chi_\zeta \mid l_z \mid \varphi_v \rangle = 2iS_\pi\,. \tag{10.106c}$$

The matrix element of l_z between χ's are calculated by expressing l_z as follows:

$$l_z = (\mathbf{r} \times \mathbf{p})_z = [(\mathbf{R}_k + \mathbf{r}_k) \times \mathbf{p}]_z = l_{kz} - i(\mathbf{R}_k \times \nabla_k)_z\,, \tag{10.107}$$

in which $\mathbf{r}_k$ is the electron coordinate whose origin is the kth ligand nuclear position, $l_{kz} = (\mathbf{r}_k \times \mathbf{p})_z$ is the z component of the angular momentum around the kth ligand nucleus, $\mathbf{R}_k$ is the position vector of the kth ligand, and

$$\nabla_k = \hat{\mathbf{i}}\,\frac{\partial}{\partial x_k} + \hat{\mathbf{j}}\,\frac{\partial}{\partial y_k} + \hat{\mathbf{k}}\,\frac{\partial}{\partial z_k}\,,$$

where $\hat{\mathbf{i}}$, $\hat{\mathbf{j}}$, and $\hat{\mathbf{k}}$ are the unit vectors along the x-, y-, and z-axes. By the relations,

$$l_{kz}\varphi_{kx} = i\varphi_{ky}\,, \tag{10.108a}$$

$$l_{kz}\varphi_{ky} = -i\varphi_{kx}\,, \tag{10.108b}$$

$$l_{kz}\varphi_{kz} = l_{kz}\varphi_{ks} = 0, \tag{10.108c}$$

and assuming for simplicity that any integral involving atomic orbitals of different ligands is zero,[‡] the use of (10.107) gives

$$\langle \chi_\xi \mid l_z \mid \chi_\eta \rangle = i/2, \tag{10.109a}$$

$$\langle \chi_\zeta \mid l_z \mid \chi_{sv} \rangle = -iA, \tag{10.109b}$$

$$\langle \chi_\zeta \mid l_z \mid \chi_{\sigma v} \rangle = -i, \tag{10.109c}$$

‡ This approximation is examined in detail by A. A. Misetich and R. E. Watson, *Phys. Rev.* **143**, 335 (1966).

in which

$$A = R_0\langle \varphi_x \mid \partial/\partial x \mid \varphi_s\rangle. \tag{10.110}$$[‡]

In (10.110) R_0 is the atomic distance between the metal ion and the ligand. From (10.104), (10.106), and (10.109) one finally obtains

$$\begin{aligned}\langle t_2\xi \mid l_z \mid t_2\eta\rangle &= i(N_t^a)^{-1}(1 - 2\lambda_\pi S_\pi + \lambda_\pi^2/2)\\ &\approx i(1 - \lambda_\pi^2/2),\end{aligned} \tag{10.111}$$

$$\begin{aligned}\langle t_2\zeta \mid l_z \mid ev\rangle &= 2i(N_t^a N_e^a)^{-1/2}[1 - \lambda_s S_s - \lambda_\sigma S_\sigma - \lambda_\pi S_\pi - \lambda_\pi(\lambda_\sigma + A\lambda_s)/2]\\ &\approx i[2 - (\lambda_s^2 + \lambda_\sigma^2 + \lambda_\pi^2) - \lambda_\pi(\lambda_\sigma + A\lambda_s)].\end{aligned} \tag{10.112}$$

Comparing (10.111) and (10.112) with (10.100) and (10.101), one arrives at the expression of k and k' as

$$\begin{aligned}k &\approx 1 - \tfrac{1}{2}\lambda_\pi^2\\ k' &\approx 1 - \tfrac{1}{2}(\lambda_s^2 + \lambda_\sigma^2 + \lambda_\pi^2) - \tfrac{1}{2}\lambda_\pi(\lambda_\sigma + A\lambda_s),\end{aligned} \tag{10.113}$$

in which the higher-order small quantities are neglected by assuming that the covalency parameters and the overlap integrals are small. As seen from the example of Section 8.4.2, these parameters appear in the g-values, often together with the spin-orbit coupling constant in the g-values of ground states. Therefore, the detailed examination of the g-values may provide the useful information about the covalency.

Problem 10.3. Confirm the results in (10.109). ◇

10.4.5 Spin-Orbit Coupling Constant

In Section 7.4.1 the matrix elements of the spin-orbit interaction have been expressed by using two parameters ζ and ζ', defined as

$$\langle t_2 \| v(1T_1) \| t_2\rangle = 6\langle t_2\, \tfrac{1}{2}\xi \mid v_{0\gamma}(1T_1) \mid t_2\, \tfrac{1}{2}\eta\rangle = 3i\zeta, \tag{10.114}$$

$$\begin{aligned}\langle t_2 \| v(1T_1) \| e\rangle &= -3\sqrt{2}\,\langle t_2\, \tfrac{1}{2}\zeta \mid v_{0\gamma}(1T_1) \mid e\tfrac{1}{2}v\rangle\\ &= -3\sqrt{2}\, i\zeta';\end{aligned} \tag{10.115}$$

to the d-function approximation one has the relation, $\zeta = \zeta'$. Here, we examine the effect of the covalency on these parameters.

According to Misetich and Buch,[§] the spin-orbit interaction for a single electron in our polycentric system is approximately given as

$$\mathscr{H}_{SO} = \xi_0(\mathbf{r})\, \mathbf{l}\cdot\mathbf{s} + \sum_k \xi_k(\mathbf{r}_k)\, \mathbf{l}_k\cdot\mathbf{s}_k \tag{10.116}$$

‡ Here, A/R_0 has been calculated by Misetich and Watson to be -0.431 for F^-.

§ A. A. Misetich and T. Buch, *J. Chem. Phys.* **41**, 2524 (1964).

instead of (7.1). In (10.116) $\mathbf{r}_k$ and $\mathbf{l}_k$ have already been introduced in (10.107). We assume that $\xi_0(\mathbf{r})$ and $\xi_k(\mathbf{r}_k)$ are extremely localized functions around the origins of their variables so that any integral involving $\xi_i(\mathbf{r}_i)$ and the atomic orbital localized at $j(i \neq j)$ is zero; to the first approximation $\xi_0(\mathbf{r})$ and $\xi_k(\mathbf{r}_k)$ are considered to be those for a free metal ion and a free ligand ion. Then, by using the explicit forms of the MO's given in (10.67), it is straightforward to obtain the following expressions for the matrix elements of the spin-orbit interaction:

$$\langle t_2 \tfrac{1}{2}\xi \mid v_{0\gamma}(1T_1) \mid t_2 \tfrac{1}{2}\eta \rangle = (2N_t^a)^{-1} \Big[\langle \varphi_\xi \mid \xi_0(\mathbf{r})\, l_z \mid \varphi_\eta \rangle + \lambda_\pi^2 \langle \chi_\xi \mid \sum_k \xi_k(\mathbf{r}_k)\, l_{kz} \mid \chi_\eta \rangle \Big], \tag{10.117}$$

$$\langle t_2 \tfrac{1}{2}\zeta \mid v_{0\gamma}(1T_1) \mid e \tfrac{1}{2}v \rangle = \tfrac{1}{2}(N_t^a N_e^a)^{-1/2} \Big[\langle \varphi_\zeta \mid \xi_0(\mathbf{r})\, l_z \mid \varphi_v \rangle + \lambda_\sigma \lambda_\pi \langle \chi_\zeta \mid \sum_k \xi_k(\mathbf{r}_k)\, l_{kz} \mid \chi_{\sigma v} \rangle \Big], \tag{10.118}$$

in which the obvious relation $l_{kz}\chi_{sv} = 0$ is already used. It is evident that

$$\begin{aligned} \langle \varphi_\xi \mid \xi_0(\mathbf{r})\, l_z \mid \varphi_\eta \rangle &= i\zeta_d\,, \\ \langle \varphi_\zeta \mid \xi_0(\mathbf{r})\, l_z \mid \varphi_v \rangle &= 2i\zeta_d\,, \end{aligned} \tag{10.119}$$

where ζ_d is considered to the first approximation to be the spin-orbit coupling constant for a d-electron in a free-transition metal ion. By using (10.64), (10.65), and (10.108), one may calculate the remaining matrix elements as

$$\begin{aligned} \langle \chi_\xi \mid \sum_k \xi_k(\mathbf{r}_k)\, l_{kz} \mid \chi_\eta \rangle &= \tfrac{1}{2} i\zeta_{Lp}\,, \\ \langle \chi_\zeta \mid \sum_k \xi_k(\mathbf{r}_k)\, l_{kz} \mid \chi_{\sigma v} \rangle &= -i\zeta_{Lp}\,, \end{aligned} \tag{10.120}$$

in which ζ_{Lp} is considered to the first approximation to be the spin-orbit coupling constant for a p-electron in a free ligand ion. Now, by using (10.119) and (10.120), (10.117) and (10.118) are expressed as

$$\begin{aligned} \langle t_2 \tfrac{1}{2}\xi \mid v_{0\gamma}(1T_1) \mid t_2 \tfrac{1}{2}\eta \rangle &= i(2N_t^a)^{-1}(\zeta_d + \tfrac{1}{2}\lambda_\pi^2 \zeta_{Lp}) \\ &\approx \tfrac{1}{2} i\{[1 + (S_\pi^2 - \gamma_\pi^2)]\, \zeta_d + \tfrac{1}{2}\lambda_\pi^2 \zeta_{Lp}\}, \end{aligned} \tag{10.121a}$$

$$\begin{aligned} \langle t_2 \tfrac{1}{2}\zeta \mid v_{0\gamma}(1T_1) \mid e \tfrac{1}{2}v \rangle &= i(N_e^a N_t^a)^{-1/2}(\zeta_d - \tfrac{1}{2}\lambda_\sigma \lambda_\pi \zeta_{Lp}) \\ &\approx i\{[1 + \tfrac{1}{2}(S_s^2 - \gamma_s^2) + \tfrac{1}{2}(S_\sigma^2 - \gamma_\sigma^2) + \tfrac{1}{2}(S_\pi^2 - \gamma_\pi^2)]\, \zeta_d - \tfrac{1}{2}\lambda_\sigma \lambda_\pi \zeta_{Lp}\}. \end{aligned} \tag{10.121b}$$

Then, comparing (10.121) with (10.114) and (10.115), one finally obtains

$$\begin{aligned}\zeta &\approx [1 + (S_\pi^2 - \gamma_\pi^2)]\,\zeta_d + \tfrac{1}{2}\lambda_\pi^2\zeta_{Lp}\,,\\ \zeta' &\approx [1 + \tfrac{1}{2}(S_s^2 - \gamma_s^2) + \tfrac{1}{2}(S_\sigma^2 - \gamma_\sigma^2) + \tfrac{1}{2}(S_\pi^2 - \gamma_\pi^2)]\,\zeta_d - \tfrac{1}{2}\lambda_\sigma\lambda_\pi\zeta_{Lp}\,.\end{aligned} \tag{10.122}$$

When the ligand ion is a light element such as F^-, ζ_{Lp} ($\sim$220 cm^{-1} for F^-) is much smaller than ζ_d ($\sim$640 cm^{-1} for Ni^{2+}) so that the terms involving ζ_{Lp} are almost negligible. Then, (10.122) shows that, if the covalency parameters are larger than the overlap integrals, both ζ and ζ' are smaller than ζ_d.

10.4.6 Slater Integrals

In Section 5.3.3 we showed that the experimentally determined values of Slater integrals or Racah parameters appearing in the ligand-field theory are smaller than those of the free ions in many cases. In principle, it is possible to calculate these Slater integrals by using the MO's given in (10.67). However, such a calculation involves the evaluation of many-center integrals, and therefore is very difficult. Any reliable calculation has not been done so far. The only conceivable way of explaining the reduction of the Slater integrals is to assume that any term involving the ligand orbitals is zero.[‡] Then the reduction factor is simply given by the normalization factors of the MO's. Since the integrals contain four MO's, the reduction factor is given by $(N_e^a)^{-n/2} \times (N_t^a)^{-2+n/2}$ in which n is the number of the e_g orbitals contained in the integral. As seen from the relations

$$\begin{aligned}(N_e^a)^{-1/2} &\approx 1 - \tfrac{1}{2}(\gamma_s^2 - S_s^2) - \tfrac{1}{2}(\gamma_\sigma^2 - S_\sigma^2),\\ (N_t^a)^{-1/2} &\approx 1 - \tfrac{1}{2}(\gamma_\pi^2 - S_\pi^2),\end{aligned} \tag{10.123}$$

the reduction could occur if the covalency parameters are larger than the corresponding overlap integrals.

This type of argument was first given by Koide and Pryce.[§] To introduce the covalency effect in the analysis of the optical spectra, they used covalency parameter ϵ. To our approximation, this parameter may be interpreted as the parameter given by

$$1 - \epsilon = N_t^a / N_e^a\,. \tag{10.124}$$

‡ Actually the value of each of these terms is very small. However, the number of these terms is very large so that this assumption cannot be justified without a detailed calculation.

§ S. Koide and M. H. L. Pryce, *Phil. Mag.* **3**, 607 (1958).

The values of parameter ϵ for Mn^{2+} ions have been experimentally found in the range 0.03–0.05.

10.5 Calculation of Covalency

The nonempirical calculations‡ of the covalency parameters γ_σ and γ_π, which employ (10.63), have been performed for $KNiF_3$. The calculations use the HF $3d$-wavefunction of a free Ni^{2+} ion for φ and the HF $2s$- and $2p$-wavefunctions of a free F^- ion for the ligand atomic orbitals. The calculated values of the covalency parameters turned out to be very small in disagreement with the experimental ones:

	Calc	Exp
γ_σ	0.074	0.23
γ_π	0.030	—

(10.125)

The failure of these calculations have been ascribed to the nature of the approximation characteristic of the simple MO theory. This point will be discussed in some detail in this section.

For this purpose we first examine the denominator of the expression for γ in (10.63). As shown in (10.62), $\bar{h}$ to be used for calculating $\bar{A}$ and $\bar{C}$ is the HF Hamiltonian for a down-spin bonding electron. Therefore, to the first approximation in which both the covalency parameters and the overlap integrals are assumed to be zero in $\bar{h}$, $\bar{A} = \langle \varphi \mid \bar{h} \mid \varphi \rangle$ may be regarded in the problem of $KNiF_3$ as the energy of a down-spin electron in φ in the ionic configuration $Ni^+(F_6)^{5-}$ and $\bar{C} = \langle \chi \mid \bar{h} \mid \chi \rangle$ as that in χ in the ionic configuration $Ni^{2+}(F_6)^{6-}$. This is clearly seen in the example of the three-electron system described in Section 10.3. To the first approximation $\bar{h}$ in (10.62) is given as

$$\bar{h}_0 = f + \langle \varphi \| \varphi \rangle + \langle \chi \| \chi \rangle, \tag{10.126}$$

and by using (10.126), one may show that

$$\begin{aligned}\langle \varphi \mid \bar{h}_0 \mid \varphi \rangle - \langle \chi \mid \bar{h}_0 \mid \chi \rangle = {} & [\langle \varphi \mid f \mid \varphi \rangle + \langle \varphi\varphi \| \varphi\varphi \rangle + \langle \varphi\chi \| \varphi\chi \rangle] \\ & - [\langle \chi \mid f \mid \chi \rangle + \langle \chi\chi \| \chi\chi \rangle + \langle \chi\varphi \| \chi\varphi \rangle].\end{aligned} \tag{10.127}$$

The three terms in the first square bracket represent the energy of a down-spin electron in φ in the configuration $\varphi\,\bar{\varphi}\,\chi$ and the three terms

‡ R. E. Watson and A. J. Freeman, *Phys. Rev.* **134**, A1526 (1964). S. Sugano and Y. Tanabe, *J. Phys. Soc. Japan* **20**, 1155 (1965).

in the second bracket that of a down-spin electron in χ in the configuration $\varphi\chi\bar{\chi}$. Actually in the HL scheme with the excited covalent configuration, where the total wavefunction is assumed to be

$$\Psi = \Psi_{\text{ion}} + \gamma\Psi_{\text{cov}}, \tag{10.128}$$

the denominator of the expression determining γ is given by the energy difference of the excited covalent configuration and the ionic configuration. The excited covalent configuration is $Ni^+(F_6)^{5-}$ for $KNiF_3$. Since this HL scheme is equivalent to the MO scheme if the same φ and χ are used in Ψ_{ion} and Ψ_{cov} as shown in (10.56), the denominator $\bar{A} - \bar{C}$ in our simple MO scheme may be interpreted as the energy required for the transfer of the down-spin ligand electron to the metal ion without readjustment of the distribution of the electrons. From this physical interpretation of the denominator, one sees that the failure of the simple MO treatment would be due to the neglect of the readjustment effect in the electron-transferred configuration, $Ni^+(F_6)^{5-}$ in the case of $KNiF_3$.

Although it is very difficult to take into account the readjustment effect on a purely theoretical basis, the HL picture may provide us an intuitively reasonable way of estimating the readjustment effect. According to Hubbard *et al.*,‡ the main readjustment effects are as follows:

(i) The readjustment of the *d*-electron distribution (the expansion of the *d*-orbitals);

(ii) The polarization of the ligand electron clouds.

By using the observed ionization potential of a Ni^+ ion, they have estimated the first effect to reduce the denominator by 0.29 a.u. and by using optical polarizability data of the ions in $KNiF_3$ the second effect to reduce the denominator by 0.17 a.u. Since the denominator calculated without taking into account the readjustment effect is 0.76 a.u., the denominator including the readjustment effect is now about 0.3 a.u., ($\sim$8 eV). They also have estimated the numerator in (10.63) and found that the readjustment effects in the numerator are not as important as in the denominator. The covalency parameters thus calculated are

$$\gamma_\sigma = 0.215, \qquad \gamma_\pi = 0.132. \tag{10.129}$$

The calculated γ_σ is in fair agreement with the experimental one, $\gamma_\sigma(\text{exp}) = 0.23$.

‡ J. Hubbard, D. E. Rimmer, and F. R. A. Hopgood, *Proc. Phys. Soc.* (*London*) **88**, 13 (1966).

Quite recently this intuitive method of taking into account the readjustment or correlation effects has been criticized by Šimanek *et al.*,[‡] who have used the HL wavefunction containing two excited configurations as

$$\Psi = a\Psi_{\text{ion}} + b\Psi_{\text{cov}} + c\Psi'_{\text{cov}}\,, \tag{10.130}$$

and applied the variation principle to determine b and c. In (10.130) Ψ_{cov} is the state in which a down-spin electron is transferred to the empty $3d$-orbital of the metal ion without the readjustment of the distribution of the other electrons and Ψ'_{cov} an excited state of Ψ_{cov}, i.e., it contains one-electron excitation in addition to the electron transfer. The correlation effect may be considered to be accounted for to some extent by the inclusion of Ψ'_{cov}. For example, the expansion of the d-orbital is accounted for to some extent by the inclusion of Ψ'_{cov} in which the $3d \rightarrow 4d$ excitation is considered in addition to the electron transfer, and the polarization of the ligands by the inclusion of Ψ'_{cov} in which the $2s \rightarrow 3p$ excitation is included. Solving the secular equation by assuming b and c to be small, they have found that the method of estimating the correlation effect by Hubbard, *et al.*, is questionable. For example, the effect of the ligand polarization cannot be calculated by using the polarizability of the ligand, which measures the degree of mixing Ψ'_{cov} into Ψ_{cov} due to the electric-field perturbation when the lifetime of the Ψ_{cov} state is longer than the time needed for the virtual electron excitation, $\Psi_{\text{cov}} \rightarrow \Psi'_{\text{cov}}$. This time is inversely proportional to the energy difference ($\sim$10 eV) of the Ψ_{cov} and Ψ'_{cov} states. In terms of the time-independent perturbuation theory, the polarizability in this case is proportional to $H'^2/(E'_{\text{cov}} - E_{\text{cov}})$ where H' is the matrix element of the electric field perturbation and E_{cov} and E'_{cov} are the energies of the Ψ_{cov} and Ψ'_{cov} states, respectively. However, in our problem in which Ψ_{cov} is a virtually excited state of a short lifetime, it is shown that the ligand polarization responsible for the correlation effect has to be calculated by using a different polarizability proportional to $H'^2/(E'_{\text{cov}} - E_{\text{ion}})$ where E_{ion} is the energy of the Ψ_{ion} state. Šimanek and Tachiki's interpretation[§] for this fact is that if the Ψ_{cov} state has a lifetime shorter than the time needed for the excitation $\Psi_{\text{cov}} \rightarrow \Psi'_{\text{cov}}$, the polarization process cannot be fully developed. A similar argument may be applied to the readjustment effect of a d-electron distribution, although it is uncertain if the inclusion of a single configuration Ψ'_{cov}

[‡] E. Šimanek, Z. Šroubek, and M. Tachiki, *J. Phys. Soc. Japan* **22**, 547 (1967).
[§] E. Šimanek and M. Tachiki, *Phys. Letters* **21**, 625 (1966).

with a relatively low energy is sufficient to account for this readjustment effect.

So far the readjustment effects have been discussed in the HL scheme. However, in view of the MO theoretical nature of the ligand-field theory in the strong-field scheme, it is highly desirable to make efforts along the line to take into account the readjustment effects within the MO scheme in a way as simple as possible.

Appendix I CHARACTER TABLES FOR THE THIRTY-TWO DOUBLE POINT-GROUPS, $\bar{G}$

For brevity we do not include those groups which may be obtained from those listed here by taking the direct product with group C_i consisting of two elements E and I. These groups are as shown in the tabulation. The character tables of these unlisted groups are easily obtained from those of listed ones by following the arguments given in Section 1.2.5.

$\bar{C}_{2h} = \bar{C}_2 \times C_i\,,$	$\bar{D}_{2h} = \bar{D}_2 \times C_i\,,$
$\bar{C}_{4h} = \bar{C}_4 \times C_i\,,$	$\bar{D}_{4h} = \bar{D}_4 \times C_i\,,$
$\bar{C}_{3i} = \bar{C}_3 \times C_i\,,$	$\bar{D}_{3d} = \bar{D}_3 \times C_i\,,$
$\bar{C}_{6h} = \bar{C}_6 \times C_i\,,$	$\bar{D}_{6h} = \bar{D}_6 \times C_i\,,$
$\bar{T}_h = \bar{T} \times C_i\,,$	$\bar{O}_h = \bar{O} \times C_i\,.$

In the tables the Γ_i's are Bethe's notations for irreducible representations and A, B, E, T, G with appropriate suffices and primes are Mulliken's notations. The half-integral suffices of E and G for two-valued irreducible representations come from Herzberg's book.[‡] In the text, however, we have used notations E_1, E_2, and G in place of $E_{1/2}$, $E_{5/2}$, and $G_{3/2}$, respectively, for group $\bar{O}$, and notations $\bar{E}$ and $\bar{A}_1 + \bar{A}_2$ in place of $E_{1/2}$ and $E_{3/2}$, respectively, for group $\bar{D}_3$; These notations are also popular.

‡ G. Herzberg, "Molecular Spectra and Molecular Structure III," Van Nostrand, Princeton, New Jersey, 1966.

GROUP $\bar{C}_1$

$\bar{C}_1$		E	R
A	Γ_1	1	1
$B_{1/2}$	Γ_2	1	—1

GROUP $\bar{C}_i$

$\bar{C}_i$		E	R	I	IR	
A_g	Γ_1^+	1	1	1	1	L_x, L_y, L_z
A_u	Γ_1^-	1	1	—1	—1	x, y, z
$B_{1/2,g}$	Γ_2^+	1	—1	1	—1	
$B_{1/2,u}$	Γ_2^-	1	—1	—1	1	

GROUPS $\bar{C}_2$ AND $\bar{C}_s$

	$\bar{C}_2$	$\bar{C}_2$	E	R	C_2	C_2R	$\bar{C}_2$	
$\bar{C}_s$		$\bar{C}_s$	E	R	σ	σR		$\bar{C}_s$
A'	A	Γ_1	1	1	1	1	z, L_z	x, y, L_z
A''	B	Γ_2	1	1	-1	-1	x, y, L_x, L_y	z, L_x, L_y
$E_{1/2}$	$E_{1/2}$	Γ_3	1	-1	i	$-i$		
		Γ_4	1	-1	$-i$	i		

GROUPS $\bar{D}_2$ AND $\bar{C}_{2v}$

	$\bar{D}_2$	$\bar{D}_2$	E	R	C_2 C_2R	C_2' $C_2'R$	C_2'' $C_2''R$	$\bar{D}_2$	
$\bar{C}_{2v}$		$\bar{C}_{2v}$	E	R	C_2 C_2R	σ_v σ_vR	σ_v' $\sigma_v'R$		$\bar{C}_{2v}$
A_1	A	Γ_1	1	1	1	1	1		z
A_2	B_1	Γ_2	1	1	1	-1	-1	z, L_z	L_z
B_1	B_2	Γ_3	1	1	-1	1	-1	y, L_y	x, L_y
B_2	B_3	Γ_4	1	1	-1	-1	1	x, L_x	y, L_x
$E_{1/2}$	$E_{1/2}$	Γ_5	2	-2	0	0	0		

GROUP $\bar{C}_3$

$\bar{C}_3$		E	R	C_3	C_3R	C_3^2	C_3^2R	
A	Γ_1	1	1	1	1	1	1	z, L_z
E	Γ_2	1	1	ω^2	ω^2	$-\omega$	$-\omega$	x, y, L_x, L_y
	Γ_3	1	1	$-\omega$	$-\omega$	ω^2	ω^2	
$E_{1/2}$	Γ_4	1	-1	ω	$-\omega$	ω^2	$-\omega^2$	
	Γ_5	1	-1	$-\omega^2$	ω^2	$-\omega$	ω	
$B_{3/2}$	Γ_6	1	-1	-1	1	1	-1	

$\omega = \exp(i\pi/3)$

GROUPS $\bar{D}_3$ AND $\bar{C}_{3v}$

$\bar{D}_3$			E	R	C_3, C_3^2R	C_3^2, C_3R	$3C_2'$	$3C_2'R$	$\bar{D}_3$	
$\bar{C}_{3v}$			E	R	C_3, C_3^2R	C_3^2, C_3R	$3\sigma_v$	$3\sigma_vR$		$\bar{C}_{3v}$
A_1		Γ_1	1	1	1	1	1	1		z
A_2		Γ_2	1	1	1	1	-1	-1	z, L_z	L_z
E		Γ_3	2	2	-1	-1	0	0	x, y, L_x, L_y	x, y, L_x, L_y
$E_{3/2}$	$(\bar{A}_1)$	Γ_4	1	-1	-1	1	i	$-i$		
	$(\bar{A}_2)$	Γ_5	1	-1	-1	1	$-i$	i		
$E_{1/2}$	$(\bar{E})$	Γ_6	2	-2	1	-1	0	0		

GROUPS $\bar{C}_4$ AND $\bar{S}_4$

$\bar{C}_4$		E	R	C_4	C_4R	C_4^2	C_4^2R	C_4^3	C_4^3R	$\bar{C}_4$	
$\bar{S}_4$		E	R	S_4	S_4R	C_2R	C_2	S_4^3	S_4^3R		$\bar{S}_4$
A	Γ_1	1	1	1	1	1	1	1	1	z	L_z
B	Γ_2	1	1	-1	-1	1	1	-1	-1	L_z	z
E	Γ_3	1	1	i	i	-1	-1	$-i$	$-i$	$x \pm iy$	$x \pm iy$
	Γ_4	1	1	$-i$	$-i$	-1	-1	i	i	$L_x \pm iL_y$	$L_x \pm iL_y$
$E_{1/2}$	Γ_5	1	-1	ω	$-\omega$	i	$-i$	ω^3	$-\omega^3$		
	Γ_6	1	-1	$-\omega^3$	ω^3	$-i$	i	$-\omega$	ω		
$E_{3/2}$	Γ_7	1	-1	$-\omega$	ω	i	$-i$	$-\omega^3$	ω^3		
	Γ_8	1	-1	ω^3	$-\omega^3$	$-i$	i	ω	$-\omega$		

$\omega = \exp(i\pi/4)$

GROUPS $\bar{D}_4$, $\bar{C}_{4v}$, AND $\bar{D}_{2d}$

$\bar{D}_4$	E	R	C_4 C_4^3R	C_4^3 C_4R	C_4^2 C_4^2R	$2C_2'$ $2C_2'R$	$2C_2''$ $2C_2''R$	$\bar{D}_4$		
$\bar{C}_{4v}$	E	R	C_4 C_4^3R	C_4^3 C_4R	C_4^2 C_4^2R	$2\sigma_v$ $2\sigma_vR$	$2\sigma_d$ $2\sigma_dR$		$\bar{C}_{4v}$	
$\bar{D}_{2d}$	E	R	S_4 S_4^3R	S_4^3 S_4R	C_2 C_2R	$2C_2'$ $2C_2'R$	$2\sigma_d$ $2\sigma_dR$			$\bar{D}_{2d}$
A_1 Γ_1	1	1	1	1	1	1	1		z	
A_2 Γ_2	1	1	1	1	1	−1	−1	z, L_z	L_z	L_z
B_1 Γ_3	1	1	−1	−1	1	1	−1			
B_2 Γ_4	1	1	−1	−1	1	−1	1			z
E Γ_5	2	2	0	0	−2	0	0	x, y, L_x, L_y	x, y, L_x, L_y	x, y, L_x, L_y
$E_{1/2}\Gamma_6$	2	−2	$\sqrt{2}$	$-\sqrt{2}$	0	0	0			
$E_{3/2}\Gamma_7$	2	−2	$-\sqrt{2}$	$\sqrt{2}$	0	0	0			

GROUPS $\bar{D}_6$, $\bar{C}_{6v}$, AND $\bar{D}_{3h}$

$\bar{D}_6$ $\bar{D}_6$	E	R	C_2 C_2R	C_3 C_3^2R	C_3^2 C_3R	C_6 C_6^5R	C_6^5 C_6R	$3C_2'$ $3C_2'R$	$3C_2''$ $3C_2''R$	$\bar{D}_6$		
$\bar{C}_{6v}$ $\bar{C}_{6v}$	E	R	C_2 C_2R	C_3 C_3^2R	C_3^2 C_3R	C_6 C_6^5R	C_6^5 C_6R	$3\sigma_v$ $3\sigma_vR$	$3\sigma_d$ $3\sigma_dR$		$\bar{C}_{6v}$	
$\bar{D}_{3h}$ $\bar{D}_{3h}$	E	R	σ_h σ_hR	C_3 C_3^2R	C_3^2 C_3R	S_3 S_3^2R	S_3^2 S_3R	$3C_2'$ $3C_2'R$	$3\sigma_v$ $3\sigma_vR$			$\bar{D}_{3h}$
A_1' A_1 Γ_1	1	1	1	1	1	1	1	1	1		z	
A_2' A_2 Γ_2	1	1	1	1	1	1	1	−1	−1	z, L_z	L_z	L_z
A_1'' B_1 Γ_3	1	1	−1	1	1	−1	−1	1	−1			
A_2'' B_2 Γ_4	1	1	−1	1	1	−1	−1	−1	1			z
E'' E_1 Γ_5	2	2	−2	−1	−1	1	1	0	0	x, y; L_x, L_y	x, y; L_x, L_y	L_x, L_y
E' E_2 Γ_6	2	2	2	−1	−1	−1	−1	0	0			x, y
$E_{1/2}$ $E_{1/2}$ Γ_7	2	−2	0	1	−1	$\sqrt{3}$	$-\sqrt{3}$	0	0			
$E_{3/2}$ $E_{3/2}$ Γ_8	2	−2	0	1	−1	$-\sqrt{3}$	$\sqrt{3}$	0	0			
$E_{5/2}$ $E_{5/2}$ Γ_9	2	−2	0	−2	2	0	0	0	0			

GROUPS $\bar{C}_6$ AND $\bar{C}_{3h}$

$\bar{C}_{3h}$	$\bar{C}_6$	$\bar{C}_6$ / $\bar{C}_{3h}$	E / E	R / R	C_6 / S_3	C_6R / S_3R	C_6^2 / C_3^2R	C_6^2R / C_3^2	C_6^3 / σ_h	C_6^3R / σ_hR	C_6^4 / C_3R	C_6^4R / C_3	C_6^5 / S_3^5	C_6^5R / S_3^5R	$\bar{C}_6$	$\bar{C}_{3h}$
A'	A	Γ_1	1	1	1	1	1	1	1	1	1	1	1	1	z, L_z	L_z
A''	B	Γ_2	1	1	-1	-1	1	1	-1	-1	1	1	-1	-1		z
E'	E_1	Γ_3	1	1	ω^4	ω^4	$-\omega^2$	$-\omega^2$	1	1	ω^4	ω^4	$-\omega^2$	$-\omega^2$		$x \pm iy$,
		Γ_4	1	1	$-\omega^2$	$-\omega^2$	ω^4	ω^4	1	1	$-\omega^2$	$-\omega^2$	ω^4	ω^4		
E''	E_2	Γ_5	1	1	ω^2	ω^2	ω^4	ω^4	-1	-1	$-\omega^2$	$-\omega^2$	$-\omega^4$	$-\omega^4$	$x \pm iy$,	$L_x \pm iL_y$
		Γ_6	1	1	$-\omega^4$	$-\omega^4$	$-\omega^2$	$-\omega^2$	-1	-1	ω^4	ω^4	ω^2	ω^2	$L_x \pm iL_y$	
$E_{5/2}$	$E_{1/2}$	Γ_7	1	-1	ω	$-\omega$	ω^2	$-\omega^2$	ω^3	$-\omega^3$	ω^4	$-\omega^4$	ω^5	$-\omega^5$		
		Γ_8	1	-1	$-\omega^5$	ω^5	$-\omega^4$	ω^4	$-\omega^3$	ω^3	$-\omega^2$	ω^2	$-\omega$	ω		
$E_{3/2}$	$E_{3/2}$	Γ_9	1	-1	i	$-i$	-1	1	$-i$	i	1	-1	i	$-i$		
		Γ_{10}	1	-1	$-i$	i	-1	1	i	$-i$	1	-1	$-i$	i		
$E_{1/2}$	$E_{5/2}$	Γ_{11}	1	-1	$-\omega$	ω	ω^2	$-\omega^2$	$-\omega^3$	ω^3	ω^4	$-\omega^4$	$-\omega^5$	ω^5		
		Γ_{12}	1	-1	ω^5	$-\omega^5$	$-\omega^4$	ω^4	ω^3	$-\omega^3$	$-\omega^2$	ω^2	ω	$-\omega$		

$$\omega = \exp(i\pi/6)$$

GROUP $\bar{T}$

$\bar{T}$		E	R	$3C_2$ $3C_2R$	$4C_3$	$4C_3R$	$4C_3^2$	$4C_3^2R$	
A	Γ_1	1	1	1	1	1	1	1	
E	Γ_2	1	1	1	ω	ω	ω^2	ω^2	
	Γ_3	1	1	1	ω^2	ω^2	ω	ω	
T	Γ_4	3	3	-1	0	0	0	0	x, y, z, L_x, L_y, L_z
$E_{1/2}$	Γ_5	2	-2	0	1	-1	1	-1	
$G_{3/2}$	Γ_6	2	-2	0	ω	$-\omega$	ω^2	$-\omega^2$	
	Γ_7	2	-2	0	ω^2	$-\omega^2$	ω	$-\omega$	

$\omega = \exp(2\pi i/3)$

GROUPS $\bar{O}$ AND $\bar{T}_d$

		$\bar{O}$	E	R	$4C_3$ $4C_3^2R$	$4C_3^2$ $4C_3R$	$3C_4^2$ $3C_4^2R$	$3C_4$ $3C_4^3R$	$3C_4^3$ $3C_4R$	$3C_2'$ $3C_2'R$	$\bar{O}$	
		$\bar{T}_d$	E	R	$4C_3$ $4C_3^2R$	$4C_3^2$ $4C_3R$	$3C_4^2$ $3C_4^2R$	$3S_4$ $3S_4^3R$	$3S_4^3$ $3S_4R$	$6\sigma_d$ $6\sigma_dR$		$\bar{T}_d$
	A_1	Γ_1	1	1	1	1	1	1	1	1		
	A_2	Γ_2	1	1	1	1	1	-1	-1	-1		
	E	Γ_3	2	2	-1	-1	2	0	0	0		
	T_1	Γ_4	3	3	0	0	-1	1	1	-1	x, y, z, L_x, L_y, L_z	L_x, L_y, L_z
	T_2	Γ_5	3	3	0	0	-1	-1	-1	1		x, y, z
(E_1)	$E_{1/2}$	Γ_6	2	-2	1	-1	0	$\sqrt{2}$	$-\sqrt{2}$	0		
(E_2)	$E_{5/2}$	Γ_7	2	-2	1	-1	0	$-\sqrt{2}$	$\sqrt{2}$	0		
(G)	$G_{3/2}$	Γ_8	4	-4	-1	1	0	0	0	0		

Appendix II TABLES OF CLEBSCH–GORDAN COEFFICIENTS, $\langle\Gamma_1\gamma_1\Gamma_2\gamma_2|\Gamma\gamma\rangle$, WITH CUBIC BASES‡

$A_2 \times A_2$

γ_1	γ_2	Γ / γ	A_1 / e_1
e_2	e_2		-1

$A_2 \times E$

γ_1	γ_2	Γ / γ	E / u	v
e_2	u		0	-1
	v		1	0

$A_2 \times T_1$

γ_1	γ_2	Γ / γ	T_2 / ξ	η	ζ
	α		1	0	0
e_2	β		0	1	0
	γ		0	0	1

‡ The C–G coefficients in Griffith's book (J. S. Griffith, "The Theory of Transition-Metal Ions." Cambridge Univ. Press, London and New York, 1964) are different from ours only in phase. The comparison of Griffith's phase and ours is given in the book by J. S. Griffith, "The Irreducible Tensor Methods for Molecular Symmetry Groups." Prentice-Hall, Englewood Cliffs, New Jersey, 1962.

$A_2 \times T_2$

γ_1	γ_2	Γ: T_1 α	T_1 β	T_1 γ
e_2	ξ	-1	0	0
	η	0	-1	0
	ζ	0	0	-1

$E \times E$

γ_1	γ_2	A_1 e_1	A_2 e_2	E u	E v
u	u	$1/\sqrt{2}$	0	$-1/\sqrt{2}$	0
	v	0	$1/\sqrt{2}$	0	$1/\sqrt{2}$
v	u	0	$-1/\sqrt{2}$	0	$1/\sqrt{2}$
	v	$1/\sqrt{2}$	0	$1/\sqrt{2}$	0

$E \times T_1$

γ_1	γ_2	T_1 α	T_1 β	T_1 γ	T_2 ξ	T_2 η	T_2 ζ
u	α	$-1/2$	0	0	$\sqrt{3}/2$	0	0
	β	0	$-1/2$	0	0	$-\sqrt{3}/2$	0
	γ	0	0	1	0	0	0
v	α	$\sqrt{3}/2$	0	0	$1/2$	0	0
	β	0	$-\sqrt{3}/2$	0	0	$1/2$	0
	γ	0	0	0	0	0	-1

$E \times T_2$

γ_1	γ_2	T_1 α	T_1 β	T_1 γ	T_2 ξ	T_2 η	T_2 ζ
u	ξ	$-\sqrt{3}/2$	0	0	$-1/2$	0	0
	η	0	$\sqrt{3}/2$	0	0	$-1/2$	0
	ζ	0	0	0	0	0	1
v	ξ	$-1/2$	0	0	$\sqrt{3}/2$	0	0
	η	0	$-1/2$	0	0	$-\sqrt{3}/2$	0
	ζ	0	0	1	0	0	0

$T_1 \times T_1$

γ_1	γ_2	Γ A_1	E		T_1			T_2		
		γ e_1	u	v	α	β	γ	ξ	η	ζ
	α	$-1/\sqrt{3}$	$1/\sqrt{6}$	$-1/\sqrt{2}$	0	0	0	0	0	0
α	β	0	0	0	0	0	$-1/\sqrt{2}$	0	0	$-1/\sqrt{2}$
	γ	0	0	0	0	$1/\sqrt{2}$	0	0	$-1/\sqrt{2}$	0
	α	0	0	0	0	0	$1/\sqrt{2}$	0	0	$-1/\sqrt{2}$
β	β	$-1/\sqrt{3}$	$1/\sqrt{6}$	$1/\sqrt{2}$	0	0	0	0	0	0
	γ	0	0	0	$-1/\sqrt{2}$	0	0	$-1/\sqrt{2}$	0	0
	α	0	0	0	0	$-1/\sqrt{2}$	0	0	$-1/\sqrt{2}$	0
γ	β	0	0	0	$1/\sqrt{2}$	0	0	$-1/\sqrt{2}$	0	0
	γ	$-1/\sqrt{3}$	$-2/\sqrt{6}$	0	0	0	0	0	0	0

$T_1 \times T_2$

γ_1	γ_2	Γ A_2	E		T_1			T_2		
		γ e_2	u	v	α	β	γ	ξ	η	ζ
	ξ	$-1/\sqrt{3}$	$-1/\sqrt{2}$	$-1/\sqrt{6}$	0	0	0	0	0	0
α	η	0	0	0	0	0	$1/\sqrt{2}$	0	0	$-1/\sqrt{2}$
	ζ	0	0	0	0	$1/\sqrt{2}$	0	0	$1/\sqrt{2}$	0
	ξ	0	0	0	0	0	$1/\sqrt{2}$	0	0	$1/\sqrt{2}$
β	η	$-1/\sqrt{3}$	$1/\sqrt{2}$	$-1/\sqrt{6}$	0	0	0	0	0	0
	ζ	0	0	0	$1/\sqrt{2}$	0	0	$-1/\sqrt{2}$	0	0
	ξ	0	0	0	0	$1/\sqrt{2}$	0	0	$-1/\sqrt{2}$	0
γ	η	0	0	0	$1/\sqrt{2}$	0	0	$1/\sqrt{2}$	0	0
	ζ	$-1/\sqrt{3}$	0	$2/\sqrt{6}$	0	0	0	0	0	0

$T_2 \times T_2$

γ_1	γ_2	Γ A_1	E		T_1			T_2		
		γ e_1	u	v	α	β	γ	ξ	η	ζ
	ξ	$1/\sqrt{3}$	$-1/\sqrt{6}$	$1/\sqrt{2}$	0	0	0	0	0	0
ξ	η	0	0	0	0	0	$1/\sqrt{2}$	0	0	$1/\sqrt{2}$
	ζ	0	0	0	0	$-1/\sqrt{2}$	0	0	$1/\sqrt{2}$	0
	ξ	0	0	0	0	0	$-1/\sqrt{2}$	0	0	$1/\sqrt{2}$
η	η	$1/\sqrt{3}$	$-1/\sqrt{6}$	$-1/\sqrt{2}$	0	0	0	0	0	0
	ζ	0	0	0	$1/\sqrt{2}$	0	0	$1/\sqrt{2}$	0	0
	ξ	0	0	0	0	$1/\sqrt{2}$	0	0	$1/\sqrt{2}$	0
ζ	η	0	0	0	$-1/\sqrt{2}$	0	0	$1/\sqrt{2}$	0	0
	ζ	$1/\sqrt{3}$	$2/\sqrt{6}$	0	0	0	0	0	0	0

Appendix III WIGNER COEFFICIENTS‡

$\langle j_1 m_1 j_2 m_2 \,|\, jm \rangle$

A general formula for calculating Wigner coefficients is given as follows:

$$
\begin{aligned}
\langle j_1 m_1 j_2 m_2 \,|\, jm \rangle &= \delta(m_1 + m_2\,, m)(2j+1)^{1/2} \Delta(j_1 j_2 j) \\
&\quad \times [(j_1+m_1)!\,(j_1-m_1)!\,(j_2+m_2)!\,(j_2-m_2)!\,(j+m)!\,(j-m)!]^{1/2} \\
&\quad \times \sum_z (-1)^z [z!\,(j_1+j_2-j-z)!\,(j_1-m_1-z)!\,(j_2+m_2-z)! \\
&\quad \times (j-j_2+m_1+z)!\,(j-j_1-m_2+z)!]^{-1}, \qquad \text{(AIII-1)}
\end{aligned}
$$

where

$$
\Delta(j_1 j_2 j) = [(j_1+j_2-j)!\,(j+j_1-j_2)!\,(j+j_2-j_1)!/(j_1+j_2+j+1)!]^{1/2}
$$

‡ Here we use notations j_1, j_2, j, and m in place of s_1, s_2, S, and M, respectively. Wigner coefficients $\langle j_1 m_1 j_2 m_2 \,|\, jm \rangle$ are related to the 3-j symbol

$$
\begin{pmatrix} j_1 & j_2 & j \\ m_1 & m_2 & m \end{pmatrix}
$$

as

$$
(-1)^{j_1 - j_2 - m} (2j+1)^{1/2} \langle j_1 m_1 j_2 m_2 \,|\, j\, -m \rangle = \begin{pmatrix} j_1 & j_2 & j \\ m_1 & m_2 & m \end{pmatrix}.
$$

Numerical values of the 3-j symbol with various sets of parameters can be found in the book: M. Rotenberg, R. Bivins, N. Metropolis, and J. K. Wooten, Jr., "The 3-j and 6-j Symbols." Technology Press, MIT, 1959.

In (AIII-1) summation parameter z takes on all integral values which make none of the factorials meaningless. Note that $0! = 1$.

Simplified formulas‡ for calculating Wigner coefficients with $j_2 = \frac{1}{2}$, 1, $\frac{3}{2}$, 2 are given as follows:

$j_2 = \frac{1}{2}$

$j =$	$m_2 = \frac{1}{2}$	$m_2 = -\frac{1}{2}$
$j_1 + \frac{1}{2}$	$\sqrt{\frac{j_1 + m + \frac{1}{2}}{2j_1 + 1}}$	$\sqrt{\frac{j_1 - m + \frac{1}{2}}{2j_1 + 1}}$
$j_1 - \frac{1}{2}$	$-\sqrt{\frac{j_1 - m + \frac{1}{2}}{2j_1 + 1}}$	$\sqrt{\frac{j_1 + m + \frac{1}{2}}{2j_1 + 1}}$

$j_2 = 1$

$j=$	$m_2 = 1$	$m_2 = 0$	$m_2 = -1$
j_1+1	$\sqrt{\frac{(j_1+m)(j_1+m+1)}{(2j_1+1)(2j_1+2)}}$	$\sqrt{\frac{(j_1-m+1)(j_1+m+1)}{(2j_1+1)(j_1+1)}}$	$\sqrt{\frac{(j_1-m)(j_1-m+1)}{(2j_1+1)(2j_1+2)}}$
j_1	$-\sqrt{\frac{(j_1+m)(j_1-m+1)}{2j_1(j_1+1)}}$	$\frac{m}{\sqrt{j_1(j_1+1)}}$	$\sqrt{\frac{(j_1-m)(j_1+m+1)}{2j_1(j_1+1)}}$
j_1-1	$\sqrt{\frac{(j_1-m)(j_1-m+1)}{2j_1(2j_1+1)}}$	$-\sqrt{\frac{(j_1-m)(j_1+m)}{j_1(2j_1+1)}}$	$\sqrt{\frac{(j_1+m+1)(j_1+m)}{2j_1(2j_1+1)}}$

‡ These formulas are taken from the book: E. U. Condon and G. H. Shortley, "The Theory of Atomic Spectra," pp. 76 and 77. Cambridge Univ. Press, London and New York, 1964.

$j_2 = \frac{3}{2}$

$j=$	$m_2 = \frac{3}{2}$	$m_2 = \frac{1}{2}$
$j_1+\frac{3}{2}$	$\sqrt{\frac{(j_1+m-\frac{1}{2})(j_1+m+\frac{1}{2})(j_1+m+\frac{3}{2})}{(2j_1+1)(2j_1+2)(2j_1+3)}}$	$\sqrt{\frac{3(j_1+m+\frac{1}{2})(j_1+m+\frac{3}{2})(j_1-m+\frac{3}{2})}{(2j_1+1)(2j_1+2)(2j_1+3)}}$
$j_1+\frac{1}{2}$	$-\sqrt{\frac{3(j_1+m-\frac{1}{2})(j_1+m+\frac{1}{2})(j_1-m+\frac{3}{2})}{2j_1(2j_1+1)(2j_1+3)}}$	$-(j_1-3m+\frac{3}{2})\sqrt{\frac{j_1+m+\frac{1}{2}}{2j_1(2j_1+1)(2j_1+3)}}$
$j_1-\frac{1}{2}$	$\sqrt{\frac{3(j_1+m-\frac{1}{2})(j_1-m+\frac{1}{2})(j_1-m+\frac{3}{2})}{(2j_1-1)(2j_1+1)(2j_1+2)}}$	$-(j_1+3m-\frac{1}{2})\sqrt{\frac{j_1-m+\frac{1}{2}}{(2j_1-1)(2j_1+1)(2j_1+2)}}$
$j_1-\frac{3}{2}$	$-\sqrt{\frac{(j_1-m-\frac{1}{2})(j_1-m+\frac{1}{2})(j_1-m+\frac{3}{2})}{2j_1(2j_1-1)(2j_1+1)}}$	$\sqrt{\frac{3(j_1+m-\frac{1}{2})(j_1-m-\frac{1}{2})(j_1-m+\frac{1}{2})}{2j_1(2j_1-1)(2j_1+1)}}$

$j=$	$m_2 = -\frac{1}{2}$	$m_2 = -\frac{3}{2}$
$j_1+\frac{3}{2}$	$\sqrt{\frac{3(j_1+m+\frac{3}{2})(j_1-m+\frac{1}{2})(j_1-m+\frac{3}{2})}{(2j_1+1)(2j_1+2)(2j_1+3)}}$	$\sqrt{\frac{(j_1-m-\frac{1}{2})(j_1-m+\frac{1}{2})(j_1-m+\frac{3}{2})}{(2j_1+1)(2j_1+2)(2j_1+3)}}$
$j_1+\frac{1}{2}$	$(j_1+3m+\frac{3}{2})\sqrt{\frac{j_1-m+\frac{1}{2}}{2j_1(2j_1+1)(2j_1+3)}}$	$\sqrt{\frac{3(j_1+m+\frac{3}{2})(j_1-m-\frac{1}{2})(j_1-m+\frac{1}{2})}{2j_1(2j_1+1)(2j_1+3)}}$
$j_1-\frac{1}{2}$	$-(j_1-3m-\frac{1}{2})\sqrt{\frac{j_1+m+\frac{1}{2}}{(2j_1-1)(2j_1+1)(2j_1+2)}}$	$\sqrt{\frac{3(j_1+m+\frac{1}{2})(j_1+m+\frac{3}{2})(j_1-m-\frac{1}{2})}{(2j_1-1)(2j_1+1)(2j_1+2)}}$
$j_1-\frac{3}{2}$	$-\sqrt{\frac{3(j_1+m-\frac{1}{2})(j_1+m+\frac{1}{2})(j_1-m-\frac{1}{2})}{2j_1(2j_1-1)(2j_1+1)}}$	$\sqrt{\frac{(j_1+m-\frac{1}{2})(j_1+m+\frac{1}{2})(j_1+m+\frac{3}{2})}{2j_1(2j_1-1)(2j_1+1)}}$

$j_2 = 2$

$j=$	$m_2 = 2$	$m_2 = 1$	$m_2 = 0$
j_1+2	$\sqrt{\frac{(j_1+m-1)(j_1+m)(j_1+m+1)(j_1+m+2)}{(2j_1+1)(2j_1+2)(2j_1+3)(2j_1+4)}}$	$\sqrt{\frac{(j_1-m+2)(j_1+m+2)(j_1+m+1)(j_1+m)}{(2j_1+1)(j_1+1)(2j_1+3)(j_1+2)}}$	$\sqrt{\frac{3(j_1-m+2)(j_1-m+1)(j_1+m+2)(j_1+m+1)}{(2j_1+1)(2j_1+2)(2j_1+3)(j_1+2)}}$
j_1+1	$-\sqrt{\frac{(j_1+m-1)(j_1+m)(j_1+m+1)(j_1-m+2)}{2j_1(j_1+1)(j_1+2)(2j_1+1)}}$	$-(j_1-2m+2)\sqrt{\frac{(j_1+m+1)(j_1+m)}{2j_1(2j_1+1)(j_1+1)(j_1+2)}}$	$m\sqrt{\frac{3(j_1-m+1)(j_1+m+1)}{j_1(2j_1+1)(j_1+1)(j_1+2)}}$
j_1	$\sqrt{\frac{3(j_1+m-1)(j_1+m)(j_1-m+1)(j_1-m+2)}{(2j_1-1)\,2j_1(j_1+1)(2j_1+3)}}$	$(1-2m)\sqrt{\frac{3(j_1-m+1)(j_1+m)}{(2j_1-1)\,j_1(2j_1+2)(2j_1+3)}}$	$\frac{3m^2-j_1(j_1+1)}{\sqrt{(2j_1-1)\,j_1(j_1+1)(2j_1+3)}}$
j_1-1	$-\sqrt{\frac{(j_1+m-1)(j_1-m)(j_1-m+1)(j_1-m+2)}{2(j_1-1)\,j_1(j_1+1)(2j_1+1)}}$	$(j_1+2m-1)\sqrt{\frac{(j_1-m+1)(j_1-m)}{(j_1-1)\,j_1(2j_1+1)(2j_1+2)}}$	$-m\sqrt{\frac{3(j_1-m)(j_1+m)}{(j_1-1)\,j_1(2j_1+1)(j_1+1)}}$
j_1-2	$\sqrt{\frac{(j_1-m-1)(j_1-m)(j_1-m+1)(j_1-m+2)}{(2j_1-2)(2j_1-1)\,2j_1(2j_1+1)}}$	$-\sqrt{\frac{(j_1-m+1)(j_1-m)(j_1-m-1)(j_1+m-1)}{(j_1-1)(2j_1-1)\,j_1(2j_1+1)}}$	$\sqrt{\frac{3(j_1-m)(j_1-m-1)(j_1+m)(j_1+m-1)}{(2j_1-2)(2j_1-1)\,j_1(2j_1+1)}}$

$j_2 = 2$ *(continued)*

$j=$	$m_2 = -1$	$m_2 = -2$
j_1+2	$\sqrt{\frac{(j_1-m+2)(j_1-m+1)(j_1-m)(j_1+m+2)}{(2j_1+1)(j_1+1)(2j_1+3)(j_1+2)}}$	$\sqrt{\frac{(j_1-m-1)(j_1-m)(j_1-m+1)(j_1-m+2)}{(2j_1+1)(2j_1+2)(2j_1+3)(2j_1+4)}}$
j_1+1	$(j_1+2m+2)\sqrt{\frac{(j_1-m+1)(j_1-m)}{j_1(2j_1+1)(2j_1+2)(j_1+2)}}$	$\sqrt{\frac{(j_1-m-1)(j_1-m)(j_1-m+1)(j_1+m+2)}{j_1(2j_1+1)(j_1+1)(2j_1+4)}}$
j_1	$(2m+1)\sqrt{\frac{3(j_1-m)(j_1+m+1)}{(2j_1-1)\,j_1(2j_1+2)(2j_1+3)}}$	$\sqrt{\frac{3(j_1-m-1)(j_1-m)(j_1+m+1)(j_1+m+2)}{(2j_1-1)\,j_1(2j_1+2)(2j_1+3)}}$
j_1-1	$-(j_1-2m-1)\sqrt{\frac{(j_1+m+1)(j_1+m)}{(j_1-1)\,j_1(2j_1+1)(2j_1+2)}}$	$\sqrt{\frac{(j_1-m-1)(j_1+m)(j_1+m+1)(j_1+m+2)}{(j_1-1)\,j_1(2j_1+1)(2j_1+2)}}$
j_1-2	$-\sqrt{\frac{(j_1-m-1)(j_1+m+1)(j_1+m)(j_1+m-1)}{(j_1-1)(2j_1-1)\,j_1(2j_1+1)}}$	$\sqrt{\frac{(j_1+m-1)(j_1+m)(j_1+m+1)(j_1+m+2)}{(2j_1-2)(2j_1-1)\,2j_1(2j_1+1)}}$

Appendix IV MATRIX ELEMENTS OF COULOMB INTERACTION

In order to satisfy the phase relation between the L- and R-states as given in (4.29), the phases of the base functions for the states indicated below are chosen to be opposite to those in the reference cited:‡

$$t_2^4\,{}^3T_1\,,\quad t_2^4\,{}^1T_2\,,\quad t_2^4\,{}^1E,$$

$$t_2^4({}^3T_1)e\,{}^2T_2\,,\quad t_2^4({}^1T_2)e\,{}^2T_2\,,\quad t_2^4({}^3T_1)e\,{}^2T_1\,,$$

$$t_2^4({}^1T_2)e\,{}^2T_1\,,\quad t_2^4({}^1E)e\,{}^2E,\quad t_2^4({}^1E)e\,{}^2A_1\,,$$

$$t_2^4({}^1E)e\,{}^2A_2\,,\quad t_2^4({}^3T_1)e\,{}^4T_1\,,\quad t_2^4({}^3T_1)e\,{}^4T_2\,.$$

Accordingly, the signs of some matrix elements listed here differ from those in the original table.‡

When each matrix listed here is diagonalized, its eigenvalues give the energies of the terms of the free ion indicated in brackets above the corresponding matrix.

‡ Y. Tanabe and S. Sugano, *J. Phys. Soc. Japan* **9**, 766 (1954). According to such phase change of the base functions, we should reverse the signs of the CFP involving $t_2^4S\Gamma\,(S\Gamma = {}^3T_1\,, {}^1T_2\,, {}^1E)$ which are tabulated in Table II of this paper. Furthermore, we should reverse the signs of the reduced matrix elements $\langle t_2^4\,{}^1A_1 \| X(\bar{\Gamma}) \| t_2^4S\Gamma\rangle$ $(S\Gamma = {}^3T_1\,, {}^1T_2\,, {}^1E)$ which are tabulated in Table Ia of the paper by Y. Tanabe and H. Kamimura, *J. Phys. Soc. Japan* **13**, 394 (1958). The signs of the reduced matrix elements of the spin-orbit interaction are corrected in Appendix VII.

(i) d^2

$^1A_1(^1G, {}^1S)$

	t_2^2	e^2
	$10B+5C$	$\sqrt{6}(2B+C)$
		$8B+4C$

$^1E(^1D, {}^1G)$

	t_2^2	e^2
	$B+2C$	$-2\sqrt{3}\,B$
		$2C$

$^1T_2(^1D, {}^1G)$

	t_2^2	t_2e
	$B+2C$	$2\sqrt{3}\,B$
		$2C$

$^3T_1(^3F, {}^3P)$

	t_2^2	t_2e
	$-5B$	$6B$
		$4B$

$t_2e\ {}^1T_1$	(^1G)	$4B+2C$
$t_2e\ {}^3T_2$	(^3F)	$-8B$
$e^2\ {}^3A_2$	(^3F)	$-8B$

(ii) d^3

$^2T_2(a\,^2D, b\,^2D, ^2F, ^2G, ^2H)$

t_2^3	$t_2^2(^3T_1)e$	$t_2^2(^1T_2)e$	$t_2e^2(^1A_1)$	$t_2e^2(^1E)$
$5C$	$-3\sqrt{3}\,B$	$-5\sqrt{3}\,B$	$4B+2C$	$2B$
	$-6B+3C$	$3B$	$-3\sqrt{3}\,B$	$-3\sqrt{3}\,B$
		$4B+3C$	$-\sqrt{3}\,B$	$\sqrt{3}\,B$
			$6B+5C$	$10B$
				$-2B+3C$

$^2T_1(^2P, ^2F, ^2G, ^2H)$

t_2^3	$t_2^2(^3T_1)e$	$t_2^2(^1T_2)e$	$t_2e^2(^3A_2)$	$t_2e^2(^1E)$
$-6B+3C$	$-3B$	$3B$	0	$-2\sqrt{3}\,B$
	$3C$	$-3B$	$3B$	$3\sqrt{3}\,B$
		$-6B+3C$	$-3B$	$-\sqrt{3}\,B$
			$-6B+3C$	$2\sqrt{3}\,B$
				$-2B+3C$

$^2E(a\,^2D, b\,^2D, ^2G, ^2H)$

t_2^3	$t_2^2(^1A_1)e$	$t_2^2(^1E)e$	e^3
$-6B+3C$	$-6\sqrt{2}\,B$	$-3\sqrt{2}\,B$	0
	$8B+6C$	$10B$	$\sqrt{3}(2B+C)$
		$-B+3C$	$2\sqrt{3}\,B$
			$-8B+4C$

$^4T_1(^4P, ^4F)$

$t_2^2(^3T_1)e$	$t_2e^2(^3A_2)$
$-3B$	$6B$
	$-12B$

$t_2^3\ ^4A_2$	(^4F)	$-15B$,
$t_2^2(^3T_1)e\ ^4T_2$	(^4F)	$-15B$,
$t_2^2(^1E)e\ ^2A_1$	(^2G)	$-11B+3C$,
$t_2^2(^1E)e\ ^2A_2$	(^2F)	$9B+3C$,

(iii) d^4

$^3T_1(a\,^3P,\, b\,^3P,\, a\,^3F,\, b\,^3F,\, ^3G,\, ^3H)$

$t_2{}^4$	$t_2{}^3(^2T_1)e$	$t_2{}^3(^2T_2)e$	$t_2{}^2(^3T_1)e^2(^1A_1)$	$t_2{}^2(^3T_1)e^2(^1E)$	$t_2{}^2(^1T_2)e^2(^3A_2)$	t_2e^3
$-15B+5C$	$\sqrt{6}\,B$	$3\sqrt{2}\,B$	$-\sqrt{2}(2B+C)$	$2\sqrt{2}\,B$	0	0
	$-11B+4C$	$5\sqrt{3}\,B$	$\sqrt{3}\,B$	$-\sqrt{3}\,B$	$3B$	$\sqrt{6}\,B$
		$-3B+6C$	$-3B$	$-3B$	$5\sqrt{3}\,B$	$\sqrt{2}(B+C)$
			$-B+6C$	$-10B$	0	$3\sqrt{2}\,B$
				$-9B+4C$	$-2\sqrt{3}\,B$	$-3\sqrt{2}\,B$
					$-11B+4C$	$\sqrt{6}\,B$
						$-16B+5C$

$^1T_2(a\,^1D,\, b\,^1D,\, a\,^1G,\, b\,^1G,\, ^1F,\, ^1I)$

$t_2{}^4$	$t_2{}^3(^2T_1)e$	$t_2{}^3(^2T_2)e$	$t_2{}^2(^3T_1)e^2(^3A_2)$	$t_2{}^2(^1T_2)e^2(^1E)$	$t_2{}^2(^1T_2)e^2(^1A_1)$	t_2e^3
$-9B+7C$	$-3\sqrt{2}\,B$	$5\sqrt{6}\,B$	0	$2\sqrt{2}\,B$	$-\sqrt{2}(2B+C)$	0
	$-9B+6C$	$-5\sqrt{3}\,B$	$3B$	$-3B$	$-3B$	$-\sqrt{6}\,B$
		$3B+8C$	$-3\sqrt{3}\,B$	$5\sqrt{3}\,B$	$-5\sqrt{3}\,B$	$\sqrt{2}(3B+C)$
			$-9B+6C$	$-6B$	0	$-3\sqrt{6}\,B$
				$-3B+6C$	$-10B$	$\sqrt{6}\,B$
					$5B+8C$	$\sqrt{6}\,B$
						$7C$

$^1A_1(a\,^1S, b\,^1S, a\,^1G, b\,^1G, ^1I)$

t_2^4	$t_2^3(^1E)e$	$t_2^2(^1A_1)e^2(^1A_1)$	$t_2^2(^1E)e^2(^1E)$	e^4
$10C$	$-12\sqrt{2}\,B$	$\sqrt{2}(4B+2C)$	$2\sqrt{2}\,B$	0
	$6C$	$-12B$	$-6B$	0
		$14B+11C$	$20B$	$\sqrt{6}(2B+C)$
			$-3B+6C$	$2\sqrt{6}\,B$
				$-16B+8C$

$^1E(a\,^1D, b\,^1D, a\,^1G, b\,^1G, ^1I)$

t_2^4	$t_2^3(^2E)e$	$t_2^2(^1E)e^2(^1A_1)$	$t_2^2(^1A_1)e^2(^1E)$	$t_2^2(^1E)e^2(^1E)$
$-9B+7C$	$-6B$	$-\sqrt{2}(2B+C)$	$2B$	$4B$
	$-6B+6C$	$-3\sqrt{2}\,B$	$-12B$	0
		$5B+8C$	$10\sqrt{2}\,B$	$-10\sqrt{2}\,B$
			$6B+9C$	0
				$-3B+6C$

$^3T_2(^3D, a\,^3F, b\,^3F, ^3G, ^3H)$

$t_2^3(^2T_1)e$	$t_2^3(^2T_2)e$	$t_2^2(^3T_1)e^2(^3A_2)$	$t_2^2(^3T_1)e^2(^1E)$	t_2e^3
$-9B+4C$	$-5\sqrt{3}\,B$	$\sqrt{6}\,B$	$\sqrt{3}\,B$	$\sqrt{6}\,B$
	$-5B+6C$	$-3\sqrt{2}\,B$	$3B$	$\sqrt{2}(3B+C)$
		$-13B+4C$	$-2\sqrt{2}\,B$	$-6B$
			$-9B+4C$	$3\sqrt{2}\,B$
				$-8B+5C$

$^1T_1(^1F, a\,^1G, b\,^1G, ^1I)$

$t_2^3(^2T_1)e$	$t_2^3(^2T_2)e$	$t_2^2(^1T_2)e^2(^1E)$	t_2e^3
$-3B+6C$	$5\sqrt{3}\,B$	$3B$	$\sqrt{6}\,B$
	$-3B+8C$	$-5\sqrt{3}\,B$	$\sqrt{2}(B+C)$
		$-3B+6C$	$-\sqrt{6}\,B$
			$-16B+7C$

$^3E(^3D, ^3G, ^3H)$

$t_2^3(^4A_2)e$	$t_2^3(^2E)e$	$t_2^2(^1E)e^2(^3A_2)$
$-13B+4C$	$-4B$	0
	$-10B+4C$	$-3\sqrt{2}\,B$
		$-11B+4C$

$^3A_2(a\,^3F, b\,^3F)$

$t_2^3(^2E)e$	$t_2^2(^1A_1)e^2(^3A_2)$
$-8B+4C$	$-12B$
	$-2B+7C$

$^1A_2(^1F, ^1I)$

$t_2^3(^2E)e$	$t_2^2(^1E)e^2(^1E)$
$-12B+6C$	$6B$
	$-3B+6C$

$t_2^3(^4A_2)e\ ^5E$	(^5D)	$-21B$,
$t_2^2(^3T_1)e^2(^3A_2)^5T_2$	(^5D)	$-21B$,
$t_2^3(^2E)e\ ^3A_1$	(^3G)	$-12B+4C$,

(iv) d^5

$^2T_2(a\,^2F, b\,^2F, a\,^2G, b\,^2G, ^2H, ^2I, a\,^2D, b\,^2D, c\,^2D)$

t_2^5	$t_2^4(^3T_1)e$	$t_2^4(^1T_2)e$	$t_2^3(^2T_1)e^2(^3A_2)$	$t_2^3(^2T_1)e^2(^1E)$
$-20B+10C$	$-3\sqrt{6}\,B$	$-\sqrt{6}\,B$	0	$-2\sqrt{3}\,B$
	$-8B+9C$	$3B$	$-\sqrt{6}\,B/2$	$3\sqrt{2}\,B/2$
		$-18B+9C$	$-3\sqrt{6}\,B/2$	$3\sqrt{2}\,B/2$
			$-16B+8C$	$2\sqrt{3}\,B$
				$-12B+8C$

$t_2^3(^2T_2)e^2(^1A_1)$	$t_2^3(^2T_2)e^2(^1E)$	$t_2^2(^1T_2)e^3$	$t_2^2(^3T_1)e^3$	t_2e^4
$4B+2C$	$2B$	0	0	0
$-3\sqrt{6}\,B/2$	$-3\sqrt{6}\,B/2$	0	$-4B-C$	0
$-5\sqrt{6}\,B/2$	$5\sqrt{6}\,B/2$	$-C$	0	0
0	0	$-3\sqrt{6}\,B/2$	$-\sqrt{6}\,B/2$	0
$-10\sqrt{3}\,B$	0	$3\sqrt{2}\,B/2$	$3\sqrt{2}\,B/2$	$-2\sqrt{3}\,B$
$2B+12C$	0	$-5\sqrt{6}\,B/2$	$-3\sqrt{6}\,B/2$	$4B+2C$
	$-6B+10C$	$-5\sqrt{6}\,B/2$	$3\sqrt{6}\,B/2$	$-2B$
		$-18B+9C$	$3B$	$-\sqrt{6}\,B$
			$-8B+9C$	$-3\sqrt{6}\,B$
				$-20B+10C$

$^2T_1(^2P, a\,^2F, b\,^2F, a\,^2G, b\,^2G, ^2H, ^2I)$

$t_2^4(^3T_1)e$	$t_2^4(^1T_2)e$	$t_2^3(^2T_1)e^2(^1A_1)$	$t_2^3(^2T_1)e^2(^1E)$	$t_2^3(^2T_2)e^2(^3A_2)$	$t_2^3(^2T_2)e^2(^1E)$	$t_2^2(^1T_2)e^3$	$t_2^2(^3T_1)e^3$
$-22B+9C$	$-3B$	$3\sqrt{2}\,B/2$	$-3\sqrt{2}\,B/2$	$3\sqrt{2}\,B/2$	$3\sqrt{6}\,B/2$	0	$-C$
	$-8B+9C$	$-3\sqrt{2}\,B/2$	$-3\sqrt{2}\,B/2$	$-15\sqrt{2}\,B/2$	$-5\sqrt{6}\,B/2$	$-4B-C$	0
		$-4B+10C$	0	0	$10\sqrt{3}\,B$	$3\sqrt{2}\,B/2$	$-3\sqrt{2}\,B/2$
			$-12B+8C$	0	0	$-3\sqrt{2}\,B/2$	$-3\sqrt{2}\,B/2$
				$-10B+10C$	$2\sqrt{3}\,B$	$15\sqrt{2}\,B/2$	$-3\sqrt{2}\,B/2$
					$-6B+10C$	$5\sqrt{6}\,B/2$	$-3\sqrt{6}\,B/2$
						$-8B+9C$	$-3B$
							$-22B+9C$

$^2E(a\,^2D, b\,^2D, c\,^2D, a\,^2G, b\,^2G, ^2H, ^2I)$

$t_2^4(^1A_1)e$	$t_2^4(^1E)e$	$t_2^3(^2E)e^2(^1A_1)$	$t_2^3(^2E)e^2(^3A_2)$	$t_2^3(^2E)e^2(^1E)$	$t_2^2(^1E)e^3$	$t_2^2(^1A_1)e^3$
$-4B+12C$	$-10B$	$6B$	$6\sqrt{3}\,B$	$6\sqrt{2}\,B$	$-2B$	$4B+2C$
	$-13B+9C$	$3B$	$-3\sqrt{3}\,B$	0	$-2B-C$	$-2B$
		$-4B+10C$	0	0	$-3B$	$-6B$
			$-16B+8C$	$2\sqrt{6}\,B$	$-3\sqrt{3}\,B$	$6\sqrt{3}\,B$
				$-12B+8C$	0	$6\sqrt{2}\,B$
					$-13B+9C$	$-10B$
						$-4B+12C$

$^2A_1(^2S, a\,^2G, b\,^2G, ^2I)$

$t_2^4(^1E)e$	$t_2^3(^2E)e^2(^1E)$	$t_2^3(^4A_2)e^2(^3A_2)$	$t_2^2(^1E)e^3$
$-3B+9C$	$3\sqrt{2}\,B$	0	$-6B-C$
	$-12B+8C$	$-4\sqrt{3}\,B$	$3\sqrt{2}\,B$
		$-19B+8C$	0
			$-3B+9C$

$^2A_2(a\,^2F, b\,^2F, ^2I)$

$t_2^4(^1E)e$	$t_2^3(^2E)e^2(^1E)$	$t_2^2(^1E)e^3$
$-23B+9C$	$-3\sqrt{2}\,B$	$2B-C$
	$-12B+8C$	$-3\sqrt{2}\,B$
		$-23B+9C$

$^4T_1(^4P, ^4F, ^4G)$

$t_2^4(^3T_1)e$	$t_2^3(^2T_2)e^2(^3A_2)$	$t_2^2(^3T_1)e^3$
$-25B+6C$	$3\sqrt{2}\,B$	$-C$
	$-16B+7C$	$-3\sqrt{2}\,B$
		$-25B+6C$

$^4T_2(^4F, ^4G, ^4D)$

$t_2^4(^3T_1)e$	$t_2^3(^2T_1)e^2(^3A_2)$	$t_2^2(^3T_1)e^3$
$-17B+6C$	$-\sqrt{6}\,B$	$-4B-C$
	$-22B+5C$	$-\sqrt{6}\,B$
		$-17B+6C$

$^4E(^4D, ^4G)$

$t_2^3(^2E)e^2(^3A_2)$	$t_2^3(^4A_2)e^2(^1E)$
$-22B+5C$	$-2\sqrt{3}\,B$
	$-21B+5C$

$t_2^3(^4A_2)e^2(^3A_2)\ ^6A_1$	(^6S)	$-35B,$
$t_2^3(^4A_2)e^2(^3A_2)\ ^4A_1$	(^4G)	$-25B+5C,$
$t_2^3(^4A_2)e^2(^1A_1)\ ^4A_2$	(^4F)	$-13B+7C,$

Appendix V COMPLEMENTARY STATES IN THE (t_2, e) SHELL

Here, we will give the proof of (4.27). First note that, by using (4.25) and (4.26), Eq. (4.24) can be reexpressed as follows:

$$\begin{aligned}
&\Psi(t_2{}^6e^4\,{}^1A_1) \\
&\quad = [{}_{10}C_N]^{-1/2} \sum_{\substack{S_1\Gamma_1S_2\Gamma_2\\ S\Gamma M\gamma\\ n',m'(n'+m'=N)}} (-1)^{n'm'} \sum_{\substack{M_1M_2\\ \gamma_1\gamma_2}} \sum_{\substack{S_1'\Gamma_1'S_2'\Gamma_2'\\ M_1'M_2'\gamma_1'\gamma_2'}} \\
&\qquad \times (-1)^{S-M}\alpha_{n'm'}(S_1\Gamma_1\,,\, S_2\Gamma_2 : S_1'\Gamma_1', S_2'\Gamma_2') \\
&\qquad \times \langle S_1M_1S_2M_2 \mid SM\rangle\langle S_1' -M_1'S_2' -M_2' \mid S -M\rangle \\
&\qquad \times \langle \Gamma_1\gamma_1\Gamma_2\gamma_2 \mid \Gamma\gamma\rangle\langle \Gamma_1'\gamma_1'\Gamma_2'\gamma_2' \mid \Gamma\gamma\rangle \\
&\qquad \times [\mathscr{A}_{n',m'}\Psi_L(t_2^{n'}S_1\Gamma_1M_1\gamma_1)\,\Psi_L(e^{m'}S_2\Gamma_2M_2\gamma_2)] \\
&\qquad \times [\mathscr{A}_{6-n',4-m'}\Psi_R(t_2^{6-n'}S_1'\Gamma_1' - M_1'\gamma_1')\,\Psi_R(e^{4-m'}S_2'\Gamma_2' - M_2'\gamma_2')],
\end{aligned} \tag{AV-1}$$

in which $\mathscr{A}_{n',m'}$ and $\mathscr{A}_{6-n',4-m'}$ are operators antisymmetrizing, respectively, the functions in the first and second square brackets with respect to the exchange of electrons in each of the t_2 and e shells. Operators $\mathscr{A}_{n',m'}$ and $\mathscr{A}_{6-n',4-m'}$ involve normalization factors

$$[{}_NC_{n'}]^{-1/2} \qquad \text{and} \qquad [{}_{10-N}C_{6-n'}]^{-1/2},$$

respectively. It should be remarked that

$$\Psi_L(t_2^{n'} S_1 \Gamma_1 M_1 \gamma_1), \qquad \Psi_L(e^{m'} S_2 \Gamma_2 M_2 \gamma_2),$$

$$\Psi_R(t_2^{6-n'} S_1' \Gamma_1' \, -M_1' \gamma_1') \quad \text{and} \quad \Psi_R(e^{4-m'} S_2' \Gamma_2' \, -M_2' \gamma_2')$$

are, respectively, the antisymmetric function of electrons $1, 2, \ldots, n'$; that of electrons $n' + 1$, $n' + 2, \ldots, N$; that of electrons $N + 1$, $N + 2, \ldots, 6 + m'$: and that of electrons $7 + m'$, $8 + m', \ldots, 10$.

On the other hand, by using (4.16), $\Psi(t_2{}^6 e^4 \, {}^1A_1)$ can also be expressed as

$$\begin{aligned}
\Psi(t_2{}^6 e^4 \, {}^1A_1) &= \mathscr{A}_{6,4} \Psi(t_2{}^6 \, {}^1A_1) \, \Psi(e^4 \, {}^1A_1) \\
&= [{}_6C_n \times {}_4C_m]^{-1/2} \sum_{\substack{S_1\Gamma_1 S_2 \Gamma_2 \\ M_1\gamma_1 M_2\gamma_2}} (-1)^{S_1+S_2-M_1-M_2} \\
&\quad \times [\mathscr{A}_{6,4} \Psi_L(t_2{}^n S_1 \Gamma_1 M_1 \gamma_1) \, \Psi_R(t_2^{6-n} S_1 \Gamma_1 \, -M_1 \gamma_1) \\
&\quad \times \Psi_L(e^m S_2 \Gamma_2 M_2 \gamma_2) \, \Psi_R(e^{4-m} S_2 \Gamma_2 \, -M_2 \gamma_2)] \\
&= [{}_6C_n \times {}_4C_m]^{-1/2} \sum_{\substack{S_1\Gamma_1 S_2 \Gamma_2 \\ M_1\gamma_1 M_2\gamma_2}} (-1)^{nm} (-1)^{S_1+S_2-M_1-M_2} \\
&\quad \times [\mathscr{A}_{6,4} \Psi_L(t_2{}^n S_1 \Gamma_1 M_1 \gamma_1) \, \Psi_L(e^m S_2 \Gamma_2 M_2 \gamma_2) \\
&\quad \times \Psi_R(t_2^{6-n} S_1 \Gamma_1 - M_1 \gamma_1) \, \Psi_R(e^{4-m} S_2 \Gamma_2 - M_2 \gamma_2)],
\end{aligned} \tag{AV-2}$$

in which $\mathscr{A}_{6,4}$ is the operator $\mathscr{A}_{n,m}$ with $n = 6$ and $m = 4$. We should mention that in the third expression $\Psi_R(t_2^{6-n} S_1 \Gamma_1 \, -M_1 \gamma_1)$ and $\Psi_L(e^m S_2 \Gamma_2 M_2 \gamma_2)$ are, respectively, the function of electrons, $n + 1$, $n + 2, \ldots, 6$, and that of electrons, $7, 8, \ldots, 6 + m$, but in the last expression they are, respectively, the function of electrons $N + 1$, $N + 2, \ldots, 6 + m'$, and that of electrons, $n + 1, n + 2, \ldots, N$.

Now, multiplying the last expressions of (AV-1) and (AV-2) by

$$\begin{aligned}
&\Psi_L(t_2{}^n S_1 \Gamma_1 M_1 \gamma_1) \, \Psi_L(e^m S_2 \Gamma_2 M_2 \gamma_2) \, \Psi_R(t_2^{6-n} S_1' \Gamma_1' \, -M_1' \gamma_1') \\
&\quad \times \Psi_R(e^{4-m} S_2' \Gamma_2' - M_2' \gamma_2'),
\end{aligned}$$

integrating over all the electron coordinates, and then using the orthogonality relations between the wavefunctions, one obtains

$$\alpha_{nm}(S_1\Gamma_1, S_2\Gamma_2 : S_1'\Gamma_1', S_2'\Gamma_2') = \alpha_{nm}^0 \, \delta(S_1 S_1') \, \delta(S_2 S_2') \, \delta(\Gamma_1 \Gamma_1') \, \delta(\Gamma_2 \Gamma_2'). \tag{AV-3}$$

Use of (AV-3) simplifies (AV-1) into the following form:

$$\Psi(t_2{}^6e^4\,{}^1A_1) = [{}_{10}C_N]^{-1/2} \sum_{\substack{S_1\Gamma_1S_2\Gamma_2 \\ M_1\gamma_1M_2\gamma_2 \\ n',m'(n'+m'=N)}} (-1)^{S_1+S_2-M_1-M_2}$$

$$\times\, \alpha^0_{n'm'}(-1)^{n'm'}[\mathscr{A}_{n',m'}\Psi_L(t_2^{n'}\,S_1\Gamma_1M_1\gamma_1)\,\Psi_L(e^{m'}\,S_2\Gamma_2M_2\gamma_2)]$$

$$\times\, [\mathscr{A}_{6-n',4-m'}\Psi_R(t_2^{6-n'}\,S_1\Gamma_1\,-M_1\gamma_1)\,\Psi_R(e^{4-m'}\,S_2\Gamma_2\,-M_2\gamma_2)]. \tag{AV-4}$$

In deriving (AV-3), use was made of the relation,

$$\langle S_1M_1S_2M_2 \mid SM\rangle = (-1)^{S_1+S_2-S}\langle S_1\,-M_1S_2\,-M_2 \mid S\,-M\rangle. \tag{AV-5}$$

Now, again multiplying the last expressions of (AV-2) and (AV-4) by

$$\Psi_L(t_2{}^nS_1\Gamma_1M_1\gamma_1)\,\Psi_L(e^mS_2\Gamma_2M_2\gamma_2)$$
$$\times\, \Psi_R(t_2^{6-n}S_1\Gamma_1\,-M_1\gamma_1)\,\Psi_R(e^{4-m}S_2\Gamma_2\,-M_2\gamma_2)$$

and integrating over all the electron coordinates, one obtains

$$\alpha^0_{nm} = 1. \tag{AV-6}$$

In deriving (AV-6) we used the relation

$${}_{10}C_N \times {}_NC_n \times {}_{10-N}C_{6-n} = {}_{10}C_6 \times {}_6C_n \times {}_4C_m\,. \tag{AV-7}$$

Appendix VI TABLES OF CLEBSCH–GORDAN COEFFICIENTS WITH TRIGONAL BASES, $\langle \Gamma_1 M_1 \Gamma_2 M_2 \mid \Gamma M \rangle = \langle \Gamma M \mid \Gamma_1 M_1 \Gamma_2 M_2 \rangle^*$

$A_2 \times E$

M_1	M_2 \ Γ M	E u_+	u_-
e_2	u_+	$-i$	0
	u_-	0	i

$A_2 \times T_1$

M_1	M_2 \ Γ M	T_2 x_+	x_-	x_0
	a_+	1	0	0
e_2	a_-	0	1	0
	a_0	0	0	1

$A_2 \times T_2$

M_1	M_2 \ Γ M	T_1 a_+	a_-	a_0
	x_+	-1	0	0
e_2	x_-	0	-1	0
	x_0	0	0	-1

$E \times E$

M_1	M_2	Γ: A_1 e_1	A_2 e_2	E u_+	E u_-
u_+	u_+	0	0	0	-1
	u_-	$-1/\sqrt{2}$	$-i/\sqrt{2}$	0	0
u_-	u_+	$-1/\sqrt{2}$	$i/\sqrt{2}$	0	0
	u_-	0	0	1	0

$E \times T_1$

M_1	M_2	T_1 a_+	T_1 a_-	T_1 a_0	T_2 x_+	T_2 x_-	T_2 x_0
u_+	a_+	0	$1/\sqrt{2}$	0	0	$i/\sqrt{2}$	0
	a_-	0	0	$-1/\sqrt{2}$	0	0	$-i/\sqrt{2}$
	a_0	$1/\sqrt{2}$	0	0	$i/\sqrt{2}$	0	0
u_-	a_+	0	0	$-1/\sqrt{2}$	0	0	$i/\sqrt{2}$
	a_-	$-1/\sqrt{2}$	0	0	$i/\sqrt{2}$	0	0
	a_0	0	$1/\sqrt{2}$	0	0	$-i/\sqrt{2}$	0

$E \times T_2$

M_1	M_2	T_1 a_+	T_1 a_-	T_1 a_0	T_2 x_+	T_2 x_-	T_2 x_0
u_+	x_+	0	$-i/\sqrt{2}$	0	0	$1/\sqrt{2}$	0
	x_-	0	0	$i/\sqrt{2}$	0	0	$-1/\sqrt{2}$
	x_0	$-i/\sqrt{2}$	0	0	$1/\sqrt{2}$	0	0
u_-	x_+	0	0	$-i/\sqrt{2}$	0	0	$-1/\sqrt{2}$
	x_-	$-i/\sqrt{2}$	0	0	$-1/\sqrt{2}$	0	0
	x_0	0	$i/\sqrt{2}$	0	0	$1/\sqrt{2}$	0

$T_1 \times T_1$

M_1	M_2	A_1 e_1	E u_+	E u_-	T_1 a_+	T_1 a_-	T_1 a_0	T_2 x_+	T_2 x_-	T_2 x_0
a_+	a_+	0	0	$-1/\sqrt{3}$	0	0	0	0	$-\sqrt{2}/\sqrt{3}$	0
	a_-	$1/\sqrt{3}$	0	0	0	0	$i/\sqrt{2}$	0	0	$-1/\sqrt{6}$
	a_0	0	$-1/\sqrt{3}$	0	$i/\sqrt{2}$	0	0	$1/\sqrt{6}$	0	0
a_-	a_+	$1/\sqrt{3}$	0	0	0	0	$-i/\sqrt{2}$	0	0	$-1/\sqrt{6}$
	a_-	0	$1/\sqrt{3}$	0	0	0	0	$\sqrt{2}/\sqrt{3}$	0	0
	a_0	0	0	$-1/\sqrt{3}$	0	$-i/\sqrt{2}$	0	0	$1/\sqrt{6}$	0
a_0	a_+	0	$-1/\sqrt{3}$	0	$-i/\sqrt{2}$	0	0	$1/\sqrt{6}$	0	0
	a_-	0	0	$-1/\sqrt{3}$	0	$i/\sqrt{2}$	0	0	$1/\sqrt{6}$	0
	a_0	$-1/\sqrt{3}$	0	0	0	0	0	0	0	$-\sqrt{2}/\sqrt{3}$

$T_1 \times T_2$

M_1	M_2 \ Γ:	A_2	E		T_1			T_2		
	M:	e_2	u_+	u_-	a_+	a_-	a_0	x_+	x_-	x_0
	x_+	0	0	$-i/\sqrt{3}$	0	$\sqrt{2}/\sqrt{3}$	0	0	0	0
a_+	x_-	$1/\sqrt{3}$	0	0	0	0	$1/\sqrt{6}$	0	0	$i/\sqrt{2}$
	x_0	0	$i/\sqrt{3}$	0	$-1/\sqrt{6}$	0	0	$i/\sqrt{2}$	0	0
	x_+	$1/\sqrt{3}$	0	0	0	0	$1/\sqrt{6}$	0	0	$-i/\sqrt{2}$
a_-	x_-	0	$-i/\sqrt{3}$	0	$-\sqrt{2}/\sqrt{3}$	0	0	0	0	0
	x_0	0	0	$-i/\sqrt{3}$	0	$-1/\sqrt{6}$	0	0	$-i/\sqrt{2}$	0
	x_+	0	$i/\sqrt{3}$	0	$-1/\sqrt{6}$	0	0	$-i/\sqrt{2}$	0	0
a_0	x_-	0	0	$-i/\sqrt{3}$	0	$-1/\sqrt{6}$	0	0	$i/\sqrt{2}$	0
	x_0	$-1/\sqrt{3}$	0	0	0	0	$\sqrt{2}/\sqrt{3}$	0	0	0

$T_2 \times T_2$

M_1	M_2 \ Γ:	A_1	E		T_1			T_2		
	M:	e_1	u_+	u_-	a_+	a_-	a_0	x_+	x_-	x_0
	x_+	0	0	$1/\sqrt{3}$	0	0	0	0	$\sqrt{2}/\sqrt{3}$	0
x_+	x_-	$-1/\sqrt{3}$	0	0	0	0	$-i/\sqrt{2}$	0	0	$1/\sqrt{6}$
	x_0	0	$1/\sqrt{3}$	0	$-i/\sqrt{2}$	0	0	$-1/\sqrt{6}$	0	0
	x_+	$-1/\sqrt{3}$	0	0	0	0	$i/\sqrt{2}$	0	0	$1/\sqrt{6}$
x_-	x_-	0	$-1/\sqrt{3}$	0	0	0	0	$-\sqrt{2}/\sqrt{3}$	0	0
	x_0	0	0	$1/\sqrt{3}$	0	$i/\sqrt{2}$	0	0	$-1/\sqrt{6}$	0
	x_+	0	$1/\sqrt{3}$	0	$i/\sqrt{2}$	0	0	$-1/\sqrt{6}$	0	0
x_0	x_-	0	0	$1/\sqrt{3}$	0	$-i/\sqrt{2}$	0	0	$-1/\sqrt{6}$	0
	x_0	$1/\sqrt{3}$	0	0	0	0	0	0	0	$\sqrt{2}/\sqrt{3}$

Appendix VII TABLES OF REDUCED MATRICES OF SPIN-ORBIT INTERACTION‡

d $\langle t_2 \| v(1T_1) \| t_2 \rangle = 3i\zeta \qquad \langle t_2 \| v(1T_1) \| e \rangle = -3\sqrt{2}i\zeta'$

d^2 $\langle t_2^2 S\Gamma \| V(1T_1) \| t_2^2 S'\Gamma' \rangle$

$S\Gamma$ \ $S'\Gamma'$	$^1A_1$	$^3T_1$	^1E	$^1T_2$
$^1A_1$		$-\sqrt{2}/\sqrt{3}$		
$^3T_1$	$\sqrt{2}/\sqrt{3}$	-1	$-1/\sqrt{3}$	$-1/\sqrt{2}$
^1E		$1/\sqrt{3}$		
$^1T_2$		$1/\sqrt{2}$		

$\times \langle t_2 \| v(1T_1) \| t_2 \rangle$

$\langle t_2 e S\Gamma \| V(1T_1) \| t_2 e S'\Gamma' \rangle$

$S\Gamma$ \ $S'\Gamma'$	$^3T_1$	$^3T_2$	$^1T_1$	$^1T_2$
$^3T_1$	$1/2$	$\sqrt{3}/2$	$1/2\sqrt{2}$	$\sqrt{3}/2\sqrt{2}$
$^3T_2$	$\sqrt{3}/2$	$-1/2$	$\sqrt{3}/2\sqrt{2}$	$-1/2\sqrt{2}$
$^1T_1$	$-1/2\sqrt{2}$	$-\sqrt{3}/2\sqrt{2}$		
$^1T_2$	$-\sqrt{3}/2\sqrt{2}$	$1/2\sqrt{2}$		

$\times \langle t_2 \| v(1T_1) \| t_2 \rangle$

‡ The signs of some reduced matrix elements in the original table [Y. Tanabe, *Progr. Theor. Phys. (Kyoto)*, Supplement No. 14, 52–65 (1960)] are changed according to the phase change of the twelve wavefunctions of the t_2^4 and t_2^4e configurations given in the note of Appendix IV. Misprints found in the original table are also corrected.

$\langle t_2^2 S\Gamma \| V(1T_1) \| t_2 e S'\Gamma' \rangle$

$S\Gamma$ \ $S'\Gamma'$	$^3T_1$	$^3T_2$	$^1T_1$	$^1T_2$
$^1A_1$	$1/\sqrt{3}$			
$^3T_1$	$-1/\sqrt{2}$	$\sqrt{3}/\sqrt{2}$	$1/2$	$-\sqrt{3}/2$
^1E	$\sqrt{2}/\sqrt{3}$			
$^1T_2$	$-1/2$	$\sqrt{3}/2$		

$\times \langle t_2 \| v(1T_1) \| e \rangle$

$\langle t_2 e S\Gamma \| V(1T_1) \| e^2 S'\Gamma' \rangle$

$S\Gamma$ \ $S'\Gamma'$	$^1A_1$	$^3A_2$	^1E
$^3T_1$	$-1/\sqrt{2}$		$-1/\sqrt{2}$
$^3T_2$		1	$1/\sqrt{2}$
$^1T_1$			
$^1T_2$		$-1/\sqrt{2}$	

$\times \langle t_2 \| v(1T_1) \| e \rangle$

d^3

(1) $\langle t_2^3 S\Gamma \| V(1T_1) \| t_2^3 S'\Gamma' \rangle$

$S\Gamma$ \ $S'\Gamma'$	$^4A_2$	^2E	$^2T_1$	$^2T_2$
$^4A_2$				$2/\sqrt{3}$
^2E				$-\sqrt{2}/\sqrt{3}$
$^2T_1$				-1
$^2T_2$	$-2/\sqrt{3}$	$-\sqrt{2}/\sqrt{3}$	-1	

$\times \langle t_2 \| v(1T_1) \| t_2 \rangle$

(2) $\langle t_2^2(S_1\Gamma_1)\, e S\Gamma \| V(1T_1) \| t_2^2(S_3\Gamma_3)\, e S'\Gamma' \rangle$

(a) $\langle t_2^2(^3T_1)\, e S\Gamma \| V(1T_1) \| t_2^2(^3T_1)\, e S'\Gamma' \rangle$

$S\Gamma$ \ $S'\Gamma'$	$^4T_1$	$^4T_2$	$^2T_1$	$^2T_2$
$^4T_1$	$\sqrt{5}/3\sqrt{2}$	$\sqrt{5}/\sqrt{6}$	$1/3\sqrt{2}$	$1/\sqrt{6}$
$^4T_2$	$\sqrt{5}/\sqrt{6}$	$-\sqrt{5}/3\sqrt{2}$	$1/\sqrt{6}$	$-1/3\sqrt{2}$
$^2T_1$	$-1/3\sqrt{2}$	$-1/\sqrt{6}$	$1/3$	$1/\sqrt{3}$
$^2T_2$	$-1/\sqrt{6}$	$1/3\sqrt{2}$	$1/\sqrt{3}$	$-1/3$

$\times \langle t_2 \| v(1T_1) \| t_2 \rangle$

(b) $\langle t_2^2(^1A_1)\, e S\Gamma \| V(1T_1) \| t_2^2(^3T_1)\, e S'\Gamma' \rangle$

$S\Gamma$ \ $S'\Gamma'$	$^4T_1$	$^4T_2$	$^2T_1$	$^2T_2$
^2E	$-2\sqrt{2}/3$	$-2\sqrt{2}/3$	$-2/3$	$-2/3$

$\times \langle t_2 \| v(1T_1) \| t_2 \rangle$

(c) $\langle t_2^2({}^3T_1)\, eS\Gamma \| V(1T_1) \| t_2^2({}^1E)\, eS'\Gamma' \rangle$

$S\Gamma$ \ $S'\Gamma'$	${}^2A_1$	${}^2A_2$	${}^2E$
${}^4T_1$	$-\sqrt{2}/3$		$\sqrt{2}/3$
${}^4T_2$		$\sqrt{2}/3$	$-\sqrt{2}/3$
${}^2T_1$	$1/3$		$-1/3$
${}^2T_2$		$-1/3$	$1/3$

$\times \langle t_2 \| v(1T_1) \| t_2 \rangle$

(d) $\langle t_2^2({}^3T_1)\, eS\Gamma \| V(1T_1) \| t_2^2({}^1T_2)\, eS'\Gamma' \rangle$

$S\Gamma$ \ $S'\Gamma'$	${}^2T_1$	${}^2T_2$
${}^4T_1$	$-1/\sqrt{2}$	$1/\sqrt{6}$
${}^4T_2$	$1/\sqrt{6}$	$1/\sqrt{2}$
${}^2T_1$	$1/2$	$-1/2\sqrt{3}$
${}^2T_2$	$-1/2\sqrt{3}$	$-1/2$

$\times \langle t_2 \| v(1T_1) \| t_2 \rangle$

(3) $\langle t_2e^2(S_2\Gamma_2)\, S\Gamma \| V(1T_1) \| t_2e^2(S_2\Gamma_2)\, S'\Gamma' \rangle$

(a) $\langle t_2e^2({}^1A_1){}^2T_2 \| V(1T_1) \| t_2e^2({}^1A_1){}^2T_2 \rangle = \langle t_2 \| v(1T_1) \| t_2 \rangle$

(b) $\langle t_2e^2({}^3A_2)\, S\Gamma \| V(1T_1) \| t_2e^2({}^3A_2)\, S'\Gamma' \rangle$

$S\Gamma$ \ $S'\Gamma'$	${}^4T_1$	${}^2T_1$
${}^4T_1$	$-\sqrt{10}/3$	$-2\sqrt{2}/3$
${}^2T_1$	$2\sqrt{2}/3$	$1/3$

$\times \langle t_2 \| v(1T_1) \| t_2 \rangle$

(c) $\langle t_2e^2({}^1E)\, S\Gamma \| V(1T_1) \| t_2e^2({}^1E)\, S'\Gamma' \rangle$

$S\Gamma$ \ $S'\Gamma'$	${}^2T_1$	${}^2T_2$
${}^2T_1$	$1/2$	$\sqrt{3}/2$
${}^2T_2$	$\sqrt{3}/2$	$-1/2$

$\times \langle t_2 \| v(1T_1) \| t_2 \rangle$

(4) $\langle t_2^3 S\Gamma \| V(1T_1) \| t_2^2(S_3\Gamma_3)\, eS'\Gamma' \rangle$

(a) $\langle t_2^3\, {}^2T_2 \| V(1T_1) \| t_2^2({}^1A_1)\, e\, {}^2E \rangle = (\sqrt{2}/\sqrt{3}) \langle t_2 \| v(1T_1) \| e \rangle$

(b) $\langle t_2^3 S\Gamma \| V(1T_1) \| t_2^2({}^3T_1)\, eS'\Gamma' \rangle$

$S\Gamma$ \ $S'\Gamma'$	${}^4T_1$	${}^4T_2$	${}^2T_1$	${}^2T_2$
${}^4A_2$		$\sqrt{10}/3$		$-2\sqrt{2}/3$
${}^2E$		$2\sqrt{2}/3$		$-1/3$
${}^2T_1$	-1	$-1/\sqrt{3}$	$1/2\sqrt{2}$	$1/2\sqrt{6}$
${}^2T_2$	1	$1/\sqrt{3}$	$-1/2\sqrt{2}$	$-1/2\sqrt{6}$

$\times \langle t_2 \| v(1T_1) \| e \rangle$

(c) $\langle t_2^3 S\Gamma \| V(1T_1) \| t_2^2(^1E)\, eS'\Gamma' \rangle$

$S\Gamma$ \ $S'\Gamma'$	$^2A_1$	$^2A_2$	^2E
$^2T_1$	$-1/\sqrt{2}$		$-1/\sqrt{2}$
$^2T_2$		$1/\sqrt{6}$	$-1/\sqrt{6}$

$\times \langle t_2 \| v(1T_1) \| e \rangle$

(d) $\langle t_2^3 S\Gamma \| V(1T_1) \| t_2^2(^1T_2)\, eS'\Gamma' \rangle$

$S\Gamma$ \ $S'\Gamma'$	$^2T_1$	$^2T_2$
^2E	1	
$^2T_1$	$-1/2\sqrt{2}$	$\sqrt{3}/2\sqrt{2}$
$^2T_2$	$1/2\sqrt{2}$	$-\sqrt{3}/2\sqrt{2}$

$\times \langle t_2 \| v(1T_1) \| e \rangle$

(5) $\langle t_2^2(S_1\Gamma_1)\, eS\Gamma \| V(1T_1) \| t_2e^2(S_4\Gamma_4)\, S'\Gamma' \rangle$

	$S_4\Gamma_4$	$^1A_1$	$^3A_2$		^1E	
$S_1\Gamma_1$	$S\Gamma$ \ $S'\Gamma'$	$^2T_2$	$^4T_1$	$^2T_1$	$^2T_1$	$^2T_2$
$^1A_1$	^2E	$-1/\sqrt{6}$	$2/3$	$1/3\sqrt{2}$	$1/\sqrt{6}$	$-1/\sqrt{6}$
$^3T_1$	$^4T_1$	-1	$-\sqrt{5}/3$	$1/3$	$2/\sqrt{3}$	
	$^4T_2$	$-1/\sqrt{3}$	$\sqrt{5}/\sqrt{3}$	$-1/\sqrt{3}$		$2/\sqrt{3}$
	$^2T_1$	$-1/2\sqrt{2}$	$1/3$	$-5/6\sqrt{2}$	$1/\sqrt{6}$	
	$^2T_2$	$-1/2\sqrt{6}$	$-1/\sqrt{3}$	$5/2\sqrt{6}$		$1/\sqrt{6}$
^1E	$^2A_1$		$2/3$	$1/3\sqrt{2}$	$1/\sqrt{6}$	
	$^2A_2$	$1/\sqrt{6}$				$1/\sqrt{6}$
	^2E	$-1/\sqrt{6}$	$-2/3$	$-1/3\sqrt{2}$	$-1/\sqrt{6}$	$-1/\sqrt{6}$
$^1T_2$	$^2T_1$	$1/2\sqrt{2}$	-1	$-1/2\sqrt{2}$		$-1/\sqrt{2}$
	$^2T_2$	$-\sqrt{3}/2\sqrt{2}$	$-1/\sqrt{3}$	$-1/2\sqrt{6}$	$1/\sqrt{2}$	

$\times \langle t_2 \| v(1T_1) \| e \rangle$

(6) $\langle t_2e^2(S_2\Gamma_2)\, S\Gamma \| V(1T_1) \| e^3\, {}^2E \rangle$

$S_2\Gamma_2$	$S\Gamma$	^2E
$^1A_1$	$^2T_2$	$1/\sqrt{2}$
$^3A_2$	$^4T_1$	$-2/\sqrt{3}$
	$^2T_1$	$1/\sqrt{6}$
^1E	$^2T_1$	$1/\sqrt{2}$
	$^2T_2$	$-1/\sqrt{2}$

$\times \langle t_2 \| v(1T_1) \| e \rangle$

d^4

(1) $\langle t_2^4 S\Gamma \| V(1T_1) \| t_2^4 S'\Gamma' \rangle$

$S\Gamma$ \ $S'\Gamma'$	$^1A_1$	$^3T_1$	^1E	$^1T_2$
$^1A_1$		$\sqrt{2}/\sqrt{3}$		
$^3T_1$	$-\sqrt{2}/\sqrt{3}$	1	$1/\sqrt{3}$	$1/\sqrt{2}$
^1E		$-1/\sqrt{3}$		
$^1T_2$		$-1/\sqrt{2}$		

$\times \langle t_2 \| v(1T_1) \| t_2 \rangle$

(2) $\langle t_2^3(S_1\Gamma_1)\, eS\Gamma \| V(1T_1) \| t_2^3(S_3\Gamma_3)\, eS'\Gamma' \rangle$

(a) $\langle t_2^3(^4A_2)\, eS\Gamma \| V(1T_1) \| t_2^3(^2T_2)\, eS'\Gamma' \rangle$

$S\Gamma$ \ $S'\Gamma'$	$^3T_1$	$^3T_2$	$^1T_1$	$^1T_2$
^5E	$\sqrt{5}/\sqrt{3}$	$\sqrt{5}/\sqrt{3}$		
^3E	$-1/\sqrt{3}$	$-1/\sqrt{3}$	$\sqrt{2}/\sqrt{3}$	$\sqrt{2}/\sqrt{3}$

$\times \langle t_2 \| v(1T_1) \| t_2 \rangle$

(b) $\langle t_2^3(^2E)\, eS\Gamma \| V(1T_1) \| t_2^3(^2T_2)\, eS'\Gamma' \rangle$

$S\Gamma$ \ $S'\Gamma'$	$^3T_1$	$^3T_2$	$^1T_1$	$^1T_2$
$^3A_1$	$1/\sqrt{3}$		$1/\sqrt{6}$	
$^3A_2$		$-1/\sqrt{3}$		$-1/\sqrt{6}$
^3E	$1/\sqrt{3}$	$-1/\sqrt{3}$	$1/\sqrt{6}$	$-1/\sqrt{6}$
$^1A_1$	$-1/\sqrt{6}$			
$^1A_2$		$1/\sqrt{6}$		
^1E	$-1/\sqrt{6}$	$1/\sqrt{6}$		

$\times \langle t_2 \| v(1T_1) \| t_2 \rangle$

(c) $\langle t_2^3(^2T_1)\, eS\Gamma \| V(1T_1) \| t_2^3(^2T_2)\, eS'\Gamma' \rangle$

$S\Gamma$ \ $S'\Gamma'$	$^3T_1$	$^3T_2$	$^1T_1$	$^1T_2$
$^3T_1$	$-\sqrt{3}/2$	$1/2$	$-\sqrt{3}/2\sqrt{2}$	$1/2\sqrt{2}$
$^3T_2$	$1/2$	$\sqrt{3}/2$	$1/2\sqrt{2}$	$\sqrt{3}/2\sqrt{2}$
$^1T_1$	$\sqrt{3}/2\sqrt{2}$	$-1/2\sqrt{2}$		
$^1T_2$	$-1/2\sqrt{2}$	$-\sqrt{3}/2\sqrt{2}$		

$\times \langle t_2 \| v(1T_1) \| t_2 \rangle$

(3) $\langle t_2^2(S_1\Gamma_1)\, e^2(S_2\Gamma_2)\, S\Gamma \| V(1T_1) \| t_2^2(S_3\Gamma_3)\, e^2(S_2\Gamma_2)\, S'\Gamma' \rangle$

(a) $\langle t_2^2(S_1\Gamma_1)\, e^2(^1A_1)\, S_1\Gamma_1 \| V(1T_1) \| t_2^2(S_3\Gamma_3)\, e^2(^1A_1)\, S_3\Gamma_3 \rangle$

	$S_3\Gamma_3$	$^3T_1$	^1E	$^1T_2$
	$S'\Gamma'$	$^3T_1$	^1E	$^1T_2$
$S_1\Gamma_1$	$S\Gamma$			
$^1A_1$	$^1A_1$	$-\sqrt{2}/\sqrt{3}$		
$^3T_1$	$^3T_1$	-1	$-1/\sqrt{3}$	$-1/\sqrt{2}$
^1E	^1E	$1/\sqrt{3}$		
$^1T_2$	$^1T_2$	$1/\sqrt{2}$		

$\times \langle t_2 \| v(1T_1) \| t_2 \rangle$

(b) $\langle t_2^2(S_1\Gamma_1)\, e^2(^3A_2)\, S\Gamma \| V(1T_1) \| t_2^2(S_3\Gamma_3)\, e^2(^3A_2)\, S'\Gamma' \rangle$

$S_1\Gamma_1$	$S\Gamma$ \ $S_3\Gamma_3$	$^1A_1$	$^3T_1$			^1E	$^1T_2$
	$S'\Gamma'$	$^3A_2$	$^5T_2$	$^3T_2$	$^1T_2$	^3E	$^3T_1$
$^1A_1$	$^3A_1$		$-\sqrt{10}/3$	$-\sqrt{2}/\sqrt{3}$	$-\sqrt{2}/3$		
$^3T_1$	$^5T_2$	$\sqrt{10}/3$	$\sqrt{5}/2$	$\sqrt{5}/2\sqrt{3}$		$-\sqrt{5}/3$	$-\sqrt{5}/\sqrt{6}$
	$^3T_2$	$-\sqrt{2}/\sqrt{3}$	$-\sqrt{5}/2\sqrt{3}$	$1/2$	$1/\sqrt{3}$	$1/\sqrt{3}$	$1/\sqrt{2}$
	$^1T_2$	$\sqrt{2}/3$		$-1/\sqrt{3}$		$-1/\sqrt{3}$	$-1/\sqrt{6}$
^1E	^3E		$\sqrt{5}/3$	$1/\sqrt{3}$	$1/3$		
$^1T_2$	$^3T_1$		$\sqrt{5}/\sqrt{6}$	$1/\sqrt{2}$	$1/\sqrt{6}$		

$\times \langle t_2 \| v(1T_1) \| t_2 \rangle$

(c) $\langle t_2^2(S_1\Gamma_1)\, e^2(^1E)\, S\Gamma \| V(1T_1) \| t_2^2(S_3\Gamma_3)\, e^2(^1E) S'\Gamma' \rangle$

$S_1\Gamma_1$	$S\Gamma$ \ $S_3\Gamma_3$	$^1A_1$	$^3T_1$		^1E			$^1T_2$	
	$S'\Gamma'$	^1E	$^3T_1$	$^3T_2$	$^1A_1$	$^1A_2$	^1E	$^1T_1$	$^1T_2$
$^1A_1$	^1E		$-\sqrt{2}/\sqrt{3}$	$-\sqrt{2}/\sqrt{3}$					
$^3T_1$	$^3T_1$	$\sqrt{2}/\sqrt{3}$	$1/2$	$\sqrt{3}/2$	$-1/\sqrt{6}$		$1/\sqrt{6}$	$-\sqrt{3}/2\sqrt{2}$	$1/2\sqrt{2}$
	$^3T_2$	$\sqrt{2}/\sqrt{3}$	$\sqrt{3}/2$	$-1/2$		$1/\sqrt{6}$	$-1/\sqrt{6}$	$1/2\sqrt{2}$	$\sqrt{3}/2\sqrt{2}$
^1E	$^1A_1$		$1/\sqrt{6}$						
	$^1A_2$			$-1/\sqrt{6}$					
	^1E		$-1/\sqrt{6}$	$1/\sqrt{6}$					
$^1T_2$	$^1T_1$		$\sqrt{3}/2\sqrt{2}$	$-1/2\sqrt{2}$					
	$^1T_2$		$-1/2\sqrt{2}$	$-\sqrt{3}/2\sqrt{2}$					

$\times \langle t_2 \| v(1T_1) \| t_2 \rangle$

(4) $\langle t_2 e^3 S\Gamma \| V(1T_1) \| t_2 e^3 S'\Gamma' \rangle$

$S\Gamma$ \ $S'\Gamma'$	$^3T_1$	$^3T_2$	$^1T_1$	$^1T_2$
$^3T_1$	$1/2$	$\sqrt{3}/2$	$1/2\sqrt{2}$	$\sqrt{3}/2\sqrt{2}$
$^3T_2$	$\sqrt{3}/2$	$-1/2$	$\sqrt{3}/2\sqrt{2}$	$-1/2\sqrt{2}$
$^1T_1$	$-1/2\sqrt{2}$	$-\sqrt{3}/2\sqrt{2}$		
$^1T_2$	$-\sqrt{3}/2\sqrt{2}$	$1/2\sqrt{2}$		

$\times \langle t_2 \| v(1T_1) \| t_2 \rangle$

(5) $\langle t_2^4 S\Gamma \| V(1T_1) \| t_2^3(S_3\Gamma_3)\, e S'\Gamma' \rangle$

(a) $\langle t_2^4 S\Gamma \| V(1T_1) \| t_2^3(^4A_2)\, e S'\Gamma' \rangle$

$S\Gamma$ \ $S'\Gamma'$	^5E	^3E
$^3T_1$	$-\sqrt{5}/\sqrt{3}$	$1/\sqrt{3}$

$\times \langle t_2 \| v(1T_1) \| e \rangle$

(b) $\langle t_2{}^4 S\Gamma \,\|\, V(1T_1) \,\|\, t_2{}^3({}^2E)\, eS'\Gamma' \rangle$

$S\Gamma$ \ $S'\Gamma'$	${}^3A_1$	${}^3A_2$	${}^3E$	${}^1A_1$	${}^1A_2$	${}^1E$
${}^3T_1$	$-1/\sqrt{3}$		$-1/\sqrt{3}$	$1/\sqrt{6}$		$1/\sqrt{6}$
${}^1T_2$		$1/\sqrt{2}$	$-1/\sqrt{2}$			

$\times \langle t_2 \,\|\, v(1T_1) \,\|\, e \rangle$

(c) $\langle t_2{}^4 S\Gamma \,\|\, V(1T_1) \,\|\, t_2{}^3({}^2T_1)\, eS'\Gamma' \rangle$

$S\Gamma$ \ $S'\Gamma'$	${}^3T_1$	${}^3T_2$	${}^1T_1$	${}^1T_2$
${}^3T_1$	$-\sqrt{3}/2$	$-1/2$	$\sqrt{3}/2\sqrt{2}$	$1/2\sqrt{2}$
${}^1E$		-1		
${}^1T_2$	$-\sqrt{3}/2\sqrt{2}$	$-1/2\sqrt{2}$		

$\times \langle t_2 \,\|\, v(1T_1) \,\|\, e \rangle$

(d) $\langle t_2{}^4 S\Gamma \,\|\, V(1T_1) \,\|\, t_2{}^3({}^2T_2)\, eS'\Gamma' \rangle$

$S\Gamma$ \ $S'\Gamma'$	${}^3T_1$	${}^3T_2$	${}^1T_1$	${}^1T_2$
${}^1A_1$	$\sqrt{2}/\sqrt{3}$			
${}^3T_1$	$1/2$	$-\sqrt{3}/2$	$-1/2\sqrt{2}$	$\sqrt{3}/2\sqrt{2}$
${}^1E$	$-1/\sqrt{3}$			
${}^1T_2$	$1/2\sqrt{2}$	$-\sqrt{3}/2\sqrt{2}$		

$\times \langle t_2 \,\|\, v(1T_1) \,\|\, e \rangle$

(6) $\langle t_2{}^3(S_1\Gamma_1)\, eS\Gamma \,\|\, V(1T_1) \,\|\, t_2{}^2(S_3\Gamma_3)\, e^2(S_4\Gamma_4)\, S'\Gamma' \rangle$

(a) $\langle t_2{}^3({}^4A_2)\, eS\Gamma \,\|\, V(1T_1) \,\|\, t_2{}^2({}^3T_1)\, e^2(S_4\Gamma_4)\, S'\Gamma' \rangle$

$S_4\Gamma_4$	${}^1A_1$	${}^3A_2$			${}^1E$	
$S\Gamma$ \ $S'\Gamma'$	${}^3T_1$	${}^5T_2$	${}^3T_2$	${}^1T_2$	${}^3T_1$	${}^3T_2$
${}^5E$	$-\sqrt{5}/\sqrt{6}$	$\sqrt{5}/2$	$-\sqrt{5}/2\sqrt{3}$		$-\sqrt{5}/\sqrt{6}$	$\sqrt{5}/\sqrt{6}$
${}^3E$	$-1/\sqrt{6}$	$-\sqrt{5}/6$	$\sqrt{3}/2$	$-2/3$	$-1/\sqrt{6}$	$1/\sqrt{6}$

$\times \langle t_2 \,\|\, v(1T_1) \,\|\, e \rangle$

(b) $(t_2{}^3({}^2E)\, eS\Gamma \,\|\, V(1T_1) \,\|\, t_2{}^2({}^3T_1)\, e^2(S_4\Gamma_4)\, S'\Gamma')$

$S_4\Gamma_4$	${}^1A_1$	${}^3A_2$			${}^1E$	
$S\Gamma$ \ $S'\Gamma'$	${}^3T_1$	${}^5T_2$	${}^3T_2$	${}^1T_2$	${}^3T_1$	${}^3T_2$
${}^3A_1$	$1/\sqrt{6}$				$1/\sqrt{6}$	
${}^3A_2$		$\sqrt{5}/3$		$-1/6$		$1/\sqrt{6}$
${}^3E$	$1/\sqrt{6}$	$\sqrt{5}/3$		$-1/6$	$1/\sqrt{6}$	$1/\sqrt{6}$
${}^1A_1$	$1/2\sqrt{3}$				$1/2\sqrt{3}$	
${}^1A_2$			$1/\sqrt{6}$			$1/2\sqrt{3}$
${}^1E$	$1/2\sqrt{3}$		$1/\sqrt{6}$		$1/2\sqrt{3}$	$1/2\sqrt{3}$

$\times \langle t_2 \,\|\, v(1T_1) \,\|\, e \rangle$

(c) $\langle t_2^3(^2E)\, eS\Gamma \| V(1T_1) \| t_2^2(^1T_2)\, e^2(S_4\Gamma_4)\, S'\Gamma' \rangle$

$S_4\Gamma_4$	$^1A_1$	$^3A_2$	^1E	
$S\Gamma$ \ $S'\Gamma'$	$^1T_2$	$^3T_1$	$^1T_1$	$^1T_2$
$^3A_1$		$1/\sqrt{2}$	$1/2$	
$^3A_2$	$1/2$			$1/2$
^3E	$-1/2$	$-1/\sqrt{2}$	$-1/2$	$-1/2$
$^1A_1$		$-1/2$		
$^1A_2$				
^1E		$1/2$		

$\times \langle t_2 \| v(1T_1) \| e \rangle$

(d) $\langle t_2^3(^2T_1)\, eS\Gamma \| V(1T_1) \| t_2^2(^3T_1)\, e^2(S_4\Gamma_4)\, S'\Gamma' \rangle$

$S_4\Gamma_4$	$^1A_1$	$^3A_2$			^1E	
$S\Gamma$ \ $S'\Gamma'$	$^3T_1$	$^5T_2$	$^3T_2$	$^1T_2$	$^3T_1$	$^3T_2$
$^3T_1$	$\sqrt{3}/2\sqrt{2}$	$-\sqrt{5}/2\sqrt{3}$		$1/4\sqrt{3}$		$1/\sqrt{2}$
$^3T_2$	$1/2\sqrt{2}$	$\sqrt{5}/2$		$-1/4$	$-1/\sqrt{2}$	
$^1T_1$	$\sqrt{3}/4$		$-1/2\sqrt{2}$			$1/2$
$^1T_2$	$1/4$		$\sqrt{3}/2\sqrt{2}$		$-1/2$	

$\times \langle t_2 \| v(1T_1) \| e \rangle$

(e) $\langle t_2^3(^2T_1)\, eS\Gamma \| V(1T_1) \| t_2^2(^1E)\, e^2(S_4\Gamma_4)\, S'\Gamma' \rangle$

$S_4\Gamma_4$	$^1A_1$	$^3A_2$	^1E		
$S\Gamma$ \ $S'\Gamma'$	^1E	^3E	$^1A_1$	$^1A_2$	^1E
$^3T_1$		1	$-1/2$		$-1/2$
$^3T_2$	$-1/\sqrt{2}$			$-1/2$	$-1/2$
$^1T_1$		$-1/\sqrt{2}$			
$^1T_2$					

$\times \langle t_2 \| v(1T_1) \| e \rangle$

(f) $\langle t_2^3(^2T_1)\, eS\Gamma \| V(1T_1) \| t_2^2(^1T_2)\, e^2(S_4\Gamma_4)\, S'\Gamma' \rangle$

$S_4\Gamma_4$	$^1A_1$	$^3A_2$	^1E	
$S\Gamma$ \ $S'\Gamma'$	$^1T_2$	$^4T_1$	$^1T_1$	$^1T_2$
$^3T_1$	$-\sqrt{3}/4$	$-1/2\sqrt{2}$	$1/2$	
$^3T_2$	$-1/4$	$\sqrt{3}/2\sqrt{2}$		$1/2$
$^1T_1$		$1/4$		
$^1T_2$		$-\sqrt{3}/4$		

$\times \langle t_2 \| v(1T_1) \| e \rangle$

(g) $\langle t_2^3(^2T_2)\, eS\Gamma \| V(1T_1) \| t_2^2(^1A_1)\, e^2(S_4\Gamma_4)\, S'\Gamma' \rangle$

	$S_4\Gamma_4$	$^1A_1$	$^3A_2$	^1E
	$S'\Gamma'$	$^1A_1$	$^3A_2$	^1E
$S\Gamma$	$^3T_1$	$-1/\sqrt{3}$		$-1/\sqrt{3}$
	$^3T_2$		$\sqrt{2}/\sqrt{3}$	$1/\sqrt{3}$
	$^1T_1$			
	$^1T_2$		$-1/\sqrt{3}$	

$\times \langle t_2 \| v(1T_1) \| e \rangle$

(h) $\langle t_2^3(^2T_2)\, eS\Gamma \| V(1T_1) \| t_2^2(^3T_1)\, e^2(S_4\Gamma_4)\, S'\Gamma' \rangle$

	$S_4\Gamma_4$	$^1A_1$	$^3A_2$			^1E	
	$S'\Gamma'$	$^3T_1$	$^5T_2$	$^3T_2$	$^1T_2$	$^3T_1$	$^3T_2$
$S\Gamma$	$^3T_1$	$1/2\sqrt{2}$	$\sqrt{5}/2$		$-1/4$	$-1/\sqrt{2}$	
	$^3T_2$	$-\sqrt{3}/2\sqrt{2}$	$\sqrt{5}/2\sqrt{3}$		$-1/4\sqrt{3}$		$-1/\sqrt{2}$
	$^1T_1$	$1/4$		$\sqrt{3}/2\sqrt{2}$		$-1/2$	
	$^1T_2$	$-\sqrt{3}/4$		$1/2\sqrt{2}$			$-1/2$

$\times \langle t_2 \| v(1T_1) \| e \rangle$

(i) $\langle t_2^3(^2T_2)\, eS\Gamma \| V(1T_1) \| t_2^2(^1E)\, e^2(S_4\Gamma_4)\, S'\Gamma' \rangle$

	$S_4\Gamma_4$	$^1A_1$	$^3A_2$	^1E		
	$S'\Gamma'$	^1E	^3E	$^1A_1$	$^1A_2$	^1E
$S\Gamma$	$^3T_1$	$1/\sqrt{6}$		$1/2\sqrt{3}$		$-1/2\sqrt{3}$
	$^3T_2$		$-1/\sqrt{3}$		$1/2\sqrt{3}$	$-1/2\sqrt{3}$
	$^1T_1$					
	$^1T_2$		$1/\sqrt{6}$			

$\times \langle t_2 \| v(1T_1) \| e \rangle$

(j) $\langle t_2^3(^2T_2)\, eS\Gamma \| V(1T_1) \| t_2^2(^1T_2)\, e^2(S_4\Gamma_4)\, S'\Gamma' \rangle$

	$S_4\Gamma_4$	$^1A_1$	$^3A_2$	^1E	
	$S'\Gamma'$	$^1T_2$	$^3T_1$	$^1T_1$	$^1T_2$
$S\Gamma$	$^3T_1$	$-1/4$	$\sqrt{3}/2\sqrt{2}$		$1/2$
	$^3T_2$	$\sqrt{3}/4$	$1/2\sqrt{2}$	$-1/2$	
	$^1T_1$		$-\sqrt{3}/4$		
	$^1T_2$		$-1/4$		

$\times \langle t_2 \| v(1T_1) \| e \rangle$

(7) $\langle t_2^2(S_1\Gamma_1)\, e^2(S_2\Gamma_2)\, S\Gamma \| V(1T_1) \| t_2e^3S'\Gamma' \rangle$

(a) $\langle t_2^2(^1A_1)\, e^2(S_2\Gamma_2)\, S_2\Gamma_2 \| V(1T_1) \| t_2e^3S'\Gamma' \rangle$

	$S'\Gamma'$	$^3T_1$	$^3T_2$	$^1T_1$	$^1T_2$
$S\Gamma$	$^1A_1$	$1/\sqrt{6}$			
	$^3A_2$		$-1/\sqrt{3}$		$-1/\sqrt{6}$
	^1E	$-1/\sqrt{6}$	$1/\sqrt{6}$		

$\times \langle t_2 \| v(1T_1) \| e \rangle$

(b) $\langle t_2^2(^3T_1)\, e^2(S_2\Gamma_2)\, S\Gamma \| V(1T_1) \| t_2e^3S'\Gamma' \rangle$

$S_2\Gamma_2$	$S\Gamma$ \ $S'\Gamma'$	$^3T_1$	$^3T_2$	$^1T_1$	$^1T_2$
$^1A_1$	$^3T_1$	$-1/2$	$\sqrt{3}/2$	$1/2\sqrt{2}$	$-\sqrt{3}/2\sqrt{2}$
$^3A_2$	$^5T_2$	$-\sqrt{5}/\sqrt{2}$	$-\sqrt{5}/\sqrt{6}$		
	$^3T_2$			$-\sqrt{3}/2$	$-1/2$
	$^1T_2$	$1/2\sqrt{2}$	$1/2\sqrt{6}$		
^1E	$^3T_1$	-1		$1/\sqrt{2}$	
	$^3T_2$		-1		$1/\sqrt{2}$

$\times \langle t_2 \| v(1T_1) \| e \rangle$

(c) $\langle t_2^2(^1E)\, e^2(S_2\Gamma_2)\, S\Gamma \| V(1T_1) \| t_2e^3S'\Gamma' \rangle$

$S_2\Gamma_2$	$S\Gamma$ \ $S'\Gamma'$	$^3T_1$	$^3T_2$	$^1T_1$	$^1T_2$
$^1A_1$	^1E	$1/\sqrt{3}$			
$^3A_2$	^3E		$-\sqrt{2}/\sqrt{3}$		$-1/\sqrt{3}$
^1E	$^1A_1$	$-1/\sqrt{6}$			
	$^1A_2$		$-1/\sqrt{6}$		
	^1E	$1/\sqrt{6}$	$1/\sqrt{6}$		

$\times \langle t_2 \| v(1T_1) \| e \rangle$

(d) $\langle t_2^2(^1T_2)\, e^2(S_2\Gamma_2)\, S\Gamma \| V(1T_1) \| t_2e^3S'\Gamma' \rangle$

$S_2\Gamma_2$	$S\Gamma$ \ $S'\Gamma'$	$^3T_1$	$^3T_2$	$^1T_1$	$^1T_2$
$^1A_1$	$^1T_2$	$-1/2\sqrt{2}$	$\sqrt{3}/2\sqrt{2}$		
$^3A_2$	$^3T_1$	$\sqrt{3}/2$	$1/2$	$\sqrt{3}/2\sqrt{2}$	$1/2\sqrt{2}$
^1E	$^1T_1$		$1/\sqrt{2}$		
	$^1T_2$	$-1/\sqrt{2}$			

$\times \langle t_2 \| v(1T_1) \| e \rangle$

(8) $\langle t_2e^3\ ^3T_1 \| V(1T_1) \| e^4\ ^1A_1 \rangle = -\langle t_2 \| v(1T_1) \| e \rangle$

d^5

(1) $\langle t_2^5 \| V(1T_1) \| t_2^5 \rangle = -\langle t_2 \| v(1T_1) \| t_2 \rangle$

(2) $\langle t_2e^4 \| V(1T_1) \| t_2e^4 \rangle = \langle t_2 \| v(1T_1) \| t_2 \rangle$

(3) $\langle t_2^4(S_1\Gamma_1)\, eS\Gamma \| V(1T_1) \| t_2^4(S_3\Gamma_3)\, eS'\Gamma' \rangle$
$= -\langle t_2^2(S_1\Gamma_1)\, eS\Gamma \| V(1T_1) \| t_2^2(S_3\Gamma_3)\, eS'\Gamma' \rangle$

(4) $\langle t_2^2(S_1\Gamma_1)\, e^3S\Gamma \| V(1T_1) \| t_2^2(S_3\Gamma_3)\, e^3S'\Gamma' \rangle$
$= \langle t_2^2(S_1\Gamma_1)\, eS\Gamma \| V(1T_1) \| t_2^2(S_3\Gamma_3)\, eS'\Gamma' \rangle$

(5) $\langle t_2^3(S_1\Gamma_1)\, e^2(S_2\Gamma_2)\, S\Gamma \,\|\, V(1T_1) \,\|\, t_2^3(S_3\Gamma_3)\, e^2(S_2\Gamma_2)\, S'\Gamma' \rangle$

(a) $\langle t_2^3(^4A_2)\, e^2(S_2\Gamma_2)\, S\Gamma \,\|\, V(1T_1) \,\|\, t_2^3(^2T_2)\, e^2(S_2\Gamma_2)\, S'\Gamma' \rangle$

	$S_2\Gamma_2$	$^1A_1$	$^3A_2$		^1E	
	$S'\Gamma'$	$^2T_2$	$^4T_1$	$^2T_1$	$^2T_1$	$^2T_2$
$S_2\Gamma_2$	$S\Gamma$					
$^1A_1$	$^4A_2$	$2/\sqrt{3}$				
$^3A_2$	$^6A_1$		$\sqrt{2}$			
	$^4A_1$		$-4/3\sqrt{3}$	$2\sqrt{5}/3\sqrt{3}$		
	$^2A_1$		$\sqrt{2}/3\sqrt{3}$	$-4/3\sqrt{3}$		
^1E	^4E				$2/\sqrt{3}$	$2/\sqrt{3}$

$\times \langle t_2 \,\|\, v(1T_1) \,\|\, t_2 \rangle$

(b) $\langle t_2^3(^2E)\, e^2(S_2\Gamma_2)\, S\Gamma \,\|\, V(1T_1) \,\|\, t_2^3(^2T_2)\, e^2(S_2\Gamma_2)\, S'\Gamma' \rangle$

	$S_2\Gamma_2$	$^1A_1$	$^3A_2$		^1E	
	$S'\Gamma'$	$^2T_2$	$^4T_1$	$^2T_1$	$^2T_1$	$^2T_2$
$S_2\Gamma_2$	$S\Gamma$					
$^1A_1$	^2E	$-\sqrt{2}/\sqrt{3}$				
$^3A_2$	^4E		$-2\sqrt{5}/3\sqrt{3}$	$-4/3\sqrt{3}$		
	^2E		$4/3\sqrt{3}$	$\sqrt{2}/3\sqrt{3}$		
^1E	$^2A_1$				$1/\sqrt{3}$	
	$^2A_2$					$-1/\sqrt{3}$
	^2E				$1/\sqrt{3}$	$-1/\sqrt{3}$

$\times \langle t_2 \,\|\, v(1T_1) \,\|\, t_2 \rangle$

(c) $\langle t_2^3(^2T_1)\, e^2(S_2\Gamma_2)\, S\Gamma \,\|\, V(1T_1) \,\|\, t_2^3(^2T_2)\, e^2(S_2\Gamma_2)\, S\Gamma \rangle$

	$S_2\Gamma_2$	$^1A_1$	$^3A_2$		^1E	
	$S'\Gamma'$	$^2T_2$	$^4T_1$	$^2T_1$	$^2T_1$	$^2T_2$
$S_2\Gamma_2$	$S\Gamma$					
$^1A_1$	$^2T_1$	-1				
$^3A_2$	$^4T_2$		$-\sqrt{10}/3$	$-2\sqrt{2}/3$		
	$^2T_2$		$2\sqrt{2}/3$	$1/3$		
^1E	$^2T_1$				$-\sqrt{3}/2$	$1/2$
	$^2T_2$				$1/2$	$\sqrt{3}/2$

$\times \langle t_2 \,\|\, v(1T_1) \,\|\, t_2 \rangle$

(6) $\langle t_2^5 \,\|\, V(1T_1) \,\|\, t_2^4(S_3\Gamma_3)\, eS'\Gamma' \rangle$

$S_3\Gamma_3$	$^1A_1$
$S'\Gamma'$	^2E
$^2T_2$	$1/\sqrt{3}$

$\times \langle t_2 \,\|\, v(1T_1) \,\|\, e \rangle$

$S_3\Gamma_3$	$^3T_1$			
$S'\Gamma'$	$^4T_1$	$^4T_2$	$^2T_1$	$^2T_2$
$^2T_2$	$-\sqrt{2}$	$-\sqrt{2}/\sqrt{3}$	$1/2$	$1/2\sqrt{3}$

$\times \langle t_2 \,\|\, v(1T_1) \,\|\, e \rangle$

$S_3\Gamma_3$	^1E		
$S'\Gamma'$	$^2A_1$	$^2A_2$	^2E
$^2T_2$		$-1/\sqrt{3}$	$1/\sqrt{3}$

$\times \langle t_2 \,\|\, v(1T_1) \,\|\, e \rangle$

$S_3\Gamma_3$	$^1T_2$	
$S'\Gamma'$	$^2T_1$	$^2T_2$
$^2T_2$	$-1/2$	$\sqrt{3}/2$

$\times \langle t_2 \,\|\, v(1T_1) \,\|\, e \rangle$

(7) $\langle t_2^2(S_1\Gamma_1)\, e^3 S\Gamma \,\|\, V(1T_1) \,\|\, t_2 e^4 \rangle$

$S_1\Gamma_1$	$S\Gamma$	$^2T_2$
$^1A_1$	^2E	$-1/\sqrt{3}$

$\times \langle t_2 \,\|\, v(1T_1) \,\|\, e \rangle$

$S_1\Gamma_1$	$S\Gamma$	$^2T_2$
$^3T_1$	$^4T_1$	$-\sqrt{2}$
	$^4T_2$	$-\sqrt{2}/\sqrt{3}$
	$^2T_1$	$-1/2$
	$^2T_2$	$-1/2\sqrt{3}$

$\times \langle t_2 \,\|\, v(1T_1) \,\|\, e \rangle$

$S_1\Gamma_1$	$S\Gamma$	$^2T_2$
^1E	$^2A_1$	
	$^2A_2$	$1/\sqrt{3}$
	^2E	$-1/\sqrt{3}$

$\times \langle t_2 \,\|\, v(1T_1) \,\|\, e \rangle$

$S_1\Gamma_1$	$S\Gamma$	$^2T_2$
$^1T_2$	$^2T_1$	$1/2$
	$^2T_2$	$-\sqrt{3}/2$

$\times \langle t_2 \,\|\, v(1T_1) \,\|\, e \rangle$

(8) $\langle t_2^4(S_1\Gamma_1)\, e S\Gamma \,\|\, V(1T_1) \,\|\, t_2^3(S_3\Gamma_3)\, e^2(S_4\Gamma_4)\, S'\Gamma' \rangle$

	$S_3\Gamma_3$	$^2T_2$				
	$S_4\Gamma_4$	$^1A_1$	$^3A_2$		^1E	
	$S'\Gamma'$	$^2T_2$	$^4T_1$	$^2T_1$	$^2T_1$	$^2T_2$
$S_1\Gamma_1$	$S\Gamma$					
$^1A_1$	^2E	$-1/\sqrt{3}$	$2\sqrt{2}/3$	$1/3$	$1/\sqrt{3}$	$-1/\sqrt{3}$

$\times \langle t_2 \,\|\, v(1T_1) \,\|\, e \rangle$

	$S_3\Gamma_3$	$^2T_1$				
	$S_4\Gamma_4$	$^1A_1$	$^3A_2$		^1E	
	$S'\Gamma'$	$^2T_1$	$^4T_2$	$^2T_2$	$^2T_1$	$^2T_2$
$S_1\Gamma_1$	$S\Gamma$					
^1E	$^2A_1$	$-1/2$			$-1/2$	
	$^2A_2$		$-\sqrt{2}/\sqrt{3}$	$-1/2\sqrt{3}$		$-1/2$
	^2E	$-1/2$	$-\sqrt{2}/\sqrt{3}$	$-1/2\sqrt{3}$	$-1/2$	$-1/2$

$\times \langle t_2 \,\|\, v(1T_1) \,\|\, e \rangle$

	$S_3\Gamma_3$	$^2T_2$				
	$S_4\Gamma_4$	$^1A_1$	$^3A_2$		^1E	
	$S'\Gamma'$	$^2T_2$	$^4T_1$	$^2T_1$	$^2T_1$	$^2T_2$
$S_1\Gamma_1$	$S\Gamma$					
^1E	$^2A_1$		$-\sqrt{2}/3$	$-1/6$	$-1/2\sqrt{3}$	
	$^2A_2$	$-1/2\sqrt{3}$				$-1/2\sqrt{3}$
	^2E	$1/2\sqrt{3}$	$\sqrt{2}/3$	$1/6$	$1/2\sqrt{3}$	$1/2\sqrt{3}$

$\times \langle t_2 \,\|\, v(1T_1) \,\|\, e \rangle$

	$S_3\Gamma_3$	^2E					
	$S_4\Gamma_4$	$^1A_1$	$^3A_2$		^1E		
	$S'\Gamma'$	^2E	^4E	^2E	$^2A_1$	$^2A_2$	^2E
$S_1\Gamma_1$	$S\Gamma$						
$^1T_2$	$^2T_1$	$1/\sqrt{2}$			$1/2$		$-1/2$
	$^2T_2$		$-2/\sqrt{3}$	$-1/\sqrt{6}$		$1/2$	$-1/2$

$\times \langle t_2 \,\|\, v(1T_1) \,\|\, e \rangle$

	$S_3\Gamma_3$	$^2T_1$				
	$S_4\Gamma_4$	$^1A_1$	$^3A_2$		^1E	
	$S'\Gamma'$	$^2T_1$	$^4T_2$	$^2T_2$	$^2T_1$	$^2T_2$
$S_1\Gamma_1$	$S\Gamma$					
$^1T_2$	$^2T_1$	$-1/4$	$-1/\sqrt{2}$	$-1/4$	$1/2$	
	$^2T_2$	$\sqrt{3}/4$	$-1/\sqrt{6}$	$-1/4\sqrt{3}$		$1/2$

$\times \langle t_2 \| v(1T_1) \| e \rangle$

	$S_3\Gamma_3$	$^2T_2$				
	$S_4\Gamma_4$	$^1A_1$	$^3A_2$		^1E	
	$S'\Gamma'$	$^2T_2$	$^4T_1$	$^2T_1$	$^2T_1$	$^2T_2$
$S_1\Gamma_1$	$S\Gamma$					
$^1T_2$	$^2T_1$	$-1/4$	$1/\sqrt{2}$	$1/4$		$1/2$
	$^2T_2$	$\sqrt{3}/4$	$1/\sqrt{6}$	$1/4\sqrt{3}$	$-1/2$	

$\times \langle t_2 \| v(1T_1) \| e \rangle$

	$S_3\Gamma_3$	$^4A_2$				
	$S_4\Gamma_4$	$^1A_1$	$^3A_2$			^1E
	$S'\Gamma'$	$^4A_2$	$^6A_1$	$^4A_1$	$^2A_1$	^4E
$S_1\Gamma_1$	$S\Gamma$					
$^3T_1$	$^4T_1$		$-\sqrt{2}$	$1/3\sqrt{3}$	$\sqrt{2}/3\sqrt{3}$	$-\sqrt{5}/3$
	$^4T_2$	$\sqrt{5}/3$				$\sqrt{5}/3$
	$^2T_1$			$-2\sqrt{5}/3\sqrt{3}$	$2/3\sqrt{3}$	$-2/3$
	$^2T_2$	$2/3$				$2/3$

$\times \langle t_2 \| v(1T_1) \| e \rangle$

	$S_3\Gamma_3$	^2E					
	$S_4\Gamma_4$	$^1A_1$	$^3A_2$		^1E		
	$S'\Gamma'$	^2E	^4E	^2E	$^2A_1$	$^2A_2$	^2E
$S_1\Gamma_1$	$S\Gamma$						
$^3T_1$	$^4T_1$		$2\sqrt{5}/3\sqrt{3}$	$-2/3\sqrt{3}$	$-\sqrt{2}/3$		$-\sqrt{2}/3$
	$^4T_2$	$-2/3$				$-\sqrt{2}/3$	$-\sqrt{2}/3$
	$^2T_1$		$-2/3\sqrt{3}$	$5/3\sqrt{6}$	$-1/6$		$-1/6$
	$^2T_2$	$-1/3\sqrt{2}$				$-1/6$	$-1/6$

$\times \langle t_2 \| v(1T_1) \| e \rangle$

	$S_3\Gamma_3$	$^2T_1$				
	$S_4\Gamma_4$	$^1A_1$	$^3A_2$		^1E	
	$S'\Gamma'$	$^2T_1$	$^4T_2$	$^2T_2$	$^2T_1$	$^2T_2$
$S_1\Gamma_1$	$S\Gamma$					
$^3T_1$	$^4T_1$	$1/\sqrt{2}$	$-\sqrt{5}/3\sqrt{2}$	$1/3\sqrt{2}$		$\sqrt{2}/\sqrt{3}$
	$^4T_2$	$1/\sqrt{6}$	$\sqrt{5}/\sqrt{6}$	$-1/\sqrt{6}$	$-\sqrt{2}/\sqrt{3}$	
	$^2T_1$	$1/4$	$1/3\sqrt{2}$	$-5/12$		$1/2\sqrt{3}$
	$^2T_2$	$1/4\sqrt{3}$	$-1/\sqrt{6}$	$5/4\sqrt{3}$	$-1/2\sqrt{3}$	

$\times \langle t_2 \| v(1T_1) \| e \rangle$

	$S_3\Gamma_3$	$^2T_2$				
	$S_4\Gamma_4$	$^1A_1$	$^3A_2$		^1E	
	$S'\Gamma'$	$^2T_2$	$^4T_1$	$^2T_1$	$^2T_1$	$^2T_2$
$S_1\Gamma_1$	$S\Gamma$					
$^3T_1$	$^4T_1$	$1/\sqrt{2}$	$\sqrt{5}/3\sqrt{2}$	$-1/3\sqrt{2}$	$-\sqrt{2}/\sqrt{3}$	
	$^4T_2$	$1/\sqrt{6}$	$-\sqrt{5}/\sqrt{6}$	$1/\sqrt{6}$		$-\sqrt{2}/\sqrt{3}$
	$^2T_1$	$1/4$	$-1/3\sqrt{2}$	$5/12$	$-1/2\sqrt{3}$	
	$^2T_2$	$1/4\sqrt{3}$	$1/\sqrt{6}$	$-5/4\sqrt{3}$		$-1/2\sqrt{3}$

$\times \langle t_2 \| v(1T_1) \| e \rangle$

(9) $\langle t_2^3(S_1\Gamma_1)\, e^2(S_2\Gamma_2)\, S\Gamma \| V(1T_1) \| t_2^2(S_3\Gamma_3)\, e^3 S'\Gamma' \rangle$

		$S_3\Gamma_3$	$^1A_1$
		$S'\Gamma'$	^2E
$S_1\Gamma_1$	$S_2\Gamma_2$	$S\Gamma$	
$^2T_2$	$^1A_1$	$^2T_2$	$1/\sqrt{3}$
	$^3A_2$	$^4T_1$	$-2\sqrt{2}/3$
		$^2T_1$	$1/3$
	^1E	$^2T_1$	$1/\sqrt{3}$
		$^2T_2$	$-1/\sqrt{3}$

$\times \langle t_2 \| v(1T_1) \| e \rangle$

		$S_3\Gamma_3$	^1E		
		$S'\Gamma'$	$^2A_1$	$^2A_2$	^2E
$S_1\Gamma_1$	$S_2\Gamma_2$	$S\Gamma$			
$^1T_1$	$^1A_1$	$^2T_1$	$-1/2$		$-1/2$
	$^3A_2$	$^4T_2$		$-\sqrt{2}/\sqrt{3}$	$-\sqrt{2}/\sqrt{3}$
		$^2T_2$		$1/2\sqrt{3}$	$1/2\sqrt{3}$
	^1E	$^2T_1$	$1/2$		$1/2$
		$^2T_2$		$1/2$	$1/2$
$^2T_2$	$^1A_1$	$^2T_2$		$1/2\sqrt{3}$	$-1/2\sqrt{3}$
	$^3A_2$	$^4T_1$	$\sqrt{2}/3$		$-\sqrt{2}/3$
		$^2T_1$	$-1/6$		$1/6$
	^1E	$^2T_1$	$-1/2\sqrt{3}$		$1/2\sqrt{3}$
		$^2T_2$		$-1/2\sqrt{3}$	$1/2\sqrt{3}$

$\times \langle t_2 \| v(1T_1) \| e \rangle$

		$S_3\Gamma_3$	$^1T_2$	
		$S'\Gamma'$	$^2T_1$	$^2T_2$
$S_1\Gamma_1$	$S_2\Gamma_2$	$S\Gamma$		
^2E	$^1A_1$	^2E	$1/\sqrt{2}$	
	$^3A_2$	^4E		$-2/\sqrt{3}$
		^2E		$1/\sqrt{6}$
	^1E	$^2A_1$	$-1/2$	
		$^2A_2$		$-1/2$
		^2E	$1/2$	$1/2$

$\times \langle t_2 \| v(1T_1) \| e \rangle$

$S_1\Gamma_1$	$S_2\Gamma_2$	$S\Gamma$ \ $S_3\Gamma_3$: ${}^1T_2$; $S'\Gamma'$:	${}^2T_1$	${}^2T_2$
${}^2T_1$	${}^1A_1$	${}^2T_1$	$-1/4$	$\sqrt{3}/4$
	${}^3A_2$	${}^4T_2$	$-1/\sqrt{2}$	$-1/\sqrt{6}$
		${}^2T_2$	$1/4$	$1/4\sqrt{3}$
	${}^1E$	${}^2T_1$	$-1/2$	
		${}^2T_2$		$-1/2$
${}^2T_2$	${}^1A_1$	${}^2T_2$	$1/4$	$-\sqrt{3}/4$
	${}^3A_2$	${}^4T_1$	$-1/\sqrt{2}$	$-1/\sqrt{6}$
		${}^2T_1$	$1/4$	$1/4\sqrt{3}$
	${}^1E$	${}^2T_1$		$-1/2$
		${}^2T_2$	$1/2$	

$\times \langle t_2 \| v(1T_1) \| e \rangle$

$S_1\Gamma_1$	$S_2\Gamma_2$	$S\Gamma$ \ $S_3\Gamma_3$: ${}^3T_1$; $S'\Gamma'$:	${}^4T_1$	${}^4T_2$	${}^2T_1$	${}^2T_2$
${}^4A_2$	${}^1A_1$	${}^4A_2$		$\sqrt{5}/3$		$-2/3$
	${}^3A_2$	${}^6A_1$	$-\sqrt{2}$			
		${}^4A_1$	$-1/3\sqrt{3}$		$-2\sqrt{5}/3\sqrt{3}$	
		${}^2A_1$	$\sqrt{2}/3\sqrt{3}$		$-2/3\sqrt{3}$	
	${}^1E$	${}^4E$	$\sqrt{5}/3$	$-\sqrt{5}/3$	$-2/3$	$2/3$
${}^2E$	${}^1A_1$	${}^2E$		$2/3$		$-1/3\sqrt{2}$
	${}^3A_2$	${}^4E$	$-2\sqrt{5}/3\sqrt{3}$		$-2/3\sqrt{3}$	
		${}^2E$	$-2/3\sqrt{3}$		$-5/3\sqrt{6}$	
	${}^1E$	${}^2A_1$	$-\sqrt{2}/3$		$1/6$	
		${}^2A_2$		$-\sqrt{2}/3$		$1/6$
		${}^2E$	$-\sqrt{2}/3$	$-\sqrt{2}/3$	$1/6$	$1/6$

$\times \langle t_2 \| v(1T_1) \| e \rangle$

$S_1\Gamma_1$	$S_2\Gamma_2$	$S\Gamma$ \ $S_3\Gamma_3$: ${}^3T_1$; $S'\Gamma'$:	${}^4T_1$	${}^4T_2$	${}^2T_1$	${}^2T_2$
${}^2T_1$	${}^1A_1$	${}^2T_1$	$-1/\sqrt{2}$	$-1/\sqrt{6}$	$1/4$	$1/4\sqrt{3}$
	${}^3A_2$	${}^4T_2$	$\sqrt{5}/3\sqrt{2}$	$-\sqrt{5}/\sqrt{6}$	$1/3\sqrt{2}$	$-1/\sqrt{6}$
		${}^2T_2$	$1/3\sqrt{2}$	$-1/\sqrt{6}$	$5/12$	$-5/4\sqrt{3}$
	${}^1E$	${}^2T_1$		$-\sqrt{2}/\sqrt{3}$		$1/2\sqrt{3}$
		${}^2T_2$	$\sqrt{2}/\sqrt{3}$		$-1/2\sqrt{3}$	
${}^2T_2$	${}^1A_1$	${}^2T_2$	$1/\sqrt{2}$	$1/\sqrt{6}$	$-1/4$	$-1/4\sqrt{3}$
	${}^3A_2$	${}^4T_1$	$\sqrt{5}/3\sqrt{2}$	$-\sqrt{5}/\sqrt{6}$	$1/3\sqrt{2}$	$-1/\sqrt{6}$
		${}^2T_1$	$1/3\sqrt{2}$	$-1/\sqrt{6}$	$5/12$	$-5/4\sqrt{3}$
	${}^1E$	${}^2T_1$	$\sqrt{2}/\sqrt{3}$		$-1/2\sqrt{3}$	
		${}^2T_2$		$\sqrt{2}/\sqrt{3}$		$-1/2\sqrt{3}$

$\times \langle t_2 \| v(1T_1) \| e \rangle$

Appendix VIII CALCULATION OF $\langle \alpha S\Gamma \| L \| \alpha' S\Gamma' \rangle$

The calculations of matrix elements of the orbital angular momentum $L_{\bar{\gamma}}$ ($\bar{\gamma} = \alpha(x), \beta(y), \gamma(z)$) are reduced to those of reduced matrices $\langle \alpha S\Gamma \| \mathbf{L} \| \alpha' S\Gamma' \rangle$ by using the relation

$$\langle \alpha S\Gamma M\gamma \mid L_{\bar{\gamma}} \mid \alpha' S'\Gamma' M'\gamma' \rangle = \delta(SS')\,\delta(MM') \times (\Gamma)^{-1/2} \langle \alpha S\Gamma \| \mathbf{L} \| \alpha' S\Gamma' \rangle \langle \Gamma\gamma \mid \Gamma'\gamma' T_1 \bar{\gamma} \rangle, \tag{AVIII-1}$$

as $L_{\bar{\gamma}}$ transforms like the $\bar{\gamma}$ base of the T_1 irreducible representation of the O-group. Since $\mathbf{L}$ is a one-electron operator, it is evident that, corresponding to (6.117), (6.118), and (6.119), respectively, one has the relations

$$\langle t_2{}^n e^m S\Gamma \| \mathbf{L} \| t_2^{n-k} e^{m+k} S\Gamma' \rangle = 0 \qquad \text{for} \quad |k| \geqslant 2, \tag{AVIII-2}$$

$$\langle t_2{}^n e^m S\Gamma \| \mathbf{L} \| t_2^{n-1} e^{m+1} S\Gamma' \rangle = C_0 \langle t_2 \| \mathbf{l} \| e \rangle, \tag{AVIII-3}$$

and

$$\langle t_2{}^n e^m S\Gamma \| \mathbf{L} \| t_2{}^n e^m S\Gamma' \rangle = C_1 \langle t_2 \| \mathbf{l} \| t_2 \rangle. \tag{AVIII-4}$$

In (AVIII-4), we used the fact that the orbital angular momentum is completely quenched in the e-state. In (AVIII-3) and (AVIII-4) C_0 and C_1 are numerical coefficients depending upon the states of interest, and are calculated by using the explicit forms of the wavefunctions in the same way as was done in the case of low-symmetry fields. It is convenient to express $\langle t_2 \| \mathbf{l} \| t_2 \rangle$ and $\langle t_2 \| \mathbf{l} \| e \rangle$ in terms of k and k' defined as

$$\langle t_2 \| \mathbf{l} \| t_2 \rangle = \sqrt{6}\,\langle t_2\xi \mid l_z \mid t_2\eta \rangle = \sqrt{6}\, ik \tag{AVIII-5}$$

and

$$\langle t_2 \| \mathbf{l} \| e \rangle = -\sqrt{3}\, \langle t_2 \zeta \mid l_z \mid ev \rangle = -2\sqrt{3}\, ik'. \qquad \text{(AVIII-6)}$$

Comparing (AVIII-5) and (AVIII-6) with (7.5), one sees that $k = k' = 1$ in the d-function approximation.

Now let us examine the reduced matrices of $\mathbf{L}$ in the complementary states. Since the matrices of $\mathbf{L}$ is Hermitian and purely imaginary for real bases, the matrix elements of $L_{\bar{\gamma}}$ in the complementary states are related to each other as

$$\begin{aligned} &\langle t_2^n(S_1\Gamma_1)\, e^m(S_2\Gamma_2)\, S\Gamma M\gamma \mid L_{\bar{\gamma}} \mid t_2^{n'}(S_1'\Gamma_1')\, e^{m'}(S_2'\Gamma_2')\, S\Gamma' M\gamma' \rangle \\ &\quad = \langle t_2^{6-n}(S_1\Gamma_1)\, e^{4-m}(S_2\Gamma_2)\, S\Gamma M\gamma \mid L_{\bar{\gamma}} \mid t_2^{6-n'}(S_1'\Gamma_1')\, e^{4-m'}(S_2'\Gamma_2')\, S\Gamma' M\gamma' \rangle \\ &\qquad\qquad (n + m = n' + m' \neq 5). \qquad \text{(AVIII-7)} \end{aligned}$$

Note that all the diagonal elements are zero. In deriving (AVIII-7) we used the fact that, if the one-electron operator is $\mathbf{L}$, (4.48) should be replaced by

$$F_{kk'}^{10-N} = F_{kk'}^{N}\,. \qquad \text{(AVIII-8)}$$

This is because (4.44) is now given as

$$F_{kk'}^{10-N} = -\langle \alpha_{N+q} \mid l_{\bar{\gamma}} \mid \alpha_p \rangle = \langle \alpha_p \mid l_{\bar{\gamma}} \mid \alpha_{N+q} \rangle. \qquad \text{(AVIII-9)}$$

Equation (AVIII-7) tells us that

$$\begin{aligned} &\langle t_2^n(S_1\Gamma_1)\, e^m(S_2\Gamma_2)\, S\Gamma \| \mathbf{L} \| t_2^{n'}(S_1'\Gamma_1')\, e^{m'}(S_2'\Gamma_2')\, S\Gamma' \rangle \\ &\quad = \langle t_2^{6-n}(S_1\Gamma_1)\, e^{4-m}(S_2\Gamma_2)\, S\Gamma \| \mathbf{L} \| t_2^{6-n'}(S_1'\Gamma_1')\, e^{4-m'}(S_2'\Gamma_2')\, S\Gamma' \rangle \\ &\qquad\qquad (n + m = n' + m' \neq 5). \qquad \text{(AVIII-10)} \end{aligned}$$

For the states of a half-filled subshell configurations, it would not be difficult to show by following the arguments given in Section 6.3.2 that

$$\langle t_2^3 S\Gamma \| \mathbf{L} \| t_2^3 S\Gamma' \rangle = 0 \qquad \text{(AVIII-11a)}$$

for the combinations

$$S\Gamma = {}^2E,\, {}^2T_1\,, \quad S\Gamma' = {}^2T_2\,, \qquad \text{(AVIII-11b)}$$

and vice versa. Equation (AVIII-10) can be shown to hold even for $n + m = n' + m' = 5$ if $n \neq 3$ and $n' \neq 3$, and, if $n = n' = 3$, one has

$$\langle t_2^3(S_1\Gamma_1)\, e^2(S_2\Gamma_2)\, S\Gamma \| \mathbf{L} \| t_2^3(S_1'\Gamma_1')\, e^2(S_2'\Gamma_2')\, S\Gamma' \rangle = 0 \qquad \text{(AVIII-12)}$$

for $S_1\Gamma_1$, $S_2\Gamma_2$, $S_1'\Gamma_1'$, and $S_2'\Gamma_2'$ giving $\mu_1\mu_2\mu_1'\mu_2' = -1$, etc.

Appendix IX SYMMETRIC AND ANTISYMMETRIC PRODUCT REPRESENTATIONS

Let $\varphi_i(\Gamma\gamma)$ $(i = 1, 2)$ be the γ base of irreducible representation Γ of group G. We consider function $\psi([\Gamma \times \Gamma]\,\gamma\gamma')$ given by

$$\psi([\Gamma \times \Gamma]\gamma\gamma') = \varphi_1(\Gamma\gamma)\,\varphi_2(\Gamma\gamma') + \varphi_1(\Gamma\gamma')\,\varphi_2(\Gamma\gamma). \qquad \text{(AIX-1)}$$

For symmetry operation R of group G, this function transforms as follows:

$$\begin{aligned} R\psi([\Gamma \times \Gamma]\gamma\gamma') &= \sum_{\gamma_1\gamma_2} \varphi_1(\Gamma\gamma_1)\,\varphi_2(\Gamma\gamma_2) \\ &\qquad \times [D^{(\Gamma)}_{\gamma_1\gamma}(R)\,D^{(\Gamma)}_{\gamma_2\gamma'}(R) + D^{(\Gamma)}_{\gamma_1\gamma'}(R)\,D^{(\Gamma)}_{\gamma_2\gamma}(R)] \\ &= \sum_{\gamma_1\gamma_2} [\varphi_1(\Gamma\gamma_1)\,\varphi_2(\Gamma\gamma_2) + \varphi_1(\Gamma\gamma_2)\,\varphi_2(\Gamma\gamma_1)] \\ &\qquad \times \tfrac{1}{2}[D^{(\Gamma)}_{\gamma_1\gamma}(R)\,D^{(\Gamma)}_{\gamma_2\gamma'}(R) + D^{(\Gamma)}_{\gamma_1\gamma'}(R)\,D^{(\Gamma)}_{\gamma_2\gamma}(R)] \\ &= \sum_{\gamma_1\gamma_2} \psi([\Gamma \times \Gamma]\gamma\gamma')\,D^{[\Gamma\times\Gamma]}_{\gamma_1\gamma_2,\gamma\gamma'}(R), \end{aligned} \qquad \text{(AIX-2)}$$

where

$$D^{[\Gamma\times\Gamma]}_{\gamma_1\gamma_2,\gamma\gamma'}(R) = \tfrac{1}{2}[D^{(\Gamma)}_{\gamma_1\gamma}(R)\,D^{(\Gamma)}_{\gamma_2\gamma'}(R) + D^{(\Gamma)}_{\gamma_1\gamma'}(R)\,D^{(\Gamma)}_{\gamma_2\gamma}(R)]. \qquad \text{(AIX-3)}$$

Therefore, $\psi([\Gamma \times \Gamma]\,\gamma\gamma')$ is the base of representation $D^{[\Gamma\times\Gamma]}$ whose matrix elements are given by (AIX-3). We call $D^{[\Gamma\times\Gamma]}$ the *symmetric product representation* of $D^{(\Gamma)}$.

Similarly, one can show that function $\psi(\{\Gamma \times \Gamma\}\gamma\gamma')$ defined as

$$\psi(\{\Gamma \times \Gamma\}\gamma\gamma') = \varphi_1(\Gamma\gamma)\,\varphi_2(\Gamma\gamma') - \varphi_1(\Gamma\gamma')\,\varphi_2(\Gamma\gamma) \qquad \text{(AIX-4)}$$

is the base of representation $D^{\{\Gamma\times\Gamma\}}$ whose matrix elements are given by

$$D^{\{\Gamma\times\Gamma\}}_{\gamma_1\gamma_2,\gamma\gamma'}(R) = \tfrac{1}{2}[D^{(\Gamma)}_{\gamma_1\gamma}(R)\,D^{(\Gamma)}_{\gamma_2\gamma'}(R) - D^{(\Gamma)}_{\gamma_1\gamma'}(R)\,D^{(\Gamma)}_{\gamma_2\gamma}(R)]. \qquad \text{(AIX-5)}$$

Representation $D^{\{\Gamma\times\Gamma\}}$ is called the *antisymmetric product representation* of $D^{(\Gamma)}$.

The characters of $D^{[\Gamma\times\Gamma]}$ and $D^{\{\Gamma\times\Gamma\}}$ are obtained as follows:

$$\begin{aligned}\chi^{[\Gamma\times\Gamma]}(R) &= \sum_{\gamma\gamma'} D^{[\Gamma\times\Gamma]}_{\gamma\gamma',\gamma\gamma'}(R)\\ &= \frac{1}{2}\Big[\sum_{\gamma} D^{(\Gamma)}_{\gamma\gamma}(R)\sum_{\gamma'} D^{(\Gamma)}_{\gamma'\gamma'}(R) + \sum_{\gamma} D^{(\Gamma)}_{\gamma\gamma}(R^2)\Big]\\ &= \frac{1}{2}\,[\chi^{(\Gamma)}(R)^2 + \chi^{(\Gamma)}(R^2)],\end{aligned} \qquad \text{(AIX-6)}$$

and

$$\chi^{\{\Gamma\times\Gamma\}}(R) = \tfrac{1}{2}[\chi^{(\Gamma)}(R)^2 - \chi^{(\Gamma)}(R^2)]. \qquad \text{(AIX-7)}$$

From the relation

$$\chi^{(\Gamma\times\Gamma)}(R) = \chi^{[\Gamma\times\Gamma]}(R) + \chi^{\{\Gamma\times\Gamma\}}(R), \qquad \text{(AIX-8)}$$

we see that a product representation can always be decomposed into a symmetric and an antisymmetric representation. By using (AIX-6) and Table 1.7, (9.89) can easily be derived.

SUBJECT INDEX

PURE AND APPLIED PHYSICS

A Series of Monographs and Textbooks

Consulting Editors

H. S. W. Massey
University College, London, England

Keith A. Brueckner
University of California, San Diego
La Jolla, California

1. F. H. Field and J. L. Franklin, Electron Impact Phenomena and the Properties of Gaseous Ions.
2. H. Kopfermann, Nuclear Moments, English Version Prepared from the Second German Edition by E. E. Schneider.
3. Walter E. Thirring, Principles of Quantum Electrodynamics. Translated from the German by J. Bernstein. With Corrections and Additions by Walter E. Thirring.
4. U. Fano and G. Racah, Irreducible Tensorial Sets.
5. E. P. Wigner, Group Theory and Its Application to the Quantum Mechanics of Atomic Spectra. Expanded and Improved Edition. Translated from the German by J. J. Griffin.
6. J. Irving and N. Mullineux, Mathematics in Physics and Engineering.
7. Karl F. Herzfeld and Theodore A. Litovitz, Absorption and Dispersion of Ultrasonic Waves.
8. Leon Brillouin, Wave Propagation and Group Velocity.
9. Fay Ajzenberg-Selove (ed.), Nuclear Spectroscopy. Parts A and B.
10. D. R. Bates (ed.), Quantum Theory. In three volumes.
11. D. J. Thouless, The Quantum Mechanics of Many-Body Systems.
12. W. S. C. Williams, An Introduction to Elementary Particles.
13. D. R. Bates (ed.), Atomic and Molecular Processes.
14. Amos de-Shalit and Igal Talmi, Nuclear Shell Theory.
15. Walter H. Barkas. Nuclear Research Emulsions. Part I.
    Nuclear Research Emulsions. Part II. *In preparation*
16. Joseph Callaway, Energy Band Theory.
17. John M. Blatt, Theory of Superconductivity.
18. F. A. Kaempffer, Concepts in Quantum Mechanics.
19. R. E. Burgess (ed.), Fluctuation Phenomena in Solids.
20. J. M. Daniels, Oriented Nuclei: Polarized Targets and Beams.
21. R. H. Huddlestone and S. L. Leonard (eds.), Plasma Diagnostic Techniques.
22. Amnon Katz, Classical Mechanics, Quantum Mechanics, Field Theory.
23. Warren P. Mason, Crystal Physics in Interaction Processes.
24. F. A. Berezin, The Method of Second Quantization.
25. E. H. S. Burhop (ed.), High Energy Physics. In four volumes.

26. L. S. Rodberg and R. M. Thaler, Introduction to the Quantum Theory of Scattering.

27. R. P. Shutt (ed.), Bubble and Spark Chambers. In two volumes.

28. Geoffrey V. Marr, Photoionization Processes in Gases.

29. J. P. Davidson, Collective Models of the Nucleus.

30. Sydney Geltman, Topics in Atomic Collision Theory.

31. Eugene Feenberg, Theory of Quantum Fluids.

32. Robert T. Beyer and Stephen V. Letcher, Physical Ultrasonics.

33. S. Sugano, Y. Tanabe, and H. Kamimura, Multiplets of Transition-Metal Ions in Crystals.

In preparation

Walter T. Grandy, Jr., Introduction to Electrodynamics and Radiation.

J. Killingbeck and G. H. A. Cole, Physical Applications of Mathematical Techniques.

Herbert Uberall, Electron Scattering from Complex Nuclei. Parts A and B.